Bayesian Statistics
for Beginners

Bayesian Statistics for Beginners

A Step-by-Step Approach

THERESE M. DONOVAN
RUTH M. MICKEY

OXFORD
UNIVERSITY PRESS

OXFORD
UNIVERSITY PRESS

Great Clarendon Street, Oxford, OX2 6DP,
United Kingdom

Oxford University Press is a department of the University of Oxford.
It furthers the University's objective of excellence in research, scholarship,
and education by publishing worldwide. Oxford is a registered trade mark of
Oxford University Press in the UK and in certain other countries

First Edition published in 2019

Impression: 7

Published in the United States of America by Oxford University Press
198 Madison Avenue, New York, NY 10016, United States of America

British Library Cataloguing in Publication Data

Data available

Library of Congress Control Number: 2019934655

ISBN 978–0–19–884129–6 (hbk.)
ISBN 978–0–19–884130–2 (pbk.)

DOI: 10.1093/oso/9780198841296.001.0001

Printed and bound by
CPI Group (UK) Ltd, Croydon, CR0 4YY

To our parents, Thomas and Earline Donovan and Ray and Jean Mickey,
for inspiring a love of learning.

To our mentors, some of whom we've met only by their written words,
for teaching us ways of knowing.

To Peter, Evan, and Ana—for everything.

Preface

Greetings. This book is our attempt at gaining membership to the Bayesian Conspiracy. You may ask, "What is the Bayesian Conspiracy?" The answer is provided by Eliezer Yudkowsky (http://yudkowsky.net/rational/bayes): "The Bayesian Conspiracy is a multi national, interdisciplinary, and shadowy group of scientists that controls publication, grants, tenure, and the illicit traffic in grad students. The best way to be accepted into the Bayesian Conspiracy is to join the Campus Crusade for Bayes in high school or college, and gradually work your way up to the inner circles. It is rumored that at the upper levels of the Bayesian Conspiracy exist nine silent figures known only as the Bayes Council."

Ha ha! Bayes' Theorem, also called Bayes' Rule, was published posthumously in 1763 in the Philosophical Transactions of the Royal Society. In The Theory That Would Not Die, author Sharon Bertsch McGrayne aptly describes how "Bayes' rule cracked the enigma code, hunted down Russian submarines, and emerged triumphant from two centuries of controversy." In short, Bayes' Rule has vast application, and the number of papers and books that employ it is growing exponentially.

Inspired by the knowledge that a Bayes Council actually exists, we began our journey by enrolling in a 5-day 'introductory' workshop on Bayesian statistics a few years ago. On Day 1, we were introduced to a variety of Bayesian models, and, on Day 2, we sheepishly had to inquire what Bayes' Theorem was and what it had to do with MCMC. In other words, the material was way over our heads. With tails between our legs, we slunk back home and began trying to sort out the many different uses of Bayes' Theorem.

As we read more and more about Bayes' Theorem, we started noting our own questions as they arose and began narrating the answers as they became more clear. The result is this strange book, cast as a series of questions and answers between reader and author. In this prose, we make heavy use of online resources such as the Oxford Dictionary of Statistics (Upton and Cook, 2014), Wolfram Mathematics, and the Online Statistics Education: An Interactive Multimedia Course for Study (Rice University, University of Houston Clear Lake, and Tufts University). We also provide friendly links to online encyclopedias such as Wikipedia and Encyclopedia Britannica. Although these should not be considered definitive, original works, we have included the links to provide readers with a readily accessible source of information and are grateful to the many authors who have contributed entries.

We are not experts in Bayesian statistics and make no claim as such. Therese Donovan is a biologist for the U. S. Geological Survey Vermont Cooperative Fish and Wildlife Research Unit, and Ruth Mickey is a statistician in the Department of Mathematics and Statistics at the University of Vermont. We were raised on a healthy dose of "frequentist" and maximum likelihood methods but have begun only recently to explore Bayesian methods. We have intentionally avoided controversial topics and comparisons between Bayesian and frequentist approaches and encourage the reader to dig deeper—much deeper—than we have here. Fortunately, a great number of experts have paved the way, and we relied heavily on the following books while writing our own:

- N. T. Hobbs and M. B. Hooten. Bayesian Models: A Statistical Primer for Ecologists. Princeton University Press, 2015.

- J. Kruschke. Doing Bayesian Data Analysis: A Tutorial with R, JAGS, and Stan. Elsevier, 2015.
- A. Gelman, J. B. Carlin, H. S. Stern, and D. B. Rubin. Bayesian Data Analysis. Chapman & Hall, 2004.
- J. V. Stone. Bayes' Rule: a Tutorial Introduction to Bayesian Analysis. Sebtel Press, 2014.
- H. Raiffa and R. Schlaifer. Applied Statistical Decision Theory. Division of Research, Graduate School of Business Administration, Harvard University, 1961.
- P. Goodwin and G. Wright. Decision Analysis for Management Judgment. John Wiley & Sons, 2014.

Although we relied on these sources, any mistakes of interpretation are our own.

Our hope is that Bayesian Statistics for Beginners is a "quick read" for the uninitiated and that, in one week or less, we could find a reader happily ensconced in a book written by one of the experts. Our goal in writing the book was to keep Bayes' Theorem front and center in each chapter for a beginning audience. As a result, Bayes' Theorem makes an appearance in every chapter. We frequently bring back past examples and explain what we did "back then," allowing the reader to slowly broaden their understanding and sort out what has been learned in order to relate it to new material. For the most part, our reviewers liked this approach. However, if this is annoying to you, you can skim over the repeated portions.

If this book is useful to you, it is due in no small part to a team of stellar reviewers. We owe a great deal of gratitude to George Allez, Cathleen Balantic, Barry Hall, Mevin Hooten, Peter Jones, Clint Moore, Ben Staton, Sheila Weaver, and Robin White. Their enthusiasm, questions, and comments have improved the narrative immensely. We offer a heartfelt thank you to Gary Bishop, Renee Westland, Melissa Murphy, Kevin Roark, John Bell, and Stuart Geman for providing pictures for this book.

Therese Donovan
Ruth Mickey

October 2018
Burlington, VT

Contents

SECTION 6 **Applications**

Appendices

SECTION 1

Basics of Probability

Overview

And so we begin. This first section deals with basic concepts in probability theory, and consists of two chapters.

- In Chapter 1, the concept of probability is introduced. Using an example, the chapter focuses on a single characteristic and introduces basic vocabulary associated with probability.
- Chapter 2 introduces additional terms and concepts used in the study of probability. The chapter focuses on two characteristics observed at the same time, and introduces the important concepts of joint probability, marginal probability, and conditional probability.

After covering these basics, your Bayesian journey will begin.

CHAPTER 1

Introduction to Probability

In this chapter, we'll introduce some basic terms used in the study of probability. By the end of this chapter, you will be able to define the following:

- Sample space
- Outcome
- Discrete outcome
- Event
- Probability
- Probability distribution
- Uniform distribution
- Trial
- Empirical distribution
- Law of Large Numbers

To begin, let's answer a few questions...

 ?? **What is probability?**

Answer: The best way to introduce probability is to discuss an example. Imagine you're a gambler, and you can win $1,000,000 if a single roll of a die turns up four. You get only one roll, and the entry fee to play this game is $10,000. If you win, you're a millionaire. If you lose, you're out ten grand.

 ?? **Should you play?**

Answer: It's up to you!

If the roll always comes up a four, you should play! If it never comes up four, you'd be foolish to play! Thus, it's helpful to know something about the die. Is it fair? That is, is each face equally likely to turn up? How likely are you to roll a four?

This type of question was considered by premier mathematicians Gerolamo Cardano (1501–1576), Pierre de Fermat (1601–1665), Blaise Pascal (1623–1662), and others. These brilliant minds created a branch of mathematics known as **probability theory**.

The rolling of a die is an example of a **random process**: the face that comes up is subject to chance. In probability, our goal is to **quantify** a random process, such as rolling a die. That is, we want to assign a number to it. If we roll a die, there are 6 possible **outcomes** (possible results), namely, one, two, three, four, five, or six. The set of all possible outcomes is called the sample space.

Bayesian Statistics for Beginners: A Step-by-Step Approach. Therese M. Donovan and Ruth M. Mickey, Oxford University Press (2019). © Ruth M. Mickey 2019.
DOI: 10.1093/oso/9780198841296.001.0001

Let's call the number of possible outcomes N, so $N = 6$. These outcomes are **discrete**, because each result can take on only one of these values. Formally, the word "discrete" is defined as "individually separate and distinct." In addition, in this example, there is a **finite** number of possible outcomes, which means "there are limits or bounds." In other words, the number of possible outcomes is not infinite.

If we believe that each and every outcome is just as likely to result as every other outcome (i.e., the die is fair), then the probability of rolling a four is $1/N$ or $1/6$. We can then say, "In 6 rolls of the die, we would expect 1 roll to result in a four," and write that as Pr(four) = 1/6.

> Here, the notation Pr means "Probability," and we will use this notation throughout this book.

?? How can we get a good estimate of Pr(four) for this particular die?

Answer: You collect some data.

Before you hand over the $10,000 entry, you ask the gamemaster if you could run an "experiment" and roll the die a few times before you make the decision to play. In this experiment, which consists of tossing a die many times, your goal is to get a rough estimate of the probability of rolling a four compared to what you expect. To your amazement, the gamemaster complies.

You start with one roll, which represents a single "trial" of an experiment. Suppose you roll a three and give the gamemaster a sneer. Table 1.1 shows you rolled 1 three, and 0 for the rest.

Table 1.1

Outcome	Frequency	Probability
One	0	0
Two	0	0
Three	1	1
Four	0	0
Five	0	0
Six	0	0
Sum	1	1

In this table, the column called **Frequency** gives the number of times each outcome was observed. The column called **Probability** is the frequency divided by the sum of the frequencies over all possible outcomes, a proportion. The probability of an event of four is the frequency of the observed number of occurrences of four (which is 0) divided by the total throws (which is 1). We can write this as:

$$Pr(four) = \frac{|\text{number of fours}|}{|\text{total trials}|} = \frac{0}{1} = 0. \tag{1.1}$$

This can be read as, "The probability of a four is the number of "four" events divided by the number of total trials." Here, the vertical bars indicate a number rather than an absolute value. This is a **frequentist** notion of probability, because we estimate Pr(four) by asking "How frequently did we observe the outcome that interests us out of the total?"

A probability distribution that is based on **raw data** is called an **empirical probability distribution**. Our empirical probability distribution so far looks like the one in Figure 1.1:

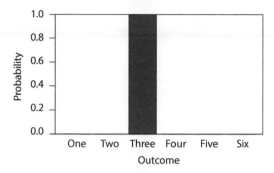

Figure 1.1 Empirical probability distribution for die outcomes, given 1 roll.

 ?? Is one roll good enough?

Answer: No.

By now, you should realize that one roll will never give us a good estimate of Pr(four). (Sometimes, though, that's all we have... we have only one Planet Earth, for example).

Next, you roll the die 9 more times, and summarize the results of the 10 total rolls (i.e., 10 trials or 10 experiments) in Table 1.2. The number of fours is 2, which allows us to estimate Pr(four) as 2/10, or 0.20 (see Figure 1.2).

Table 1.2

Outcome	Frequency	Probability
One	0	0.0
Two	2	0.2
Three	5	0.5
Four	2	0.2
Five	0	0.0
Six	1	0.1
Sum	10	1.0

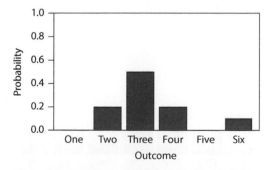

Figure 1.2 Empirical probability distribution for die outcomes, given 10 rolls.

What can you conclude from these results? The estimate of Pr(four) = 0.2 seems to indicate that the die may be in your favor! (Remember, you expect Pr(four) = 0.1667 if the die was fair). But $10,000 is a lot of money, and you decide that you should keep test rolling until the gamemaster shouts "Enough!" Amazingly, you are able to squeeze in 500 rolls, and you obtain the results shown in Table 1.3.

Table 1.3

Outcome	Frequency	Probability
One	88	0.176
Two	91	0.182
Three	94	0.188
Four	41	0.082
Five	99	0.198
Six	87	0.174
Sum	**500**	**1.000**

The plot of the frequency results in Figure 1.3 is called a **frequency histogram**. Notice that frequency, not probability, is on the y-axis. We see that a four was rolled 41 times. Notice also that the sum of the frequencies is 500. The frequency distribution is an example of an **empirical distribution**: It is constructed from raw data.

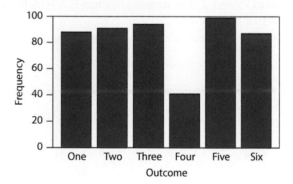

Figure 1.3 Frequency distribution of 500 rolls.

We can now estimate Pr(four) as 41/500 = 0.082. We can calculate the probability estimates for the other outcomes as well and then plot them as the probability distribution in Figure 1.4.

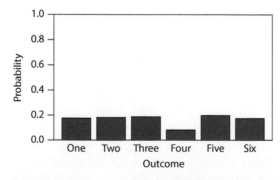

Figure 1.4 Empirical probability distribution for 500 rolls.

According to the Law of Large Numbers in probability theory, the formula:

$$\text{probability} = \frac{|\text{number of observed outcomes of interest}|}{|\text{total trials}|} \tag{1.2}$$

yields an estimate that is closer and closer to the true probability as the number of trials increases. In other words, your estimate of Pr(four) gets closer and closer to or approaches the true probability when you use more trials (rolls) in your calculations.

 ?? ## What would we expect if the die were fair?

Table 1.4 lists the six possible outcomes, and the probability of each event ($1/6 = 0.167$). Notice that the sum of the probabilities across the events is 1.0.

Table 1.4

Outcome	Probability
One	0.167
Two	0.167
Three	0.167
Four	0.167
Five	0.167
Six	0.167
Sum	1

Figure 1.5 shows exactly the same information as Table 1.4; both are examples of a probability distribution. On the horizontal axis, we list each of the possible outcomes. On the vertical axis is the probability. The height of each bar provides the probability of observing each outcome. Since each outcome has an equal chance of being rolled, the heights of the bars are all the same and show as 0.167, which is $1/N$. Note that this is not an empirical distribution, because we did not generate it from an experiment. Rather, it was based on the assumption that all outcomes are equally likely.

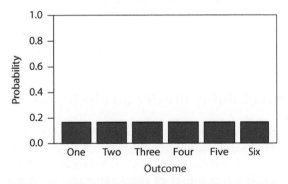

Figure 1.5 Probability distribution for rolling a fair die.

Again, this is an example of a **discrete uniform probability distribution**: discrete because there are a discrete number of separate and distinct outcomes; uniform because each and every event has the same probability.

?? **How would you change the table and probability distribution if the die were loaded in favor of a four?**

Answer: Your answer here!

There are many ways to do this. Suppose the probability of rolling a four is 0.4. Since all the probabilities have to add up to 1.0, this leaves 0.6 to distribute among the remaining five outcomes, or 0.12 for each (assuming these five are equally likely to turn up). If you roll this die, it's not a sure thing that you'll end up with a four, but getting an outcome of four is more likely than, say, getting a three (see Table 1.5).

Table 1.5

Outcome	Probability
One	0.12
Two	0.12
Three	0.12
Four	0.4
Five	0.12
Six	0.12
Sum	1

The probabilities listed in the table sum to 1.0 just as before, as do the heights of the corresponding blue bars in Figure 1.6.

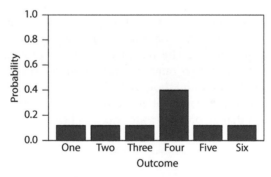

Figure 1.6 Probability distribution for rolling a loaded die.

 ?? **What would the probability distribution be for the bet?**

Answer: The bet is that if you roll a four, you win $1,000,000, and if you **don't** roll a four, you lose $10,000. It would be useful to group our 6 possible outcomes into one of two **events**. As the Oxford Dictionary of Statistics explains, "An event is a particular collection of outcomes, and is a subset of the sample space." Probabilities are assigned to events.

Our two events are E1 = {four} and E2 = {one, two, three, five, six}. The brackets { } indicate the set of outcomes that belong in each event. The first event contains one outcome, while the second event consists of five possible outcomes. Thus, in probability theory, outcomes can be grouped into new events at will. We started out by considering six outcomes, and now we have collapsed those into two events.

Now we assign a probability to each event. We know that Pr(four) = 0.4. What is the probability of NOT rolling a four? That is the probability of rolling one OR two OR three OR five OR six. Note that these events **cannot occur simultaneously**. Thus, we can write Pr($\sim$four) as the SUM of the probabilities of events one, two, three, five, and six (which is $0.12 + 0.12 + 0.12 + 0.12 + 0.12 = 0.6$). Incidentally, the $\sim$ sign means "complement of." If A is an event, $\sim$A is its complement (i.e., everything but A). This is sometimes written as A^c. The word **OR** is a tip that you **ADD** the individual probabilities together to get your answer as long as the events are mutually exclusive (i.e., cannot occur at the same time).

This is an example of a fundamental rule in probability theory: if two or more events are mutually exclusive, then the probability of any occurring is the sum of the probabilities of each occurring. Because the different outcomes of each roll (i.e., rolling a one, two, three, five, or six) are mutually exclusive, the probability of getting any outcome other than four is the sum of the probability of each one occurring (see Table 1.6).

Table 1.6

Event	Probability
Four	0.4
Not Four	0.6
Sum	1

Note that the probabilities of these two possible events sum to 1.0. Because of that, if we know that Pr(four) is 0.4, we can quickly compute the Pr($\sim$four) as $1 - 0.4 = 0.6$ and save a few mental calculations.

The probability distribution looks like the one shown in Figure 1.7:

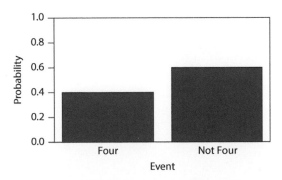

Figure 1.7 Probability distribution for rolling a four or not.

Remember: The SUM of the probabilities across all the different outcomes MUST EQUAL 1! In the discrete probability distributions above, this means that the heights of the bars summed across all discrete outcomes (i.e., all bars) totals 1.

Do you still want to play? At the end of the book, we will introduce decision trees, an analytical framework that employs Bayes' Theorem to aid in decision-making. But that chapter is a long way away.

 ?? **Do Bayesians think of probability as long-run averages?**

Answer: We're a few chapters away from hitting on this very important topic. You'll see that Bayesians think of probability in a way that allows the testing of theories and hypotheses. But you have to walk before you run. What you need now is to continue learning the basic vocabulary associated with probability theory.

 ?? **What's next?**

Answer: In Chapter 2, we'll expand our discussion of probability. See you there.

CHAPTER 2

Joint, Marginal, and Conditional Probability

Now that you've had a short introduction to probability, it's time to build on our probability vocabulary. By the end of this chapter, you will understand the following terms:

- Venn diagram
- Marginal probability
- Joint probability
- Independent events
- Dependent events
- Conditional probability

Let's start with a few questions.

 ?? ### What is an eyeball event?

A gala that celebrates the sense of vision? Nope. The eyeball event in this chapter refers to whether a person is right-eyed dominant or left-eyed dominant. You already know if you are left- or right-handed, but did you know that you are also left- or right-eyed? Here's how to tell (http://www.wikihow.com/Determine-Your-Dominant-Eye; see Figure 2.1):

Figure 2.1 Determining your dominant eye.

Bayesian Statistics for Beginners: A Step-by-Step Approach. Therese M. Donovan and Ruth M. Mickey, Oxford University Press (2019). © Ruth M. Mickey 2019.
DOI: 10.1093/oso/9780198841296.001.0001

1. Stretch your arms out in front of you and create a hole with your hands by joining your finger tips to make a triangular opening, as shown.
2. Find a small object nearby and align your hands with it so that you can see it in the triangular hole. Make sure you are looking straight at the object through your hands—cocking your head to either side, even slightly, can affect your results. Be sure to keep both eyes open!
3. Slowly move your hands toward your face to draw your viewing window toward you. As you do so, keep your head perfectly still, but keep the object lined up in the hole between your hands. Don't lose sight of it.
4. Draw your hands in until they touch your face—your hands should end up in front of your dominant eye. For example, if you find that your hands end up so you are looking through with your right eye, that eye is dominant.

The eyeball characteristic has two discrete outcomes: lefty (for left-eyed dominant people) or righty (for right-eyed dominant people). Because there are only two outcomes, we can call them **events** if we want.

Let us suppose that you ask 100 people if they are "lefties" or "righties." In this case, the 100 people represent our "universe" of interest, which we designate with the letter U. The total number of elements (individuals) in U is written $|U|$. (Once again, the vertical bars here simply indicate that U is a number; it doesn't mean the absolute value of U.)

Here, there are only two possible events: "lefty" and "righty." Together, they make up a set of possible outcomes. Let A be the event "left-eye dominant," and $\sim A$ be the event "right-eye dominant." Here, the tilde means "complement of," and here it can be interpreted as "everything but A." Notice that these two events are mutually exclusive: you cannot be both a "lefty" and a "righty." The events are also "exhaustive" because you must be either a lefty or righty.

Suppose that 70 of 100 people are lefties. These people are a subset of the larger population. The number of people in event A can be written $|A|$, and in this example $|A| = 70$. Note that $|A|$ must be less than or equal to $|A|$, which is 100. Remember that we use the vertical bars here to highlight that we are talking about a number.

Since there are only two possibilities for eye dominance type, this means that $100 - 70 = 30$ people are righties. The number of people in event $\sim A$ can be written $|\sim A|$, and in this example $|\sim A| = 30$. Note that $|\sim A|$ must be less than or equal to $|U|$.

Our universe can be summarized as shown in Table 2.1.

Table 2.1

Event	Frequency
Lefty (A)	$|A| = 70$
Righty ($\sim A$)	$|\sim A| = 30$
Universe (U)	$|U| = 100$

We can illustrate this example in a diagrammatic form, as shown in Figure 2.2. Here, our universe of 100 people is captured inside a box.

This is a Venn diagram, which shows A and $\sim A$. The universe U is 100 people and is represented by the entire box. We then allocate those 100 individuals into A and $\sim A$. The blue circle represents A; lefties stand inside this circle; righties stand outside the circle, but inside the box. You can see that A consists of 70 **elements**, and $\sim A$ consists of 30 elements.

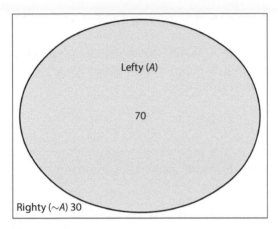

Lefty (A)

70

Righty ($\sim A$) 30

Figure 2.2

 ?? Why is it called a Venn diagram?

Answer: Venn diagrams are named for John Venn (see Figure 2.3), who wrote his seminal article in 1880.

Figure 2.3 John Venn.

According to the MacTutor History of Mathematics Archive, Venn's son described him as "of spare build, he was throughout his life a fine walker and mountain climber, a keen botanist, and an excellent talker and linguist."

 ?? **What is the probability that a person in universe U is in group A?**

Answer: We write this as $\Pr(A)$. Remember that **Pr** stands for **probability**, so $\Pr(A)$ means the probability that a person is in group A and therefore is a lefty. We can determine the probability that a person is in group A as:

$$\Pr(A) = \frac{|A|}{|U|} = \frac{70}{100} = 0.7. \tag{2.1}$$

Probability is determined as the number of persons in group A out of the total. The probability that the randomly selected person is in group A is 0.7.

 ?? **What about people who are not in group A?**

Answer: There are 30 of them ($100 - 70 = 30$), and they are righties.

$$\Pr(\sim A) = \frac{|\sim A|}{|U|} = \frac{30}{100} = 0.3. \tag{2.2}$$

With only two outcomes for the eyeball event, our notation focuses on the probability of being in a given group (A = lefties) and the probability of **not** being in the given group ($\sim A$ = righties).

 ?? **I'm sick of eyeballs. Can we consider another characteristic?**

Answer: Yes, of course. Let's probe these same 100 people and find out other details about their anatomy. Suppose we are curious about the presence or absence of Morton's toe. People with "Morton's toe" have a large second metatarsal, longer in fact than the first metatarsal (which is also known the big toe or hallux toe). Wikipedia articles suggest that this is a normal variation of foot shape in humans and that less than 20% of the human population have this condition. Now we are considering a second characteristic for our population, namely toe type.

Let's let B designate the event "Morton's toe." Let the number of people with Morton's toe be written as $|B|$. Let $\sim B$ designate the event "common toe." Suppose 15 of the 100 people have Morton's toe. This means $|B| = 15$, and $|\sim B| = 85$. The data are shown in Table 2.2, and the Venn diagram is shown in Figure 2.4.

Table 2.2

Event	Frequency
Morton's toe (B)	$|B| = 15$
Common toe ($\sim B$)	$|\sim B| = 85$
Universe (U)	$|U| = 100$

These events can be represented in a Venn diagram, where a box holds our universe of 100 people.

Note the size of this red circle is smaller than the previous example because the number of individuals with Morton's toe is much smaller.

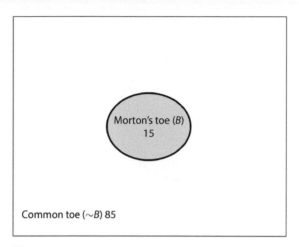

Morton's toe (*B*)
15

Common toe (∼*B*) 85

Figure 2.4

 ?? Can we look at both characteristics simultaneously?

Answer: You bet. With two characteristics, each with two outcomes, we have four possible combinations:

1. Lefty AND Morton's toe, which we write as $A \cap B$.
2. Lefty AND common toe, which we write as $A \cap \sim B$.
3. Righty AND Morton's toe, which we write as $\sim A \cap B$.
4. Righty AND common toe, which we write as $\sim A \cap \sim B$.

The upside-down ∩ is the mathematical symbol for **intersection**. Here, you can read it as "BOTH" or "AND."

The number of individuals in $A \cap B$ can be written $|A \cap B|$, where the bars indicate a number (not absolute value). Let's suppose we record the frequency of individuals in each of the four combinations (see Table 2.3).

Table 2.3

	Lefty (*A*)	Righty (∼*A*)	Sum
Morton's toe (*B*)	0	15	15
Common toe (∼*B*)	70	15	85
Sum	**70**	**30**	**100**

Let's study this table carefully.

Notice that this table has four "quadrants," so to speak. Our actual values are stored in the upper left quadrant, shaded dark blue. The upper right quadrant (shaded light blue) sums the number of people with and without Morton's toe. The lower left quadrant (shaded light blue) sums the number of people that are lefties and righties. The lower right (white) quadrant gives the grand total, or $|U|$.

Now let's plot the results in the same Venn diagram (see Figure 2.5).

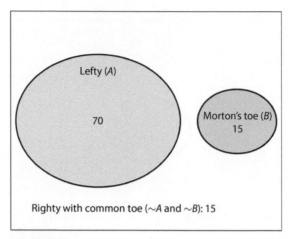

Figure 2.5

The updated Venn diagram shows the blue circle with 70 lefties, the red circle with 15 people with Morton's toe, and no overlap between the two. This means $|A \cap B| = 0$, $|A| = |A \cap \sim B| = 70$, and $|B| = |\sim A \cap B| = 15$. By subtraction, we know the number of individuals that are not in A OR B | $\sim A \cap \sim B$ | is 15 because we need to account for all 100 individuals somewhere in the diagram.

 ?? **Is it possible to have Morton's toe AND be a lefty?**

Answer: For our universe, no. If you are a person with Morton's toe, you are standing in the red circle and cannot also be standing in the blue circle. So these two events (Morton's toe and lefties) are mutually exclusive because they do not occur at the same time.

 ?? **Is it possible NOT to have Morton's toe if you are a lefty?**

Answer: You bet. All 70 lefties do not have Morton's toe. These two events are non-mutually exclusive.

 ?? **What if five lefties also have Morton's toe?**

Answer: In this case, we need to adjust the Venn diagram to show that five of the people that are lefties also have Morton's toe. These individuals are represented as the intersection between the two events (see Figure 2.6). Note that the total number of individuals is still 100; we need to account for everyone!

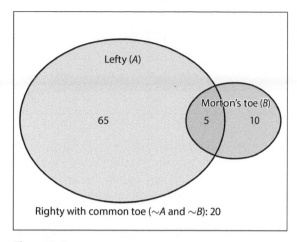

Figure 2.6

We'll run with this example for the rest of the chapter.

This Venn diagram is pretty accurate (except for the size of the box overall). There are 70 people in A, 15 people in B, and 5 people in $A \cap B$. The blue circle contains 70 elements in total, and 5 of those elements also occur in B. The red circle contains 15 elements (so is a lot smaller than the blue circle), and 5 of these are also in A. So 5/15 (33%) of the red circle overlaps with the blue circle, and 5/70 (7%) of the blue circle overlaps with the red circle.

 ?? Of the four events (A, $\sim A$, B, and $\sim B$), which are not mutually exclusive?

Answer:

- A and B are not mutually exclusive (a lefty can have Morton's toe).
- A and $\sim B$ are not mutually exclusive (a lefty can have a common toe).
- $\sim A$ and B are not mutually exclusive (a righty can have Morton's toe).
- $\sim A$ and $\sim B$ are not mutually exclusive (a righty can have a common toe).

In Venn diagrams, if two events overlap, they are not mutually exclusive.

 ?? Are any events mutually exclusive?

Answer: If we focus on each circle, A and $\sim A$ are mutually exclusive (a person cannot be a lefty and righty). B and $\sim B$ are mutually exclusive (a person cannot have Morton's toe and a common toe). Apologies for the trick question!

 ?? If you were one of the lucky 100 people included in the universe, where would you fall in this diagram?

Answer: Your answer here!

It's handy to look at the numbers in a table format too. Table 2.4 shows the same values as Figure 2.6.

Table 2.4

	Lefty (A)	Righty (~A)	Sum
Morton's toe (B)	5	10	15
Common toe (~B)	65	20	85
Sum	70	30	100

This is an important table to study. Once again, notice that there are four quadrants or sections in this table. In the upper left quadrant, the first two columns represent the two possible events for eyeball dominance: lefty and righty. The first two rows represent the two possible events for toe type: Morton's toe and common toe.

The upper left entry indicates that 5 people are members of both A and B.

- Look for the entry that indicates that 20 people are $\sim A \cap \sim B$, that is, *both* not A and not B.
- Look for the entry that indicates that 65 people are $A \cap \sim B$.
- Look for the entry that indicates that 10 people are $\sim A \cap B$.

The lower left and upper right quadrants of our table are called the **margins** of the table. They are shaded light blue. Note that the **total** number of individuals in A (regardless of B) is 70, and the total number of individuals in $\sim A$ is 30. The total number of individuals in B (regardless of A) is 15 and the total number of individuals $\sim B$ is 85. Any way you slice it, the grand total must equal 100 (the lower right quadrant).

 ?? **What does this have to do with probability?**

Answer: Well, if you are interested in determining the probability that an individual belongs to any of these four groups, you could use your universe of 100 individuals to do the calculation. Do you remember the **frequentist** way to calculate probability? We learned about that in Chapter 1.

$$\Pr = \frac{|\text{number of observed outcomes of interest}|}{|U|}. \tag{2.3}$$

Our total **universe** in this case is the 100 individuals. To get the probability that a person selected at random would belong to a particular event, we simply divide the entire table above by our total, which is 100, and we get the results shown in Table 2.5.

Table 2.5

	Lefty (A)	Righty (~A)	Sum
Morton's toe (B)	0.05	0.1	0.15
Common toe (~B)	0.65	0.2	0.85
Sum	0.7	0.3	1

We've just converted the raw numbers to probabilities by dividing the frequency table by the grand total.

Note that this differs from our die rolling exercise in Chapter 1, where you were unsure what the probability was and had to repeatedly roll a die to estimate it. By the Law of Large Numbers, the more trials you have, the more you zero in on the actual probability. In this case, however, we are given the number of people in each category, so the calculation is straightforward. These 100 people are the only people of interest. They represent our universe of interest; we are not using them to sample a larger group. If you didn't know the make-up of the universe, you could randomly select one person out of the universe over and over again to get the probabilities, where all persons are equally likely to be selected.

Let's walk through one calculation. Suppose we want to know the probability that an individual is a lefty AND has Morton's toe. The number of individuals in A and B is written $|A \cap B|$ and the probability that an individual is a lefty with Morton's toe is written:

$$\Pr(A \cap B) = \frac{|A \cap B|}{|U|} = \frac{5}{100} = 0.05. \tag{2.4}$$

This is officially called the **joint probability** and is the upper left entry in our table. The Oxford Dictionary of Statistics states that the "joint probability of a set of events is the probability that all occur simultaneously." Joint probabilities are also called **conjoint probabilities**. Incidentally, a table that lists joint probabilities such as the one above is sometimes referred to as a **conjoint table**.

When you hear the word **joint**, you should think of the word **AND** and realize that you are considering (and quantifying) more than one characteristic of the population. In this case, it indicates that someone is in A **AND** B. This is written as:

$$\Pr(A \cap B) \ldots \text{ or, equivalently,} \ldots \Pr(B \cap A). \tag{2.5}$$

 ?? What is the probability that a person selected at random is a righty and has Morton's toe?

Answer: This is equivalent to asking, what is the **joint probability** that a person is right-eye dominant AND has Morton's toe? See if you can find this entry in Table 2.5. The answer is 0.1.

In addition to the joint probabilities, the table also provides the **marginal** probabilities, which look at the probability of A or $\sim A$ (regardless of B) and the probability of B or $\sim B$ (regardless of A).

 ?? What does the word "marginal" mean?

Answer: The word **marginal** in the dictionary is defined as "pertaining to the margins; or situated on the border or edge." In our table, the marginal probabilities are just the probabilities for one characteristic of interest (e.g., A and $\sim A$) regardless of other characteristics that might be listed in the table.

Let's now label each cell in our conjoint table by its probability type (see Table 2.6).

Table 2.6

	Lefty (A)	Righty (~A)	Sum
Morton's toe (B)	Joint	Joint	Marginal
Common toe (~B)	Joint	Joint	Marginal
Sum	Marginal	Marginal	Total

Suppose you know only the following facts: the marginal probability of being a lefty is 0.7, the marginal probability of having Morton's toe is 0.15, and the joint probability of being a lefty with Morton's toe is 0.05. Also suppose that you haven't looked at Table 2.6!

 ?? Can you fill in the empty cells in Table 2.7?

Take out some scratch paper and a pencil. You can do it! Here are some hints:

- the lower right hand quadrant must equal 1.00;
- for any given characteristic, the sum of the two marginal probabilities must equal 1.00.

Table 2.7

	Lefty (A)	Righty (~A)	Sum
Morton's toe (B)	0.05	?	0.15
Common toe (~B)	?	?	?
Sum	0.7	?	?

Answer: Because the marginal probabilities for eyeballs must sum to 1.00, and the marginal probabilities for toes must sum to 1.00 (because they deal with mutually exclusive events), we can fill in the missing marginal probabilities.

The marginal probability of a lefty, $Pr(A)$, is 0.7, so the marginal of a righty, $Pr(\sim A)$, must be $1.00 - 0.7 = 0.3$.

The marginal probability of having Morton's toe, $Pr(B)$, is 0.15, so the marginal of $Pr(\sim B)$ must be $1.00 - 0.15 = 0.85$.

So far, so good. Once we know the marginals, we can calculate the joint probabilities in the upper left quadrant (see Table 2.8). For example:

- if the marginal $Pr(A) = 0.7$, then we know that $Pr(A \cap \sim B) = 0.7 - 0.05 = 0.65$;
- if the marginal $Pr(B) = 0.15$, then we know that $Pr(\sim A \cap B) = 0.15 - 0.05 = 0.1$;
- if the marginal $Pr(\sim B) = 0.3$, then we know that $Pr(\sim A \cap \sim B) = 0.3 - 0.1 = 0.2$.

Table 2.8

	Lefty (A)	Righty (~A)	Sum
Morton's toe (B)	0.05	0.1	0.15
Common toe (~B)	0.65	0.2	0.85
Sum	0.7	0.3	1

?? **Quickly: What is the marginal probability of having Morton's toe with this conjoint table?**

Answer: The marginal probability of having Morton's toe is written:

$$\Pr(B) = 0.15 \tag{2.6}$$

 ?? **Can you express the marginal probability of having Morton's toe as the sum of joint probabilities?**

Answer: Don't cheat now... try to express $\Pr(B)$ as the sum of joint probabilities before reading on! This step is essential for understanding Bayesian inference in future chapters!

How did you do?

Hint 1: We can decompose the total, 0.15, into its two pieces: $0.05 + 0.1$.

The probability that a lefty has Morton's toe can be written:

$$\Pr(A \cap B). \tag{2.7}$$

The probability that a righty has Morton's toe can be written:

$$\Pr(\sim A \cap B). \tag{2.8}$$

If we put these two terms together, we can express the marginal probability of having Morton's toe as:

$$\Pr(B) = \Pr(A \cap B) + \Pr(\sim A \cap B) \tag{2.9}$$

$$\Pr(B) = 0.05 + 0.1 = 0.15. \tag{2.10}$$

 ?? **Can we look at this problem from the Venn diagram perspective again?**

Of course! Here it is in Figure 2.7.

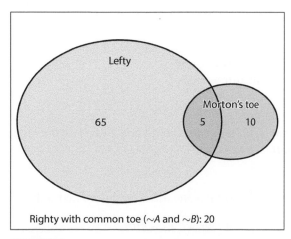

Figure 2.7

We are now poised to ask some very interesting questions.

?? **If you have Morton's toe, does that influence your probability of being a lefty?**

Answer: To answer this question, we must introduce the very important concept of **conditional probability**.

 ?? **What is conditional probability?**

Answer: Conditional probability is the probability of an event **given** that another event has occurred.

Conditional probability is written as:

- Pr($A|B$), which is read "the probability of A, given that B occurs"; in our context, Pr($A|B$) is Pr(lefty | Morton's toe);
- Pr($A|{\sim}B$), which is read "the probability of A, given that $\sim B$ occurs"; in our context, Pr($A|{\sim}B$) is Pr(lefty | common toe);
- Pr($B|{\sim}A$), which is read "the probability of B, given that $\sim A$ occurs"; in our context, Pr($B|{\sim}A$) is Pr(Morton's toe | righty);
- etc.

The vertical bar means "given."

 ?? **How exactly do you calculate the probability that a person is a lefty, given the person has Morton's toe?**

Answer: You use the following equation, which is a standard equation in probability theory:

$$\Pr(A|B) = \frac{\Pr(A \cap B)}{\Pr(B)}. \tag{2.11}$$

It's essential that you understand conditional probability, so let's look at this equation from a few different angles and, in the words of Kalid Azad, "let's build some intuition" about what it means.

Angle 1: The Venn diagram zoom
We already know that the numerator

$$\Pr(A \cap B) \tag{2.12}$$

is the intersection in the Venn diagram where A and B overlap (the probability of a lefty and Morton's toe). This can be written as

$$\Pr(B \cap A) \tag{2.13}$$

as well. The intersection of A and B is the intersection, no matter how you write it:

$$\Pr(A \cap B) = \Pr(B \cap A). \tag{2.14}$$

And we know that the denominator Pr(B) is the probability of Morton's toe.

In the Venn diagram, we can focus on the area of B and then look to see what fraction of the total B is occupied by A. In this example, we restrict our attention to the 15 people with Morton's toe, and note that 5 of them are lefties. Therefore, about 5/15 or 1/3 of the red circle is overlapped by the blue circle.

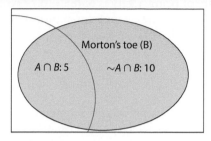

Figure 2.8

For the numbers given, we can see that $\Pr(A \mid B) = 5/15 = 1/3 = 0.333 = 33.3\%$. A general rule can help with the visualization: zoom to the denominator space B, then determine what fraction of this space is occupied by A. Similarly, $\Pr(\sim A \mid B) = 10/15 = 2/3 = 0.667 = 66.7\%$. Note that these probabilities sum to 1.

Angle 2: The table approach

We can also tackle this problem using the raw data (see Table 2.9).

Table 2.9

	Lefty (A)	Righty ($\sim A$)	Sum
Morton's toe (B)	5	10	15
Common toe ($\sim B$)	65	20	85
Sum	70	30	100

Here's the key equation again:

$$\Pr(A \mid B) = \frac{\Pr(A \cap B)}{\Pr(B)}. \tag{2.15}$$

In words, this equation says, what fraction of B consists of $A \cap B$? From our table, we calculated $\Pr(A \cap B)$ as:

$$\Pr(A \cap B) = \frac{|A \cap B|}{|U|} = \frac{5}{100}. \tag{2.16}$$

And we know that $\Pr(B)$ is:

$$\Pr(B) = \frac{|B|}{|U|} = \frac{15}{100}. \tag{2.17}$$

Now we can calculate the probability of A given B as:

$$\Pr(A \mid B) = \frac{\frac{|A \cap B|}{|U|}}{\frac{|B|}{|U|}} = \frac{|A \cap B|}{|B|} = \frac{5}{15} = 0.333. \tag{2.18}$$

?? So if you have Morton's toe, does that influence your probability of being a lefty?

Answer: If you have Morton's toe, the probability of being a lefty is 0.33. If you don't have Morton's toe, the probability of being a lefty is $65/85 = 0.77$ (you can confirm this too).

If Morton's toe does not matter, these conditional probabilities should be equal to the marginal probability, which is 0.7. This is clearly not the case here.

 ?? Does $\Pr(A \mid B) = \Pr(B \mid A)$?

Answer: In other words, is the probability of a lefty, given Morton's toe, the same thing as the probability of Morton's toe, given a lefty? Let's try it!

$$\Pr(A \mid B) = \frac{\frac{|A \cap B|}{|U|}}{\frac{|B|}{|U|}} = \frac{|A \cap B|}{|B|} = \frac{5}{15} = 0.333 \tag{2.19}$$

$$\Pr(B \mid A) = \frac{\frac{|A \cap B|}{|U|}}{\frac{|A|}{|U|}} = \frac{|A \cap B|}{|A|} = \frac{5}{70} = 0.072. \tag{2.20}$$

So the answer is **No!** These two probabilities are very different things. The first asks what is the probability of A given that event B happens (with a result of 0.333), while the second asks what is the probability of B given that A happens (with a result of 0.072).

 ?? Can you calculate the conditional probability of being a lefty, given you have Morton's toe, from our conjoint table instead of the raw numbers?

Answer: Yes...see if you can find it before looking at Table 2.10!

Table 2.10

	Lefty (A)	Righty ($\sim A$)	Sum
Morton's toe (B)	0.05	0.1	0.15
Common toe ($\sim B$)	0.65	0.2	0.85
Sum	**0.7**	**0.3**	**1**

Remember, when dealing with conditional probabilities, the key word is "zoom." Let's start with $\Pr(A \mid B)$:

$$\Pr(A \mid B) = \frac{\Pr(A \cap B)}{\Pr(B)}. \tag{2.21}$$

If B happens, we zoom to row 1 (Morton's toe), and then ask what fraction of the people with Morton's toe are lefties:

$$\Pr(A \mid B) = \frac{.05}{0.15} = 0.333. \tag{2.22}$$

?? **Can you calculate conditional probability of having Morton's toe, given you are a lefty, from our conjoint table?**

Answer: If A happens, we zoom to the first column (lefties) and then ask what fraction of the lefties have Morton's toe:

$$\Pr(B \mid A) = \frac{0.05}{0.7} = 0.072. \tag{2.23}$$

 ?? **If we know the conditional and marginal probabilities, can we calculate the joint probabilities?**

Yes! Don't forget this fundamental equation:

$$\Pr(A \mid B) = \frac{\Pr(A \cap B)}{\Pr(B)}. \tag{2.24}$$

You can rearrange this to your heart's content. For this book, the most important rearrangement is:

$$\Pr(A \cap B) = \Pr(A \mid B) * \Pr(B). \tag{2.25}$$

This formula can be used to calculate joint probability, $\Pr(A \cap B)$. Take some time to make sure this equation sinks in and makes full sense to you.

As an aside, if the occurrence of one event does not change the probability of the other occurring, the two events are said to be **independent**. This means that $\Pr(A \mid B) = \Pr(A \mid {\sim} B) = \Pr(A)$.

So, when A and B are independent:

$$\Pr(A \cap B) = \Pr(A) * \Pr(B). \tag{2.26}$$

 ?? **Are $\Pr(A \mid B)$ and $\Pr(B \mid A)$ related in some way?**

Answer: That, dear reader, is the subject of our next chapter, where we will derive Bayes' Theorem. See you there!

SECTION 2

Bayes' Theorem and Bayesian Inference

Overview

Welcome to Section 2! This section provides an introduction to Bayesian inference and provides three (hopefully) fun examples to get your feet wet.
This section consists of 5 chapters.

- In Chapter 3, Bayes' Theorem is introduced. The chapter shows its derivation and describes two ways to think about it. First, Bayes' Theorem describes the relationship between two inverse conditional probabilities, P(A|B) and P(B|A). Second, Bayes' Theorem can be used to express how a degree of belief for a given hypothesis can be updated in light of new evidence. This chapter focuses on the first interpretation.

- Chapter 4 introduces the concept of Bayesian inference. The chapter discusses the scientific method, and illustrates how Bayes' Theorem can be used for scientific inference. Bayesian Inference is the use of Bayes' Theorem to draw conclusions about a set of mutually exclusive and exhaustive alternative hypotheses by linking prior knowledge about each hypothesis with new data. The result is updated probabilities for each hypothesis of interest. The ideas of prior probabilities, likelihood, and posterior probabilities are introduced.

- Chapter 5, the "Author Problem," provides a concrete example of Bayesian inference. This chapter draws on work by Frederick Mosteller and David Wallace, who used Bayesian inference to assign authorship for unsigned Federalist Papers. The Federalist Papers were a collection of papers known to be written during the American Revolution. However, some papers were unsigned by the author, resulting in disputed authorship. The chapter provides a very basic Bayesian analysis of the unsigned "Paper 54," which was written by Alexander Hamilton or James Madison. The example illustrates the principles of Bayesian inference for two competing hypotheses.

- Chapter 6, the "Birthday Problem," is intended to highlight the decisions the analyst (you!) must make in setting the prior distribution. The "Birthday Problem" expands consideration from two hypotheses to multiple, discrete hypotheses. In this chapter, interest is in determining the posterior probability that a woman named Mary was born in a given month; there are 12 alternative hypotheses. Furthermore, consideration is given to assigning prior probabilities. The priors represent *a priori* probabilities that each alternative hypothesis is correct, where *a priori* means "prior to data collection," and can be "informative" or "non-informative." A Bayesian analysis cannot be conducted without using a prior distribution. The concept of likelihood is explored more deeply.

• Chapter 7, the "Portrait Problem," highlights the fact that multiple pieces of information can be used in a Bayesian analysis. A key concept in this chapter is that multiple sources of data can be combined in a Bayesian inference framework. The main take home point is that Bayesian analysis can be very, very flexible. A Bayesian analysis is possible as long as the likelihood of observing the data under each hypothesis can be computed.

By the end of this section, you will have a good understanding of how Bayes' Theorem is related to the scientific method.

CHAPTER 3

Bayes' Theorem

In this chapter, we're going to build on the content in Section 1 and derive Bayes' Theorem. This is what you've been waiting for!

By the end of this chapter, you will be able to derive Bayes' Theorem and explain the relationship between $\Pr(A \mid B)$ and $\Pr(B \mid A)$.

Let's begin with a few questions.

First, who is Bayes?

Answer: Thomas Bayes (1701–1761) was an English mathematician and Presbyterian minister, known for having formulated a specific case of the theorem that bears his name, Bayes' Theorem.

Is that really a picture of Thomas Bayes in Figure 3.1?

Answer: It could be, but nobody is really sure! We'll revisit this question in a future chapter.

Thomas Bayes?

REV. T. BAYES
Improver of the Columnar Method developed by Barrett.

Figure 3.1 "Thomas Bayes" (Photocopied from Terrence O'Donnell)

Bayesian Statistics for Beginners: A Step-by-Step Approach. Therese M. Donovan and Ruth M. Mickey, Oxford University Press (2019). © Ruth M. Mickey 2019.
DOI: 10.1093/oso/9780198841296.001.0001

?? Ok, what exactly is Bayes' Theorem?

Answer: There are two ways to think about Bayes' Theorem:

- It describes the relationship between $\Pr(A \mid B)$ and $\Pr(B \mid A)$.
- It expresses how a subjective degree of belief should rationally change to account for evidence.

The fact that there are two interpretations can be a source of confusion, but we hope you will fully appreciate the difference soon! In this chapter, we'll derive Bayes' Theorem and discuss the first interpretation of the theorem: the relationship between $\Pr(A \mid B)$ and $\Pr(B \mid A)$.

To help guide us, let's return to our familiar Venn diagram and conjoint table of eye dominance and toes. As before, our example will consider 100 total elements and four events. Remember that our data consisted of 70 lefties and 15 Morties (people with Morton's toe). This time, however, we will simply refer to the eye dominance events (left-eyed vs. right-eyed dominant) as A and $\sim A$, and the toe events (Morton's toe vs. common toe) as B and $\sim B$ (see Figure 3.2). In this way, we can generalize the problem so that it pertains to any two characteristics of interest.

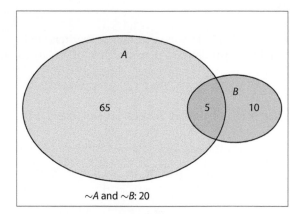

Figure 3.2

These same numbers can be expressed in tabular form (Table 3.1), or as a conjoint table (Table 3.2).

Table 3.1 Frequency table

	A	$\sim A$	Sum
B	5	10	15
$\sim B$	65	20	85
Sum	70	30	100

Table 3.2 Conjoint probability table

	A	$\sim A$	Sum
B	0.05	0.10	0.15
$\sim B$	0.65	0.20	0.85
Sum	0.70	0.30	1.00

In Chapter 2, we learned that the conditional probability

$$\Pr(A\,|\,B) = \frac{\Pr(A \cap B)}{\Pr(B)} \tag{3.1}$$

can be expressed as:

$$\Pr(A \cap B) = \Pr(A\,|\,B) * \Pr(B). \tag{3.2}$$

In words, **the joint probability of A and B is the product of the conditional probability of A, given B, and the marginal probability of B.** For the conditional probability equation (Equation 3.1), you visually 'zoom' to the denominator space, B, then ask what fraction is also occupied by A.

And we also learned that this conditional probability equation:

$$\Pr(B\,|\,A) = \frac{\Pr(B \cap A)}{\Pr(A)} \tag{3.3}$$

can be expressed as:

$$\Pr(B \cap A) = \Pr(B\,|\,A) * \Pr(A). \tag{3.4}$$

In words, **the joint probability of B and A is the product of the conditional probability of B, given A, and the marginal probability of A.** Equation 3.3 indicates that you "zoom" to the denominator space, A, and then ask what fraction is also occupied by B to get the conditional probability of B, given A.

We additionally learned that $\Pr(A\,|\,B)$ is not necessarily the same thing as $\Pr(B\,|\,A)$.

And yet, when you need to calculate the joint probability of B **and** A, we end up with the same result, no matter which approach we use.

Let's confirm this:

$$\Pr(A \cap B) = \Pr(A\,|\,B) * \Pr(B) = \frac{.05}{0.15} * 0.15 = 0.05 \tag{3.5}$$

$$\Pr(B \cap A) = \Pr(B\,|\,A) * \Pr(A) = \frac{.05}{0.7} * 0.7 = 0.05. \tag{3.6}$$

Equations 3.5 and 3.6 produce the same result. We should expect this, right? We are trying to estimate the probability of A and B occurring together, which is where the blue and red circles intersect. Although $\Pr(A\,|\,B)$ is 0.05/0.15 = 0.33, and $\Pr(B\,|\,A)$ is 0.05/0.7 = 0.071, when we multiply each by their respective marginals, we end up with the same joint probability estimate.

 ??   What does this have to do with Bayes' Theorem?

We are just a couple of steps from Bayes' Theorem. Are you ready?

We'll start with a simple reminder that the joint probability of A **and** B can be viewed in two ways:

$$\Pr(A \cap B) = \Pr(A\,|\,B) * \Pr(B) \tag{3.7}$$

$$\Pr(A \cap B) = \Pr(B\,|\,A) * \Pr(A). \tag{3.8}$$

Therefore:

$$\Pr(A \mid B) * \Pr(B) = \Pr(B \mid A) * \Pr(A). \qquad (3.9)$$

And dividing both sides by $\Pr(B)$ gives us Bayes' Theorem!

$$\Pr(A \mid B) = \frac{\Pr(B \mid A) * \Pr(A)}{\Pr(B)}. \qquad (3.10)$$

As we'll soon see, Bayes' Theorem can be expressed in other ways as well, but in this version it should be clear that Bayes' Theorem is a way of calculating conditional probability. Figure 3.3 provides a handy-dandy summary of the key concepts.

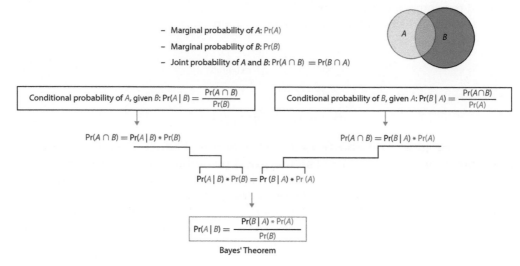

Figure 3.3

 ?? **What is so remarkable about this?**

Answer: Let's review the definitions:

- Bayes' Theorem describes the relationship between $\Pr(A \mid B)$ and $\Pr(B \mid A)$, the focus of this chapter.
- It expresses how a subjective degree of belief should rationally change to account for evidence. We will discuss this in depth in Chapter 4.

In the first interpretation, Bayes' Theorem is a fixed relationship between $\Pr(A)$, $\Pr(B)$, $\Pr(A \mid B)$, and $\Pr(B \mid A)$. Suppose we give you a problem to solve, we provide you with one type of conditional probability, $\Pr(B \mid A)$, and then we challenge you to find the reverse: $\Pr(A \mid B)$. With Bayes' Theorem, the calculations can be straightforward.

Let's look at our conjoint table, Table 3.3, which gives the joint and marginal probabilities.

Table 3.3

	A	$\sim A$	Sum
B	0.05	0.10	0.15
$\sim B$	0.65	0.20	0.85
Sum	0.70	0.30	1.00

?? If you have a member of *B*, what is the probability that he/she is also a member of *A*?

Answer: This is a problem where Bayes' Theorem could be used because you are asked to find a conditional probability: what is the probability of *A* **given** *B*. But we solved this easily in Chapter 2 by just using the joint probability table: zoom to row *B*, and ask what proportion of the total consists of *A*:

$$\Pr(A \,|\, B) = \frac{0.05}{0.15} = 0.333. \tag{3.11}$$

This example demonstrates that even if you are given a Bayes' type of problem, there are still ways to solve the problem without really using the theorem. But, just to be complete, let's use Bayes' Theorem to solve it:

$$\Pr(A \,|\, B) = \frac{\Pr(B \,|\, A) * \Pr(A)}{\Pr(B)} \tag{3.12}$$

$$\Pr(A \,|\, B) = \frac{\frac{0.05}{0.7} * 0.7}{0.15} = 0.333. \tag{3.13}$$

It works!

 ?? So, when would we need to use Bayes' Theorem?

Answer: You can use Bayes' Theorem when you are provided with one kind of conditional probability, like $\Pr(A \,|\, B)$, but are asked to find its inverse, $\Pr(B \,|\, A)$.

A notable example is posted at http://yudkowsky.net/rational/bayes. In that article, Yudkowsky poses the following challenge:

> One percent of women at age forty who participate in routine screening have breast cancer; 80% of women with breast cancer will have a positive mammogram (test), while 9.6% of women without breast cancer will also get a positive result. A woman in this age group had a positive mammogram in a routine screening. What is the probability that she actually has breast cancer?

We are given the probability that a woman with breast cancer will get a positive mammogram. But we are asked for the reverse of this: what is the probability that a woman with a positive mammogram has breast cancer?

Shall we give this a go? Let's let *A* represent women with breast cancer, and ~*A* represent women without it. And let's let *B* represent a positive test, and ~*B* represent a negative test. Bayes' Theorem would allow us to estimate the probability of cancer, given a positive test result, as:

$$\Pr(A \,|\, B) = \frac{\Pr(B \,|\, A) * \Pr(A)}{\Pr(B)}. \tag{3.14}$$

So we know $\Pr(A) = 0.01$, and we're given $\Pr(B \,|\, A) = 0.8$. To get our answer via Bayes' Theorem, all we need is to determine the denominator, $\Pr(B)$.

Let's set up a conjoint table to visualize the problem (see Table 3.4).

Table 3.4 Breast Cancer Problem

	A: Cancer	~A: No Cancer	Sum
B: Positive	?	?	?
~B: Negative	?	?	?
Sum	?	?	?

Our universe consists of women who participate in routine screenings for breast cancer. From the problem, we know that $\Pr(A) = 0.01$. This means that 1% of women have breast cancer. It also means that $\Pr(\sim A) = 0.99$, or 99% of women do not have breast cancer. These are the marginal probabilities for the cancer characteristic (see Table 3.5).

Table 3.5 Breast cancer problem

	A: Cancer	~A: No Cancer	Sum
B: Positive	?	?	?
~B: Negative	?	?	?
Sum	0.01	0.99	1.00

We can make some headway with this knowledge of $\Pr(A)$. We're also given the probability of a positive test result **given** cancer:

$$\Pr(B \mid A) = 0.8. \tag{3.15}$$

We can use that information to compute the joint probability of A *and* B (cancer and positive test). If the marginal probability of having cancer, $\Pr(A)$, is 0.01, then the probability of a positive test result and cancer is 0.008:

$$\Pr(B \cap A) = \Pr(B \mid A) * \Pr(A) = 0.8 * 0.01 = 0.008. \tag{3.16}$$

Now, we can calculate the probability of a negative test result and cancer through simple subtraction:

$$\Pr(\sim B \cap A) = 0.01 - 0.008 = 0.002. \tag{3.17}$$

Let's add these entries to our joint and marginal probability table as follows, filling in any cells we can (as shown in Table 3.6).

Table 3.6 Breast Cancer Problem

	A: Cancer	~A: No Cancer	Sum
B: Positive	0.008	?	?
~B: Negative	0.002	?	?
Sum	0.01	0.99	1.00

Notice that $\Pr(B \mid A)$, which is 0.8, is nowhere in this table. But $0.008/0.01 = 0.8$.

So far, so good. Now we just need either the joint probability of $\sim A \cap B$ or the joint probability of $\sim A \cap \sim B$, and we can our find the last missing piece to solve the problem—the marginal $\Pr(B)$.

In the problem statement, we're given the probability of a positive test result given no cancer:

$$\Pr(B \mid \sim A) = 0.096. \tag{3.18}$$

This allows us to compute the joint probability that a woman has a positive test AND does not have breast cancer:

$$\Pr(\sim A \cap B) = \Pr(B \mid \sim A) * \Pr(\sim A) = 0.096 * 0.99 = 0.095. \tag{3.19}$$

The marginal probability of testing positive is then:

$$\Pr(B) = \Pr(A \cap B) + \Pr(\sim A \cap B) = 0.008 + 0.095 = 0.103. \tag{3.20}$$

We can also compute the joint probability that a woman has a negative test AND does not have breast cancer as:

$$\Pr(\sim A \cap B) = 0.99 - 0.095 = 0.895. \tag{3.21}$$

This allows us to fill in the rest of our table (as shown in Table 3.7).

Table 3.7 Breast Cancer Problem

	A: Cancer	∼A: No Cancer	Sum
B: Positive	0.008	0.095	0.103
∼B: Negative	0.002	0.895	0.897
Sum	0.010	0.990	1.000

It may be helpful to think in terms of counts instead of probabilities. Suppose there are 1000 women. We would expect that these women would be partitioned into the four events, as shown in Table 3.8.

Table 3.8 Breast Cancer Problem

	A: Cancer	∼A: No Cancer	Sum
B: Positive	8	95	103
∼B: Negative	2	895	897
Sum	10	990	1000

Our Venn diagram would look roughly like the one in Figure 3.4, with the following results (noting that the size of the box would be much, much larger than shown):

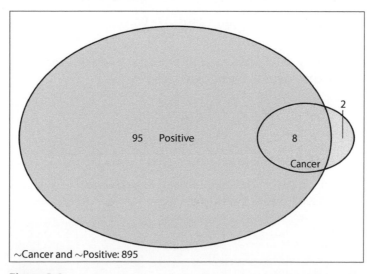

Figure 3.4

- 10 of 1000 women have breast cancer (blue circle);
- 990 of 1000 women do not have breast cancer (and are any portion of the diagram that is not blue);
- 103 of 1000 women will have a positive mammogram (red circle);
- 897 of 1000 women will have a negative mammogram (and are standing in any portion of the diagram that is not red);
- 8 of the women with cancer will have a positive mammogram; 2 will have a negative mammogram;
- 95 of the women without cancer will have a positive mammogram.

With the joint and marginal table filled in, we can now answer Yudkowsky's challenge: a woman in this age group had a positive mammogram in a routine screening. What is the probability that she actually has breast cancer?

$$\Pr(A \mid B) = \frac{\Pr(B \mid A) * \Pr(A)}{\Pr(B)} = \frac{\frac{0.008}{0.01} * 0.01}{0.103} = 0.0776. \tag{3.22}$$

She has a 7.76% chance of having cancer, given her positive test result. This probability may be surprisingly small to you. You might have noticed that you didn't need to fill out the full conjoint table to get the answer. But this little exercise was a quick review of Chapter 2.

Do you see how the Theorem allows you to switch the information around? We were provided with information about the probability of a test result, given a cancer condition. We used Bayes' Theorem to determine the probability of cancer given the test result.

This use of Bayes' Theorem relates inverse representations of the probabilities concerning two events: $\Pr(A \mid B)$ and $\Pr(B \mid A)$. Bayes noted that there is an intricate relationship between $\Pr(A \mid B)$ and $\Pr(B \mid A)$. His theorem is useful because sometimes it is far easier to estimate one of these conditional probabilities than the other.

 ?? ## Is that all there is to it?

Answer: In terms of what Bayes' Theorem is and how you can use it to calculate conditional probability, yes.

You'll often see Bayes' Theorem written in the following form:

$$\Pr(A \mid B) = \frac{\Pr(B \mid A) * \Pr(A)}{\Pr(B)}. \tag{3.23}$$

There are other, equally valid ways to express Bayes' Theorem. For example, the two equations below are equivalent to the one given above:

$$\Pr(A \mid B) = \frac{\Pr(B \mid A) * \Pr(A)}{\Pr(A \cap B) + \Pr(\sim A \cap B)} \tag{3.24}$$

$$\Pr(A \mid B) = \frac{\Pr(B \mid A) * \Pr(A)}{\Pr(B \mid A) * \Pr(A) + \Pr(B \mid \sim A) * \Pr(\sim A)}. \tag{3.25}$$

The second definition of Bayes' Theorem focuses on **inference**—a topic we'll explore in Chapter 4, which relies on this expanded version.

CHAPTER 4

Bayesian Inference

Now that you've been introduced to Bayes' Theorem, we'll focus our attention on **Bayesian inference**. By the end of this chapter, you will understand the following concepts:

- Bayesian inference
- Induction
- Deduction
- Hypothesis
- Alternative hypotheses
- Prior probability of a hypothesis
- Likelihood of the observed data
- Posterior probability of a hypothesis, given the data

Before we get started, let's quickly review Bayes' Theorem, using a Venn diagram as a visual aid (see Figure 4.1).

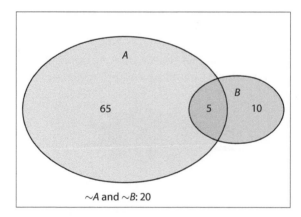

Figure 4.1

Here, A and B represent two events. Therefore, this diagram consists of four possible joint events:

- $A \cap B$
- $A \cap \sim B$
- $\sim A \cap B$
- $\sim A \cap \sim B$

In Section 1, we learned that the joint probability of A and B occurring can be expressed as:

$$\Pr(A \cap B) = \Pr(A \mid B) * \Pr(B) \tag{4.1}$$

Bayesian Statistics for Beginners: A Step-by-Step Approach. Therese M. Donovan and Ruth M. Mickey, Oxford University Press (2019). © Ruth M. Mickey 2019.
DOI: 10.1093/oso/9780198841296.001.0001

In words **the joint probability of _A_ and _B_ is the product of the conditional probability of _A_, given _B_, and the marginal probability of _B_**.

Similarly, the joint probability of _A_ and _B_ occurring can be expressed as:

$$\Pr(B \cap A) = \Pr(B \mid A) * \Pr(A). \tag{4.2}$$

Therefore:

$$\Pr(A \mid B) * \Pr(B) = \Pr(B \mid A) * \Pr(A). \tag{4.3}$$

And dividing both sides by $\Pr(B)$ gives us Bayes' Theorem:

$$\Pr(A \mid B) = \frac{\Pr(B \mid A) * \Pr(A)}{\Pr(B)}. \tag{4.4}$$

In Chapter 3, we provided the two definitions of Bayes' Theorem and discussed definition 1:

- The Theorem relates inverse representations of the probabilities concerning two events, that is, $\Pr(A|B)$ and $\Pr(B|A)$.

We walked through a few different examples to show how to use Bayes' Theorem to determine, for example, the probability that a woman has breast cancer if her mammogram came back positive.

In this chapter, we turn our attention to the second definition of Bayes' Theorem:

- The Theorem can express how a subjective degree of belief should rationally change to account for evidence.

Here's the theorem we'll be using for definition 2:

$$\Pr(A \mid B) = \frac{\Pr(B \mid A) * \Pr(A)}{\Pr(B)}. \tag{4.5}$$

Yes, the equation is the same, but in this second interpretation, Bayes' Theorem opens up entirely new possibilities for thinking about probabilities and the conduct of

Science

 ?? What exactly is science?

Answer: There are many definitions, formal and informal. Generally speaking, science refers to a system of acquiring knowledge.

Answer: (From NASA): Science is curiosity in thoughtful action about the world and how it behaves.

Answer: (From Wikipedia): Science (from Latin _scientia_, meaning "knowledge") is a systematic enterprise that builds and organizes knowledge in the form of testable explanations and predictions about the universe.

 ?? How do we go about actually conducting science?

Answer: We normally use what is called the scientific method. The Oxford English Dictionary (Stevenson, 2010) says that the scientific method is "a method or procedure that has

characterized natural science since the 17th century, consisting in systematic observation, measurement, and experiment, and the formulation, testing, and modification of hypotheses."

A key concept in scientific endeavors is formulating testable, alternative explanations about how the universe works. The scientific method actually consists of two types of inquiry: **induction** and **deduction**, which, when used in concert, produce knowledge.

The scientific process is nicely captured in the diagram in Figure 4.2 (adapted from Rao, 1997). Let's walk through this diagram, noting that it is a (diamond-shaped) circle at heart and has neither beginning nor end. It can be thought of as a race track, or a wheel.

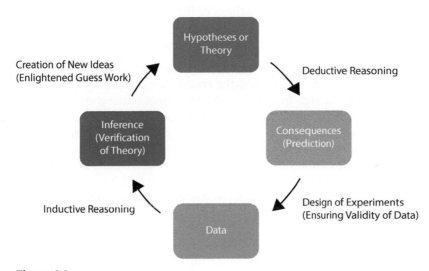

Figure 4.2

We need to start somewhere in this diagram, which contains four boxes and four arrows. Let's start with the upper box:

1. **Hypothesis or Theory Box**. A hypothesis is a proposed explanation for a phenomenon. A scientific theory is a coherent group of propositions formulated to explain a group of facts or phenomena in the natural world and repeatedly confirmed through experiment or observation. We will also note that process-based models, such as models of global climate circulation, are hypotheses at heart.
   - A theory: Darwin's Theory of Evolution
   - A hypothesis: The earth is warming due to increased CO_2 levels in the atmosphere.
2. **Deductive Reasoning Arrow**. The Oxford Reference tells us that deductive reasoning is reasoning from the general to the particular. Here, you start with a hypothesis or theory and test it from an examination of facts.
3. **Consequences or Predictions Box**. Dr. Sylvia Wassertheil-Smoller, a research professor at Albert Einstein College of Medicine, explains, "In deductive inference, we hold a theory and based on it we make a prediction of its consequences. That is, we predict what the observations should be if the theory were correct" (source: http://www.livescience.com/21569-deduction-vs-induction.html; accessed August 17, 2017). A prediction is the result of deduction.
4. **Design of Experiments Arrow**. In this step, you plan an experiment that will allow you to collect data to test your hypothesis or hypotheses. A well-designed experiment will ensure that the data you collect are valid.

5. **Data Box**. After the experiment is designed, we then collect some data. We can also use existing datasets if they are appropriate.

6. **Inductive Reasoning Arrow**. The Oxford Reference tells us that inductive reasoning involves inferring general principles from specific examples.

7. **Inference Box**. Inductive reasoning makes broad generalizations from specific observations. Dr. Sylvia Wassertheil-Smoller explains, "In inductive inference, we go from the specific to the general. We make many observations, discern a pattern, make a generalization, and infer an explanation or a theory" (source: http://www.livescience.com/ 21569-deduction-vs-induction.html; accessed August 17, 2017). At this point, we may verify the pattern, or falsify our hypothesis or theory.

8. **Creativity Arrow**. The final process involves creativity, in which we bring to bear creative ideas that may explain a pattern or a phenomenon, which brings us full circle.

C. R. Rao (1997) notes that the inference-through-consequences portion of the diagram comes under the subject of research and the creative role played by scientists, while the design of experiments through inductive reasoning comes under the realm of statistics.

Note that you can go through a portion of the scientific race track or wheel (e.g., collect data in the lower box, analyze the data to infer a generalized pattern, and then stop). However, going around the race track once or multiple times builds **knowledge** about a system. As Dr. Wassertheil-Smoller states, "In science there is a constant interplay between inductive inference (based on observations) and deductive inference (based on theory), until we get closer and closer to the 'truth,' which we can only approach but not ascertain with complete certainty."

> "Our search is only for working hypotheses which are supported by observational facts and which, in course of time, may be replaced by better working hypotheses with more supporting evidence from a wider set of data and provide wider applicability." – C.R. Rao.

There are reams of writings about the scientific process—much of it is (thankfully) beyond the scope of this book. The nice thing about Bayes' Theorem is that it can be used to address many kinds of scientific questions, but the underlying concept of Bayesian inference is the notion of **alternative hypotheses**, which is why we started in the upper box.

 ??   How on earth did Thomas Bayes make a connection between probability and scientific inference?

Answer: You'll have to read more about the story of Bayes' Theorem to learn about how he came to his great insight. It involves a thought experiment with balls rolling on a table.

Due credit should also be given to Pierre-Simon Laplace. A Wikipedia author notes "Pierre-Simon Laplace was an influential French scholar whose work was important to the development of mathematics, statistics, physics and astronomy. He summarized and extended the work of his predecessors in his five-volume Mécanique Céleste (Celestial Mechanics) (1799–1825). This work translated the geometric study of classical mechanics to one based on calculus, opening up a broader range of problems. In statistics, the Bayesian interpretation of probability was developed mainly by Laplace" (article accessed August 15, 2017).

If you'd like to learn more, a terrific read is The Theory That Would Not Die: How Bayes' Rule Cracked the Enigma Code, Hunted Down Russian Submarines, and Emerged Triumphant from Two Centuries of Controversy, by Sharon Bertsch McGrayne (2011). If you can't buy the book, at the very least, watch a lecture.

 ?? What is Bayesian inference?

Answer: Bayesian inference is the process of confronting alternative hypotheses with new data and using Bayes' Theorem to update your beliefs in each hypothesis.

The Oxford Dictionary of Statistics (Upton and Cook 2014) describes Bayesian inference as "an approach concerned with the consequences of modifying our previous beliefs as a result of receiving new data."

Wikipedia defines it this way: "Bayesian inference is a method of statistical inference in which Bayes' Theorem is used to update the probability for a hypothesis as more evidence or information becomes available" (article accessed August 15, 2017).

 ?? How does Bayesian inference work?

Answer: That is the million dollar question. Here we go!

Here's Bayes' Theorem as we derived it at the beginning of this chapter:

$$\Pr(A\,|\,B) = \frac{\Pr(B\,|\,A) * \Pr(A)}{\Pr(B)}. \tag{4.6}$$

To use Bayes' Theorem for scientific inference, it's useful (actually, essential) to replace the marginal denominator $\Pr(B)$ as the sum of the joint probabilities that make it up:

$$\Pr(A\,|\,B) = \frac{\Pr(B\,|\,A) * \Pr(A)}{\Pr(A \cap B) + \Pr(\sim A \cap B)}. \tag{4.7}$$

It's critical that you understand this: **the marginal probability of B is the sum of the joint probabilities that make it up**. Perhaps the conjoint table will help jog your memory (see Table 4.1).

Table 4.1 Conjoint table with joint and marginal probabilities.

	A	$\sim A$	Marginal
B	$\Pr(A \cap B)$	$\Pr(\sim A \cap B)$	$\Pr(B)$
$\sim B$	$\Pr(A \cap \sim B)$	$\Pr(\sim A \cap \sim B)$	$\Pr(\sim B)$
Marginal	$\Pr(A)$	$\Pr(\sim A)$	Total $= 1.00$

Remember that the marginal probabilities are stored, well, in the margins, and that you obtain them by adding the joint probabilities. So far, so good.

Now, let's tackle a problem. Returning to the Yudkowsky example in Chapter 3, suppose we were asked to find the probability that a woman has breast cancer (A), given that her mammogram test came back positive (B). Here, the results of the mammogram are the "data."

OK, we are asked to find $\Pr(A|B)$. Thus, $\Pr(B)$ is in the denominator in Bayes' Theorem, which is:

$$\Pr(A|B) = \frac{\Pr(B|A) * \Pr(A)}{\Pr(A \cap B) + \Pr(\sim A \cap B)}. \tag{4.8}$$

Let's now identify the parts of this problem in terms of the scientific method:

- We have two competing **hypotheses** regarding cancer: the woman has cancer (A) vs. she does not ($\sim A$).
- We have **data** for this problem: the test came back positive. So, B represents our observed data. Notice that $\sim B$ does not appear in the problem because we did not observe a negative test.

 ?? How can we turn this into a Bayesian inference problem?

Answer: We can make great headway if we now replace the **joint probabilities** with their conditional probability equivalents:

$$\Pr(A \cap B) = \Pr(B|A) * \Pr(A) \tag{4.9}$$

$$\Pr(\sim A \cap B) = \Pr(B| \sim A) * \Pr(\sim A). \tag{4.10}$$

Then, Bayes' Theorem looks like this:

$$\Pr(A|B) = \frac{\Pr(B|A) * \Pr(A)}{\Pr(B|A) * \Pr(A) + \Pr(B| \sim A) * \Pr(\sim A)}. \tag{4.11}$$

The denominator is still just the sum of the two joint probabilities that make up the marginal probability of B. **Never, ever, lose sight of this!** Let's look at this form of Bayes' Theorem more deeply.

 ?? Is there a pattern in the denominator of this new version?

Answer: The denominator, which is the marginal probability of B, is the probability of observing our data, B, given A multiplied by the probability of A, plus the probability of observing our data, B, given $\sim A$ multiplied by the probability of $\sim A$. This has a certain ring to it, don't you agree?

 ?? Does anything else about this equation strike you as notable?

Answer: You might have noticed that one term in the denominator is exactly the same thing as the numerator!

$$\Pr(A|B) = \frac{\Pr(B|A) * \Pr(A)}{\Pr(B|A) * \Pr(A) + \Pr(B| \sim A) * \Pr(\sim A)}. \tag{4.12}$$

Now, it should be clear that Bayes' Theorem returns a proportion, or probability. After all, we need to find $\Pr(A|B)$, which is a probability that ranges between 0 and 1.

?? **So, why all the fuss?**

Answer: Well, Bayes' great insight was that when we use the Theorem in this way, the equation itself can be used to draw inferences regarding competing hypotheses and thus is directly tied to the scientific method. Cool!

Let's revisit our scientific method diagram (see Figure 4.3).

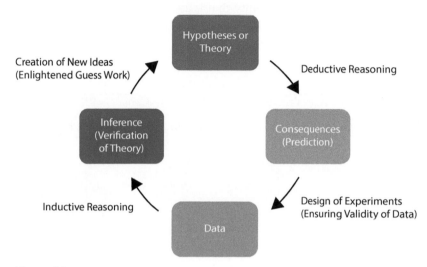

Figure 4.3

Again, the scientific "race track" has no beginning or end. But Bayesian inference begins with the notion of multiple hypotheses, so we'll start with the upper box. Our problem is to determine the probability that a woman has breast cancer, given the test result was positive. Let's focus on the four main boxes only:

1. **Hypothesis or Theory box**. Identify your hypotheses. Here, we are interested in testing two alternative hypotheses: the woman has cancer (*A*) versus she does not (~*A*). These hypotheses are mutually exclusive and exhaustive. We now assign the probability that each hypothesis is true (prior to obtaining mammogram results). The probabilities associated with each hypothesis must sum to 1.0. In Bayesian circles, these are called **prior probabilities** because they represent our current belief in each hypothesis *prior* to data collection.

2. **Consequences box**. We know that the data we will be collecting are medical tests (mammograms). So, now we need to write out equations for calculating the probability of observing the test data under each hypothesis. This probability is called **likelihood**, and figuring out how to calculate the likelihood of the data under each hypothesis is often the most challenging part of Bayesian inference.

3. **Data box**. Next, we collect data. The test came back positive.

4. **Inference Box**. With data in hand, we can now plug our data into the likelihood equations:
   • likelihood of observing the data (a positive test result) under the cancer hypothesis
   • likelihood of observing the data (a positive test result) under the no-cancer hypothesis.

Finally, we use Bayes' Theorem to determine a **posterior probability** for each hypothesis. The posterior probability represents our updated belief in each hypothesis after new data are collected:

- Probability of cancer, given the observed data
- Probability of no cancer, given the observed data.

Thus, in Bayesian inference, the posterior probability of a random event or an uncertain proposition is the conditional probability that is assigned **after** the relevant evidence or background is taken into account (Wikipedia; accessed August 15, 2017). Here's what Bayes' Theorem looks like for this problem:

$$\text{Pr(Cancer} \mid \text{Positive)}$$
$$= \frac{\text{Pr(Positive} \mid \text{Cancer)} * \text{Pr(Cancer)}}{\text{Pr(Positive} \mid \text{Cancer)} * \text{Pr(Cancer)} + \text{Pr(Positive} \mid \sim\text{Cancer)} * \text{Pr}(\sim\text{Cancer)}}. \tag{4.13}$$

Now, let's replace each term with its Bayesian inference definition (see Figure 4.4).

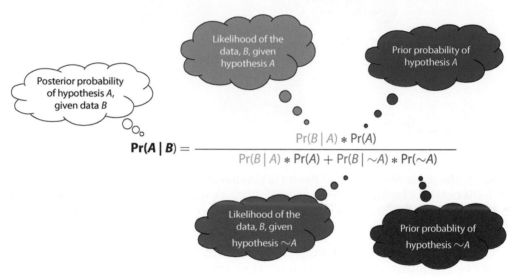

Figure 4.4

Look for the following terms in this diagram:

- Prior probability of each hypothesis, A and $\sim A$
- Likelihood of observing the data, B, under each hypothesis
- Posterior probability of hypothesis A, given B

Since there are only two hypotheses in this example, A and $\sim A$, if we use Bayes' Theorem to estimate the posterior probability of A, which is written Pr(A|data), then by subtraction we also estimate Pr($\sim A$|data) as 1 − Pr(A | data).

However, we could also solve it directly, as shown below:

$$\text{Pr}(\sim\text{Cancer} \mid \text{Positive)}$$
$$= \frac{\text{Pr(Positive} \mid \sim\text{Cancer)} * \text{Pr}(\sim\text{Cancer)}}{\text{Pr(Positive} \mid \text{Cancer)} * \text{Pr(Cancer)} + \text{Pr(Positive} \mid \sim\text{Cancer)} * \text{Pr}(\sim\text{Cancer)}}. \tag{4.14}$$

?? ## How does this relate to science?

Answer: Your answer here!

Science refers to a system of acquiring knowledge. Bayes' Theorem allows us to posit specific hypotheses for some phenomena and express our current belief that each hypothesis is true. Then, as new data become available, we update our belief in each hypothesis.

> Paul Samuelson, the Nobel laureate from the Massachusetts Institute of Technology, recalled that John Maynard Keynes once was challenged for altering his position on some economic issue. "When my information changes, I change my mind. What do you do?"

 ?? ## Ok, what exactly is the difference between the two interpretations of Bayes' Theorem?

Answer: Remember the alternative definitions:

- Bayes' Theorem describes the relationship between $\Pr(A|B)$ and $\Pr(B|A)$.
- The Theorem expresses how a subjective degree of belief should rationally change to account for evidence.

Let's return to our example problem:

> One percent of women at age forty who participate in routine screening have breast cancer. Eighty percent of women with breast cancer will get positive mammograms. In addition, 9.6% of women without breast cancer will get positive mammograms. A woman in this age group had a positive mammogram in a routine screening. What is the probability that she actually has breast cancer?

Here is our conjoint table once more:

Table 4.2 Breast Cancer Problem.

	A: Cancer	~A: No Cancer	Sum
B: Positive	0.008	0.095	0.103
~B: Negative	0.002	0.895	0.897
Sum	0.010	0.990	1.000

In Chapter 3, we solved this problem using this version of Bayes' Theorem:

$$\Pr(A|B) = \frac{\Pr(B|A) * \Pr(A)}{\Pr(B)} = \frac{\frac{0.008}{0.01} * 0.01}{0.103} = 0.0776. \tag{4.15}$$

In this chapter, we are solving the problem with the second interpretation of Bayes' Theorem:

$$\Pr(A|B) = \frac{\Pr(B|A) * \Pr(A)}{\Pr(B|A) * \Pr(A) + \Pr(B|\sim A) * \Pr(\sim A)}. \tag{4.16}$$

$$\Pr(A \mid B) = \frac{\frac{0.008}{0.01} * 0.01}{\frac{0.008}{0.01} * 0.01 + \frac{0.095}{0.99} * 0.99} = 0.0776. \tag{4.17}$$

Both return the same answer, and both are correct. The second interpretation, however, places the problem within a scientific context, where you posit hypotheses and then update your beliefs in each hypothesis after data are collected. In this particular example, our hypotheses are that the woman has cancer and that she does not. Without a test, our initial belief that a woman has breast cancer is 1% (because 1% of the population has breast cancer), and our belief that she does not have breast cancer is 99%. With the test, however, we can use Bayes' Theorem to update our beliefs. In light of the new data, the probability that a woman has breast cancer is 7.8%.

 ??   What if there are more than two hypotheses?

Now let's generalize this problem even more. Suppose there are n hypotheses. Let's number our hypotheses from $i = 1$ to n. Here, i is an **index of hypotheses**. In our example, $n = 2$ since there are two hypotheses. We could let hypothesis $i = 1$ be the "cancer hypothesis", and $i = 2$ be the "no cancer hypothesis." But we can generalize this as H_1, H_2,..., H_n hypotheses. In other words, there can be any **discrete** number of hypotheses.

Let's let **data** represent the data we collected, that is, our **observed data**. So now Bayes' Theorem can be written:

$$\Pr(H_i \mid \text{data}) = \frac{\Pr(\text{data} \mid H_i) * \Pr(H_i)}{\sum_{j=1}^{n} \Pr(\text{data} \mid H_j) * \Pr(H_j)}. \tag{4.18}$$

- The left side of the equation, $\Pr(H_i|\text{data})$, can be read: "The posterior probability of hypothesis i, given the data."
- The numerator requires the likelihood of observing the data under hypothesis i and is written $\Pr(\text{data}|H_i)$. This is then multiplied by the prior probability for hypothesis i, which is written $\Pr(H_i)$.
- In the denominator, the symbol Σ means "sum." The "j" associated with the sum symbol is an **index of summation**. Thus, in the denominator, we are summing over **all** hypotheses. So the index j goes from 1 to n. In our example, we had two hypotheses, so there must be two terms in the denominator.

We use Bayes' Theorem to compute the posterior probabilities for each and every hypothesis under consideration.

> Broadly speaking, you start with a scientific question and set forth two or more alternative hypotheses. You then assign a prior probability that each alternative hypothesis is true. Next, you collect data. Finally, you use Bayes' Theorem to update the probability for each hypothesis considered.

?? One more time...what is Bayesian inference again?

The Merriam–Webster dictionary defines "inferred" or "inferring" as "deriving conclusions from facts or premises," and "infer" as "to form an opinion or reach a conclusion through reasoning and information." The Merriam–Webster Thesaurus suggests the following synonyms for the word "infer": conclude, decide, deduce, derive, extrapolate, gather, judge, make out, reason, and understand.

Bayesian inference, then, is the use of Bayes' Theorem to draw conclusions about a set of mutually exclusive, exhaustive, alternative hypotheses by linking prior knowledge about each hypothesis with new data. The result is updated probabilities for each hypothesis of interest:

Initial Belief in Hypothesis i + New Data $\rightarrow$ Updated Belief in Hypothesis i.

 ?? What if I collect more data?

Each time you collect more data and apply Bayes' Theorem, you're updating your belief in each alternative hypothesis. The new posterior probabilities then become priors for the next analysis. By tracking our beliefs through time, we track our **learning**. The phrase "Today's posterior is tomorrow's prior" captures the process nicely, as long as the same data are not used to update the prior again the next time around.

 ?? What other sort of questions have been tackled using Bayesian inference approaches?

Answer: Loads of them! And the use of Bayes' Theorem is skyrocketing in many fields. In Chapter 5, we'll work on another problem to give you more practice.

CHAPTER 5

The Author Problem: Bayesian Inference with Two Hypotheses

In this chapter, we provide a concrete example of **Bayesian inference**. By the end of this chapter, you should have a renewed understanding of:

- Bayesian inference
- Hypothesis
- Alternative Hypothesis
- Prior probability of a hypothesis
- Prior probability distribution
- Likelihood of the observed data
- Posterior probability of a hypothesis
- Posterior probability distribution.

We'll walk through a classic example of Bayesian Inference. In 1964, Frederick Mosteller and David Wallace published an article in which they studied the disputed authorship of some of the Federalist Papers. Mosteller and Wallace introduce their paper as follows:

> "The Federalist papers were published anonymously in 1787–8 by Alexander Hamilton, John Jay, and James Madison to persuade the citizens of the State of New York to ratify the Constitution. Of the 77 essays, each 900 to 2500 words in length, that appeared in newspapers, it is generally agreed that Jay wrote five (Nos. 2, 3, 4, 5, and 64), leaving no further question about Jay's share. Hamilton is identified as the author of 43 papers, and Madison of 14. The authorship of 12 papers (Nos. 49–58, 62, and 63) is in dispute between Hamilton and Madison."

In the 12 disputed papers, either Alexander Hamilton (see Figure 5.1, left) or James Madison (see Figure 5.1, right) signed his name as "Publius" instead of using his given name. Mosteller and Wallace used a Bayesian analysis to rightfully attribute the authorship of each paper.

Let's assume we are working with **a specific** paper of unknown authorship (No. 54), and walk through a Bayesian inference analysis. This paper discusses the way in which the seats in the United States House of Representatives are apportioned among the states. It is titled "The Apportionment of Members Among the States."

First, let's recall Bayes' Theorem:

$$\Pr(A \mid B) = \frac{\Pr(B \mid A) * \Pr(A)}{\Pr(B \mid A) * \Pr(A) + \Pr(B \mid \sim A) * \Pr(\sim A)}. \tag{5.1}$$

Bayesian Statistics for Beginners: A Step-by-Step Approach. Therese M. Donovan and Ruth M. Mickey, Oxford University Press (2019). © Ruth M. Mickey 2019.
DOI: 10.1093/oso/9780198841296.001.0001

Figure 5.1 Alexander Hamilton (left) and James Madison (right).

We'll use a series of steps to conduct our analysis, which follow the scientific race track described in Chapter 4:

1. Identify your hypotheses.
2. Express your belief that each hypothesis is true in terms of prior probabilities.
3. Gather the data.
4. Determine the **likelihood** of the observed data under each hypothesis.
5. Use Bayes' Theorem to compute the posterior probabilities for each hypothesis.

We'll be using these steps throughout the book. Here we go!

 ?? **What is step 1?**

Answer: In step 1, we identify our hypotheses. Paper No. 54 is 2008 words in length, and our goal is to determine its most likely author. For this paper, we have two hypotheses for authors: Hamilton or Madison. Let's nail down our nomenclature:

- Hamilton = Hamilton hypothesis
- Madison = Madison hypothesis

Note that the hypotheses are exhaustive and mutually exclusive. Since there are only two hypotheses, it makes sense that:

$$\Pr(\text{Hamilton}) = \Pr(\sim\text{Madison}) \tag{5.2}$$

and

$$\Pr(\text{Madison}) = \Pr(\sim\text{Hamilton}). \tag{5.3}$$

In terms of our notation for Bayes' Theorem in Equation 5.1 above, A could correspond to the hypothesis that the author is Alexander Hamilton, and $\sim A$ could correspond to the hypothesis that the author is James Madison. And B in Bayes' Theorem above represents the observed data.

The posterior probability that the author was Hamilton is then:

$$\Pr(\text{Hamilton} \mid \text{data}) = \frac{\Pr(\text{data} \mid \text{Hamilton}) * \Pr(\text{Hamilton})}{\Pr(\text{data} \mid \text{Hamilton}) * \Pr(\text{Hamilton}) + \Pr(\text{data} \mid \text{Madison}) * \Pr(\text{Madison})}.$$

(5.4)

 ??   ## What is step 2?

Answer: We express our belief that each hypothesis is true in terms of prior probabilities. We do this because the paper is unsigned and we have some uncertainty about which hypothesis is correct. If the author authentically signed the paper, the probability of their authorship would be 1.0, and we wouldn't have a problem to tackle!

- Pr(Hamilton) = prior probability that the true author is Hamilton.
- Pr(Madison) = prior probability that the true author is Madison.

Remember that the sum of the prior probabilities must add to 1.0. We have several options; here are a few:

- Pr(Hamilton) = 0.1 and Pr(Madison) = 0.9
- Pr(Hamilton) = 0.5 and Pr(Madison) = 0.5
- Pr(Hamilton) = 0.7 and Pr(Madison) = 0.3
- Pr(Hamilton) = 0.75 and Pr(Madison) = 0.25

There are many possible combinations of prior probabilities to choose from. Knowing that Hamilton penned 43 papers and that Madison penned 14, it may be reasonable to assign Pr(Hamilton) a prior probability of 0.75 (the last option above) because Hamilton wrote 43 of the 57 papers that were signed by either man (43/(43+14) = 0.75). That would mean that Pr(Madison) = 0.25.

However, Mosteller and Wallace set the odds at 50:50, which is Pr(Hamilton) = 0.5 and Pr(Madison) = 0.5. This gives each hypothesis the same "weight" of belief as the other.

Let's now represent our hypotheses and prior probabilities as a graph (see Figure 5.2).

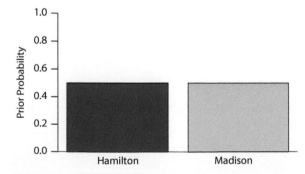

Figure 5.2 Prior probability distribution.

A graph of the prior probabilities is called the **prior probability distribution**. This distribution represents hypotheses on the x-axis and their probabilities on the y-axis.

Note that the sum of the probabilities across hypotheses must be 1.0; this means that the list of alternative hypotheses is exhaustive. In addition, the hypotheses are mutually exclusive: the paper can't be written by both men. Figure 5.2 is an example of a discrete prior probability distribution.

?? What is step 3?

Answer: Gather the data. **The data for this problem are found in Paper No. 54, and it is 2008 words long.**

Ultimately, we are asking, which hypothesis (Hamilton or Madison) is most consistent with the data we observe (the paper)?

- Pr(Hamilton | data) = what is the probability the author is Hamilton, given the paper?
- Pr(Madison | data) = what is the probability the author is Madison, given the paper?

This is the challenge that Mosteller and Wallace set out to tackle, and it is a very tough nut to crack!

Mosteller and Wallace didn't use ALL of the data at their disposal (i.e., every word in the paper), but instead focused on some signature words and phrases as a key to identify each author.

Madison tended to use the word **by** more frequently than Hamilton, whereas Hamilton tended to use the word **to** more frequently than Madison. The best single discriminator, however, was the use of the word **upon**.

Hamilton used **upon** with overwhelmingly greater frequency than Madison. To be sure, other metrics could have been used, including the length of each man's sentences. However, the two men are nearly identical in this respect, with Hamilton averaging 34.55 words per sentence, and Madison averaging 34.59 per sentence. Sentence length, therefore, would not be very helpful in separating the two hypotheses.

Our sincere apologies to Mosteller and Wallace, but to keep things simple, we will simplify their analysis tremendously and focus on the frequency of the use of **upon** by each author.

The word "upon" appeared *twice* in the paper in question. Then, the rate of **upon** is calculated as:

$$\frac{\text{\#upons}}{\text{total words}} = \frac{2}{2008} = 0.000996. \tag{5.5}$$

If we standardize this rate to 1000 words, then our manuscript has a standardized rate of **upon** of

$$0.000996 * 1000 = 0.996 \tag{5.6}$$

We need some standardization because papers vary in length: a paper with 1 **upon** in 20 words has a very different rate of **upon** than a paper with 1 **upon** in 2008 words.

Now that we have the data in hand, we need to determine the likelihood of observing a rate of 0.996 **upon**s under each hypothesis.

?? What is step 4?

Answer: Determine the **likelihood** of the observed data, assuming each hypothesis is true.

In other words, determine:

$$\Pr(0.996 \mid \text{Hamilton}) \tag{5.7}$$

and

$$\Pr(0.996 \mid \text{Madison}). \tag{5.8}$$

Here we pit the observed data against each hypothesis. This step is often the most challenging part of a Bayesian analysis. How exactly how do you compute the likelihood of the data?

Dictionary.com provides three definitions of likelihood:

1. The state of being likely or probable; probability.
2. A probability or chance of something: There is a strong likelihood of his being elected.
3. Archaic.indication of a favorable end; promise.

So, **likelihood** is another word for probability. There is a subtle difference, however, in the way the term is used in statistical analysis, where likelihood describes the probability of observing data that have already been collected. We will cover this in more detail in future chapters, but for now it's enough just to have a good feel for the concept: **likelihood involves the collection of data, and we look retrospectively at the probability of collecting those data**. If we live in the Sahara Desert, we know the probability of rain is low. Suppose it rains for seven consecutive days; this represents our data. Can we all agree that that it is very unlikely to have observed seven consecutive days of rain, given that we are in the Sahara Desert? We hope so!

Wolfram Math World differentiates likelihood and probability this way: "Likelihood is the hypothetical probability that an event that has already occurred would yield a specific outcome. The concept differs from that of a probability in that a probability refers to the occurrence of future events, while a likelihood refers to past events with known outcomes."

Let's put this in terms of authorship of the disputed paper. The paper is a past event with a known outcome: the rate of **upon** is 0.996 per 1000 words. Our next step, then, is to determine how likely it is to observe 0.996 **upon**s per thousand words under each hypothesis:

$$\Pr(0.996 \mid \text{Hamilton}). \tag{5.9}$$

- Equation 5.9 asks: "What is the likelihood of observing a rate of 0.996 **upon**s per 1000 words, given that the author is Hamilton?"

$$\Pr(0.996 \mid \text{Madison}). \tag{5.10}$$

- Equation 5.10 asks: "What is the likelihood of observing a rate of 0.996 **upon**s per 1000 words, given that the author is Madison?"

> Notice that the likelihoods are conditional for each hypothesis! In this Bayesian analysis, the likelihood is interpreted as the probability of observing the data, given the hypothesis.

?? **How exactly do we compute the likelihood?**

Answer: Computing the likelihood of the observed data is a critical part of Bayesian analysis. Throughout this book, you'll be learning about probability distributions that will enable you to estimate the likelihood of the observed data under each hypothesis. But, for now, what you need is a good, intuitive feel for what it is, and a firm understanding of how the observed data (a rate of 0.996) has a different likelihood depending on whether the author was Hamilton or Madison.

Imagine Mosteller and Wallace had a whole team of students who searched through 98 articles **known to be penned by Hamilton or Madison**, and tediously counted the number of times each author used the word **upon** and also tallied the total number of words in each document. The data might look like that shown in Table 5.1.

Table 5.1

Article	Author	Length	Upons	Rate	Standardized Rate
1	Madison	1672	1	0.0006	0.598
2	Hamilton	2196	2	0.00091	0.911
3	Madison	1690	2	0.00118	1.183
4	Hamilton	1013	3	0.00296	2.962
5	Madison	1160	1	0.00086	0.862

Our dataset consists of 98 such articles, 48 of which are **known to be penned** by Hamilton, and 50 of which are **known to be penned** by Madison. **Only the first 5 papers are shown above, and these are hypothetical, to give you a better flavor for what was involved**. Here, each article is listed in column 1. The author of each article is identified in column 2, and the word length of the article is given in column 3. The total number of **upon**s is provided in column 4, and the rate of **upon**s per 1000 words is calculated in column 5. As you can see, the first article was known to be written by Madison, and he used the word **upon** 1 time in 1672 words, for a standardized rate of 0.598 **upon**s per 1000 words.

Mosteller and Wallace summarized their standardized results as shown in Table 5.2 (Table 2.3 in their 1964 publication).

Table 5.2

Rate	Hamilton	Madison
0 (exactly)	0	41
(0,1]	1	7
(1,2]	10	2
(2,3]	11	0
(3,4]	11	0
(4,5]	10	0
(5,6]	3	0
(6,7]	1	0
(7,8]	1	0
	48	50

This table assigns each article into a "rate bin," where the bins are provided in column 1. The first bin includes all manuscripts where the word **upon** is never used (i.e., the rate of **upon** = 0). Let's talk about the notation used for the remaining bins. A closed interval such as [0,1] would include all manuscripts with a rate of 0 through 1, including the endpoints. An open interval such as (0,1) does not include the endpoints 0 or 1. A half-closed interval has the notation (0,1]. Thus, the interval (0,1] is a bin that includes all manuscripts where the rate of **upon** was > 0 but ≤ 1. For the 48 papers known to be penned by Hamilton, 0 of them had a rate of 0, 1 had a rate in the (0,1] bin, 10 had a rate in the (1,2] bin, and so on. For the 50 papers known to be penned by Madison, 41 of them had a rate of 0 (in which the word **upon** was never used), 7 had a rate > 0 and ≤ 1, and 2 had a rate > 1 and ≤ 2.

The same data can be shown as a **frequency histogram** (see Figure 5.3). The x-axis of this histogram are the bins, or the rate of **upon**s per 1000 words. The y-axis is the frequency, or number of articles, corresponding to each rate of **upon**. Frequency histograms therefore depict the raw data in graphic format.

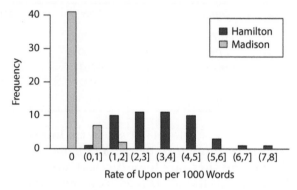

Figure 5.3 Hamilton's and Madison's rates of "upon".

Remember, for this step we are trying to get estimates of the **likelihood of the observed data under each hypothesis**. The unsigned manuscript has a rate of **upon** = 0.996, which falls into the second bin:

$$\Pr(0.996 \,|\, \text{Hamilton}) \tag{5.11}$$

- Equation 5.11 asks: "What is the probability of observing a rate of 0.996 **upon**s per 1000 words, given that the author is Hamilton?"

$$\Pr(0.996 \,|\, \text{Madison}) \tag{5.12}$$

- Equation 5.12 asks: "What is the probability observing a rate of 0.996 **upon**s per 1000 words, given that the author is Madison?"

 ?? Which of the two hypotheses more closely matches the observed rate?

Answer: Intuitively, the data are more consistent with the Madison hypothesis. Can you see that too?

?? **If likelihood is a probability, how do we quantify this "consistency" in terms of probability?**

Answer: Great question! We'll try this different ways. First, though, think about how you would approach this problem before reading on.

Looking back at Table 5.2, we can see the following:

- One of Hamilton's 48 manuscripts had a rate of **upon** greater than 0 but less than 1.0. Therefore, we can use $1/48 = 0.021$ as an estimate of the likelihood of the data under the Hamilton hypothesis.
- Seven of Madison's 50 manuscripts had a rate of **upon** greater than 0 but less than 1.0. Therefore, we can use $7/50 = 0.140$ as an estimate of the likelihood of the data under the Madison hypothesis.

These are quick and dirty estimates of each author's use of **upon**, and we'll run with this.

?? **What is step 5?**

Answer: Use Bayes' Theorem to compute the updated Pr(Hamilton) and Pr(Madison), given the data.

First, let's recall Bayes' Theorem and the familiar inferential terms (see Figure 5.4).

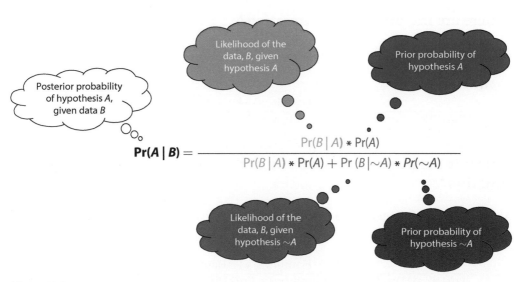

Figure 5.4

Bayes' Theorem can be used to calculate the posterior probability that the author is Hamilton:

$$\Pr(\text{Hamilton} \mid 0.996)$$

$$= \frac{\Pr(0.996 \mid \text{Hamilton}) * \Pr(\text{Hamilton})}{\Pr(0.996 \mid \text{Hamilton}) * \Pr(\text{Hamilton}) + \Pr(0.996 \mid \text{Madison}) * \Pr(\text{Madison})}. \tag{5.13}$$

In words...

- The left side of the equation can be read as "the posterior probability that Hamilton penned the paper, given the paper had a rate of 0.996 **upon** per thousand words."
- On the right, the numerator multiplies two terms: the likelihood of observing a rate of **upon**s per 1000 words of 0.996 under the Hamilton hypothesis, multiplied by the prior probability of the Hamilton hypothesis.
- The denominator repeats the numerator, but adds in the Madison hypothesis as well.

Of course, since there are only two hypotheses and they are mutually exclusive, you can get the posterior probability of Madison as 1 minus the posterior probability of Hamilton. Or you can write it out fully for the sake of completeness:

$$Pr(\text{Madison} \mid 0.996)$$
$$= \frac{Pr(0.996 \mid \text{Madison}) * Pr(\text{Madison})}{Pr(0.996 \mid \text{Hamilton}) * Pr(\text{Hamilton}) + Pr(0.996 \mid \text{Madison}) * Pr(\text{Madison})}. \tag{5.14}$$

In short, we can use Bayes' Theorem to estimate the posterior probability for each hypothesis. Let's keep our focus on the Hamilton hypothesis for now:

$$Pr(\text{Hamilton} \mid 0.996)$$
$$= \frac{Pr(0.996 \mid \text{Hamilton}) * Pr(\text{Hamilton})}{Pr(0.996 \mid \text{Hamilton}) * Pr(\text{Hamilton}) + Pr(0.996 \mid \text{Madison}) * Pr(\text{Madison})}. \tag{5.15}$$

 ?? Where are the priors in this equation?

Answer: They are shown in color below (red and purple):

$$Pr(\text{Hamilton} \mid 0.996)$$
$$= \frac{Pr(0.996 \mid \text{Hamilton}) * Pr(\text{Hamilton})}{Pr(0.996 \mid \text{Hamilton}) * Pr(\text{Hamilton}) + Pr(0.996 \mid \text{Madison}) * Pr(\text{Madison})}. \tag{5.16}$$

 ?? Where is the posterior probability of the Hamilton hypothesis in this equation?

Answer: It's highlighted in red below:

$$Pr(\text{Hamilton} \mid 0.996)$$
$$= \frac{Pr(0.996 \mid \text{Hamilton}) * Pr(\text{Hamilton})}{Pr(0.996 \mid \text{Hamilton}) * Pr(\text{Hamilton}) + Pr(0.996 \mid \text{Madison}) * Pr(\text{Madison})}. \tag{5.17}$$

 ?? Where are the likelihoods of the observed data under each hypothesis in this equation?

Answer: The likelihoods are shown in blue and green below:

$$Pr(\text{Hamilton} \mid 0.996)$$
$$= \frac{Pr(0.996 \mid \text{Hamilton}) * Pr(\text{Hamilton})}{Pr(0.996 \mid \text{Hamilton}) * Pr(\text{Hamilton}) + Pr(0.996 \mid \text{Madison}) * Pr(\text{Madison})}. \tag{5.18}$$

Don't lose track that we broke the components of Bayes' Theorem into pieces!

 ?? So, what is the posterior probability of the Hamilton hypothesis?

Remember that we set the priors to 0.5 for Hamilton and 0.5 for Madison. Also remember that the likelihood of observing a rate of **upon**s of 0.996 was 0.021 for Hamilton and 0.140 for Madison. Now let's fill in our priors and likelihood:

$$\Pr(\text{Hamilton} \,|\, 0.996) = \frac{0.021 * 0.5}{0.021 * 0.5 + 0.140 * 0.5} = \frac{0.0105}{0.0805} = 0.1304. \quad (5.19)$$

The answer is the posterior probability that the author of the paper in question was Alexander Hamilton:

$$\Pr(\text{Hamilton} \,|\, 0.996) = 0.1304. \quad (5.20)$$

Since we have only two, mutually exclusive, hypotheses, this means that our posterior probability that the author was James Madison is $1 - 0.1304 = 0.8696$:

$$\Pr(\text{Madison} \,|\, 0.996) = 0.8696. \quad (5.21)$$

These new posterior estimates can be graphed as the posterior probability distribution.

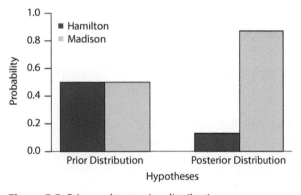

Figure 5.5 Prior and posterior distributions.

Thus, given the manuscript (and ancillary data about word use by the two authors from KNOWN data), we now increase our belief that that the author of Federalist Paper No. 54 was James Madison. You can read about it here!

 ?? How do we set the prior probabilities?

Answer: Great question! We'll cover the topic of priors in more detail in Chapter 6. But for now, let's see what happens when we use a different set of priors instead:

$$\Pr(\text{Hamilton}) = 0.75 \quad (5.22)$$

$$\Pr(\text{Madison}) = 0.25. \quad (5.23)$$

We considered these potential priors at the beginning of the chapter. Let's use Bayes' Theorem to calculate the posterior probability that the author was Alexander Hamilton with these new priors:

$$\Pr(\text{Hamilton} \mid 1.00) = \frac{0.021 * 0.75}{0.021 * 0.75 + 0.140 * 0.25} = \frac{0.01575}{0.05075} = 0.3103. \qquad (5.24)$$

The answer now is 0.3103, whereas our first result was 0.1304. In both cases, the analysis now suggests that there is a greater probability that James Madison was the author. However, the posterior probability for the Madison hypothesis is smaller when the odds are stacked against him in the second example. In other words, priors matter... a point that we will discuss in more detail in Chapter 6.

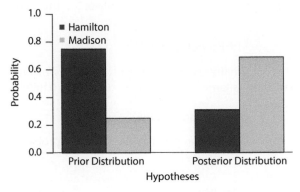

Figure 5.6 Prior and posterior distributions.

 ?? **What if we found more papers known to be authored by Hamilton and Madison?**

Answer: Fantastic! The more information you have to calculate the likelihood, the better. In this case, you would use this new information to get better estimates of the probability of each author's use of the word **upon**. Additionally, the discovery of more papers may influence your choice of priors.

 ?? **Do the likelihoods of the data have to add to 1.0?**

Answer: No, don't confuse the prior probabilities for a set of hypotheses (which must sum to 1.0) with the probability of the data. This can be confusing!

> In Bayesian analysis, the sum of the prior probabilities across hypotheses, as well as the sum of the posterior probabilities across hypotheses, must be 1.0. This is not true for the likelihoods of observing the data under each hypothesis.

This is a very important concept, so let's review our calculations to nail this point (see Table 5.3).

Table 5.3

Observed Rate	Hypothesis	Prior	Likelihood	Prior * L	Posterior
0.996	Hamilton	0.5	0.0210	0.0105	0.1304
0.996	Madison	0.5	0.1400	0.0700	0.8696
	Sum	**1.0**	**0.1610**	**0.0805**	**1.0000**

Table 5.3 shows our calculations for the analysis where the prior probabilities for the two hypotheses were equal. You've seen these calculations already, but notice that the sum of the priors is 1.000 (as required), while the sum of the likelihoods for the two hypotheses is 0.1610.

Let's have a look at our second analysis, in which the priors were unequal (see Table 5.4).

Table 5.4

Observed Rate	Hypothesis	Prior	Likelihood	Prior * L	Posterior
0.996	Hamilton	0.75	0.0210	0.0158	0.3103
0.996	Madison	0.25	0.1400	0.0350	0.6897
	Sum	**1.00**	**0.1610**	**0.0508**	**1.0000**

Once again, notice the sum of the likelihoods is 0.161, whereas the sum of the priors is 1.000.

?? Did Mosteller and Wallace really use this approach?

Not exactly. They used a Bayesian inference approach, but they calculated their likelihoods a bit differently than we did here. Typically, the analyst makes an assumption about the distribution that gives rise to the data. Mosteller and Wallace were premier statisticians and instead fit the histogram data to what is called a **Poisson distribution**. The Poisson distribution is used to estimate the probability of an event occurring (such as the number of **upon**s per 1000 words, the number of births per 1000 females, or the number of car crashes in 100 days at a corner), given the average rate of that event. We'll cover the Poisson probability distribution in depth later in the book. Here, we are simply trying to give you a **general understanding** of how to quantify likelihood.

?? Can we summarize this problem?

OK. To summarize, Mosteller and Wallace set out to determine the author of Paper No. 54. This paper had a standardized rate of **upon**s = 0.996 per 1000 words. They sought out the posterior probability associated with two hypotheses:

- Pr(Hamilton | data) = what is the probability the author is Hamilton, given the rate of **upon** is 0.996?
- Pr(Madison | data) = what is the probability the author is Madison, given the rate of **upon** is 0.996?

To answer this, they assigned each hypothesis a prior probability of being true. Then, they gathered data from manuscripts **known** to be penned by each author and used this information to calculate the likelihood of observing the data under each hypothesis:

- Pr(data | Hamilton) = what is the likelihood of the observed rate of 0.996, given the author is Hamilton?
- Pr(data | Madison) = what is the likelihood of the observed rate of 0.996, given the author is Madison?

Mosteller and Wallace used Bayes' Theorem, which says there is a relationship between the original conditional probabilities and their inversions, in order to generate updated beliefs that each hypothesis is true:

$$\text{Pr(Hamilton | 0.996)}$$
$$= \frac{\text{Pr}(0.996\,|\,\text{Hamilton}) * \text{Pr(Hamilton)}}{\text{Pr}(0.996\,|\,\text{Hamilton}) * \text{Pr(Hamilton)} + \text{Pr}(0.996\,|\,\text{Madison}) * \text{Pr(Madison)}}. \tag{5.25}$$

The use of Bayes' Theorem in this way is known as **Bayesian inference**. The Merriam–Webster dictionary defines inference as the "the act or process of inferring." It then defines "inferred" or "inferring" as "deriving conclusions from facts or premises," and "infer" as "to form an opinion or reach a conclusion through reasoning and information."

Bayesian inference is then the use of Bayes' Theorem to draw conclusions about a set of mutually exclusive, exhaustive, alternative hypotheses by linking prior knowledge about each hypothesis with new data. The result is an updated probability for each hypothesis of interest.

 ?? How does this problem differ from the Breast Cancer Problem in the last chapter?

Your answer here!

Answer: Both problems had two alternative hypotheses (e.g., Madison vs. Hamilton, and Cancer vs. No Cancer). Both used Bayesian inference. The rate of **upon**s corresponds to "testing positive." The rate of **upon**s could extend up to 8.0 per 1000 words, whereas the test result was either positive or negative. A main difference, however, is the assignment of prior probabilities for each hypothesis. In the Breast Cancer Problem, we were told that 1% of women of a particular age have breast cancer. This 1% estimate presumably comes from a different source of information. In the author problem, Mosteller and Wallace did not have an external source of information, so they set their priors at 50–50.

In Chapter 6, we will tackle a new problem that will highlight the different approaches for setting the prior distribution.

CHAPTER 6

The Birthday Problem: Bayesian Inference with Multiple Discrete Hypotheses

Now that you've had a taste of what Bayesian inference is all about, we'll start on a new example and explore two key features of Bayesian analysis: assigning priors and estimating likelihoods. One of the more controversial aspects of Bayesian inference is the use of priors. They are controversial, but they can't be ignored.

In the frequentist notion of Bayes' Theorem, the priors are just marginal probabilities. But in Bayesian inference, the priors represent *a priori* probabilities that each alternative hypothesis is correct, where *a priori* means "prior to data collection." **You cannot conduct a Bayesian analysis without using a prior distribution**. You must also collect some information to estimate the likelihood of the data given a hypothesis.

By the end of this chapter, you should be able to define:

- Informative prior distribution
- Non-informative prior distribution
- Objective priors
- Subjective priors
- Prior sensitivity analysis

Let's begin with a new example. This problem is taken from a short story called Absent Treatment by P. G. Wodehouse (author of the infamous Jeeves collection). In this story, the main character and author, Reggie Pepper, is helping his friend Bobbie Cardew. Bobbie is in a pinch because he forgot the date of his wife's birthday. The wife, Mary, has had enough and has ditched poor Bobbie. Let's pick up the story with Mary's send-off letter to Bobbie and then listen in on an exchange between Reggie and Bobbie:

> MY DEAR BOBBIE, I am going away. When you care enough about me to remember to wish me many happy returns on my birthday, I will come back. My address will be Box 341, London Morning News. – Mary

I [Reggie] read it twice, then I said, "Well, why don't you?"

"Why don't I what?"

"Why don't you wish her many happy returns? It doesn't seem much to ask."

"But she says on her birthday."

"Well, when is her birthday?"

Bayesian Statistics for Beginners: A Step-by-Step Approach. Therese M. Donovan and Ruth M. Mickey, Oxford University Press (2019). © Ruth M. Mickey 2019.
DOI: 10.1093/oso/9780198841296.001.0001

"Can't you understand?" said Bobbie. "I've forgotten."

"Forgotten!" I said.

"Yes," said Bobbie, "Forgotten."

"How do you mean, forgotten?" I said. "Forgotten whether it's the twentieth or twenty-first, or what? How near do you get to it?"

"I know it came somewhere between the first of January and the thirty-first of December. That's how near I get to it."

"Think."

"Think? What's the use of saying 'Think'? Think I haven't thought? I've been knocking sparks out of my brain ever since I've opened that letter."

"And you can't remember?"

"No."

I rang the bell and ordered restoratives.

"Well, Bobbie," I said, "it's a pretty hard case to spring on an untrained amateur like me. Suppose someone had come to Sherlock Holmes and said, 'Mr. Holmes, here's a case for you. When is my wife's birthday?' Wouldn't that have given Sherlock a jolt? However, I know enough about the game to understand that a fellow can't shoot off his deductive theories unless you start him with a clue, so rouse yourself out of that pop-eyed trance and come across with two or three. For instance, can't you remember the last time she had a birthday? What sort of weather was it? That might fix the month."

Bobbie shook his head. "It was just ordinary weather, as near as I can recollect."

"Warm?"

"Warmish."

"Or cold?"

"Well, fairly cold, perhaps. I can't remember."

I ordered two more of the same. They seemed indicated in the Young Detective's Manual. Getting ideas is like golf. Some days you're right off, others it's as easy as falling off a log. I don't suppose dear old Bobbie had ever had two ideas in the same morning before in his life; but now he did it without an effort. He just loosed another dry Martini into the undergrowth, and before you could turn round it had flushed quite a brain-wave.

Do you know the little books called "When You Were Born"? There's one for each month. They tell you your character, your talents, your strong points, and your weak points at fourpence halfpenny a go. Bobbie's idea was to buy the whole twelve, and go through them till we found out which month hit off Mary's character. That would give us the month, and narrow it down a whole lot.

A pretty hot idea for a non-thinker like dear old Bobbie. We sallied out at once. He took half and I took half, and we settled down to work. As I say, it sounded good. But when we came to go into the thing, we saw that there was a flaw. There was plenty of information all right, but there wasn't a single month that didn't have something that exactly hit off Mary. For instance, in the December book it said, "December people are apt to keep their own secrets. They are extensive travelers." Well, Mary had certainly kept her secret, and she had travelled quite extensively enough for Bobbie's needs. Then, October people were "born with original ideas" and "loved moving." You couldn't have summed up Mary's little jaunt more neatly. February people had "wonderful memories"—Mary's specialty.

We took a bit of a rest, then had another go at the thing.

Bobbie was all for May, because the book said that the women born in that month were "inclined to be capricious, which is always a barrier to a happy married life;" but I plumped for February because February women "are unusually determined to have their own way, are very earnest, and expect a full return in their companion or mates," which he owned was about as like Mary as anything could be.

In the end he tore the books up, stamped on them, burnt them, and went home.

What a great example! Here, our task is to use a Bayesian inference approach to determine the month in which Mary was born. For the purposes of this chapter, let's suppose that Bobbie at least knows that Mary was born in the year 1900.

To begin, we have 12 discrete hypotheses ($n = 12$). Remember that Bayes' Theorem can be expressed as:

$$\Pr(H_i \mid \text{data}) = \frac{\Pr(\text{data} \mid H_i) * \Pr(H_i)}{\sum_{j=1}^{n} \Pr(\text{data} \mid H_j) * \Pr(H_j)}. \tag{6.1}$$

In words, the posterior probability for hypothesis i given the observed data (data) is written $\Pr(H_i \mid \text{data})$. It is equal to a proportion. The numerator consists of the likelihood of observing the data under hypothesis i, which is written $\Pr(\text{data} \mid H_i)$, multiplied by the prior probability of hypothesis i, which is written $\Pr(H_i)$. The denominator repeats this process for all 12 hypotheses, and sums them. The symbol $\sum_{j=1}^{n}$ means "sum;" in this case, it could have been written $\sum_{j=1}^{12}$. The indices are from $j = 1$ to 12 to indicate that there are 12 pieces of information to sum together. In short, i indicates a specific hypothesis, whereas j is the index of summation and indicates a term in the denominator.

OK, here we go.

Step 1. Identify your hypotheses. There are 12 months in a year, and these represent the alternative hypotheses. Note that the hypotheses are mutually exclusive and exhaustive. They are mutually exclusive because Mary's birth month cannot occur in both October and November. They are exhaustive because Reggie and Bobbie considered every possible outcome with respect to Mary's birth month. Let's get our nomenclature down:

- January = January hypothesis
- February = February hypothesis
- March = March hypothesis
- Etc.

Step 2. Express our belief that each hypothesis is true in terms of probabilities. We do this because Bobbie and Reggie are uncertain about which hypothesis is correct.

- Pr(January) = prior probability that Mary's true birth month is January
- Pr(February) = prior probability that Mary's true birth month is February
- Pr(March) = prior probability that Mary's true birth month is March
- Etc.

Now we assign prior probabilities that each hypothesis is true. You may be wondering how to set priors for a problem in general. Let's peek in on a famous article written by Eliezer Yudkowsky for some insight:

Q. How can I find the priors for a problem?

A. Many commonly used priors are listed in the Handbook of Chemistry and Physics.

Q. Where do priors originally come from?

A. Never ask that question.

Q. Uh huh. Then where do scientists get their priors?

A. Priors for scientific problems are established by annual vote of the AAAS. In recent years the vote has become fractious and controversial, with widespread acrimony, factional polarization, and several outright assassinations. This may be a front for infighting within the Bayes Council, or it may be that the disputants have too much spare time. No one is really sure.

Q. I see. And where does everyone else get their priors?

A. They download their priors from Kazaa.

Q. What if the priors I want aren't available on Kazaa?

A. There's a small, cluttered antique shop in a back alley of San Francisco's Chinatown. Don't ask about the bronze rat.

Ha ha! Seriously, now. When Bobbie and Reggie first started out, they had absolutely no idea which month was the birth month. Consequently, they gave equal weight to each hypothesis, which is 0.083. Remember that the sum of the probabilities that make up the prior distribution must be 1.0, so each month has a probability of $1/12 = 0.083$. The chart in Figure 6.1 depicts the prior probability distribution; the prior probability of each and every alternative hypothesis is provided.

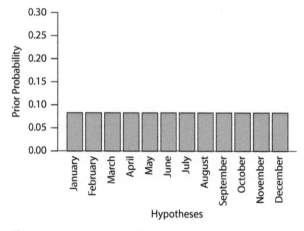

Figure 6.1 Prior probability distribution.

These are called **non-informative priors** and result in equal probabilities for all hypotheses; that is, each month has the same prior probability of being Mary's birth month. Other terms for the same concept include **vague** or **diffuse** priors. A SAS webpage on Bayesian analysis states that "a prior distribution is non-informative if the prior is 'flat' relative to the likelihood function." Wikipedia defines a non-informative prior as a prior that "expresses vague or general information about a variable…The simplest and oldest rule for determining a non-informative prior is the principle of indifference, which assigns equal probabilities to all possibilities."

A non-informative prior is a prior distribution that adds little or no information to the Bayesian inference. When an analyst uses a non-informative prior, their goal is to obtain a posterior distribution that is shaped primarily by the likelihood of the data.

However, Reggie and Bobbie might have used the information within the "When Were You Born" books to set the priors differently. In that case, they would employ an **informative**

prior distribution. A SAS webpage states that "an informative prior is a prior that is not dominated by the likelihood and that has an impact on the posterior distribution. If a prior distribution dominates the likelihood, it is clearly an informative prior." Wikipedia defines an informative prior as a prior that "expresses specific, definite information about a variable." These could arise from expert opinion or from previous study.

We will return to this topic shortly.

An informative prior is a prior distribution that adds information to the Bayesian inference. When an analyst uses an informative prior, their goal is to obtain a posterior distribution that is shaped by both the prior and the likelihood of the data.

Suppose Reggie and Bobbie believe that the February and May hypotheses are more likely than the rest based on the information in "When Were You Born." Their informative prior distribution may have looked like the one shown in Figure 6.2, with the sum of the probabilities across hypotheses equal to 1.00.

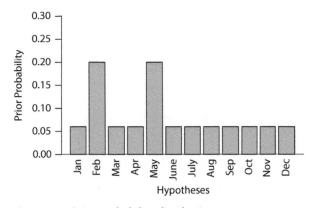

Figure 6.2 Prior probability distribution.

Usually, though, an informative prior is based on previous scientific study or expert opinion...this is probably not the case with "When You Were Born!"

 ?? **Should Bobbie and Reggie use an informative prior?**

Answer: Great question! The analyst must select a prior distribution. A colleague notes that "the most important principle of prior selection is that your prior should represent the best knowledge that you have before you look at the data." He notes that it is unjustified to use default, ignorance, or other automatic priors if you have substantial information that can affect the answer (the posterior). Of course, this assumes that your best knowledge is well founded, which perhaps is not the case here!

Step 3. Gather the data. In this case, we have one datapoint: one woman named Mary born in the year 1900.

Now that we have our data, Bayes' Theorem can provide the posterior probability of her birth month. Here's Bayes' Theorem again:

$$\Pr(H_i \,|\, \text{data}) = \frac{\Pr(\text{data} \,|\, H_i) * \Pr(H_i)}{\sum\limits_{j=1}^{n} \Pr(\text{data} \,|\, H_j) * \Pr(H_j)}. \tag{6.2}$$

For example, the posterior probability that Mary was born in January can be written:

$$\Pr(\text{January} \,|\, 1\text{Mary}) = \frac{\Pr(1\text{Mary} \,|\, \text{January}) * \Pr(\text{January})}{\sum\limits_{j=1}^{n} \Pr(1\text{Mary} \,|\, H_j) * \Pr(H_j)}. \tag{6.3}$$

Step 4. Now Reggie and Bobbie need to estimate the likelihood of observing the data (1 Mary) for each monthly hypothesis. As we've said, figuring out how to estimate the likelihood of observing the data under each hypothesis is often the trickiest part of a Bayesian inference problem.

 ?? **What data do we need then?**

Answer: For this problem, we need the rate at which girls are named Mary per month. Let's suppose that we had data on the names of girls born each month in 1900 in an English "shire" where Mary was born (see Table 6.1). These figures represent the data from which we will calculate our likelihoods. Note that Mary—the subject of our story—is included somewhere in this table, but we're not sure where! (This doesn't need to be the case, however, for Bayesian analysis.)

Table 6.1

Months	Female Births	Marys
January	1180	57
February	963	14
March	899	22
April	1190	20
May	862	20
June	976	28
July	1148	11
August	906	10
September	1147	8
October	945	80
November	907	95
December	917	100

We can represent these raw data as a **frequency histogram**, with the months on the x-axis and the total number of births on the y-axis (see Figure 6.3). Remember that a frequency histogram is a distribution of the **raw** data. You can see that, in 1900, the name Mary is a popular winter name, but not so much at other times of the year.

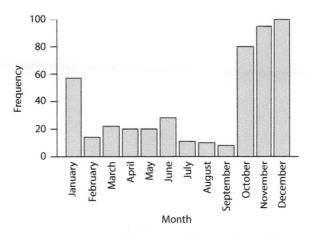

Figure 6.3 Frequency histogram of Marys by month.

Now we need to determine the likelihood of the data, assuming each hypothesis is true. In this case, we need to determine how likely it is to observe a newborn named Mary in each month. We need:

- Pr(1 Mary | January)
- Pr(1 Mary | February)
- Pr(1 Mary | March)
- Pr(1 Mary | April)
- etc.

> Notice that the likelihoods are conditional for each hypothesis! In Bayesian analyses, the likelihood is interpreted here as the probability of observing the data, given the hypothesis.

The likelihood of being named Mary in each month is just the total Marys divided by the total births. The graph in Figure 6.4 provides the likelihood of being named Mary for each month in the year 1900. Notice the y-axis is probability (or likelihood) and that the shape of this graph is similar to our frequency distribution but not exactly like it, since we had to adjust for the fact that each month had a different number of births. For instance, 20 Marys were born in both April and May, but in April there were 1190 births while in May there were only 862 births.

Now we have the likelihood of observing a baby named Mary for each month of the year. It looks like October, November, and December are more popular when it comes to "Mary."

Step 5. Use Bayes' Theorem to compute the posterior probabilities Pr(January), Pr(February), Pr(March), and so on, given the data (a baby named Mary born in the year 1900, who grew up to marry Bobbie).

At this point, we combine the priors and the likelihoods to get a posterior probability that each month is Mary's birth month with Bayes' Theorem. We'll use the informative priors based on the "When You Were Born" books (Figure 6.2).

Let's start by calculating the posterior probability of the January hypothesis:

$$\Pr(\text{January} \mid 1\text{Mary}) = \frac{\Pr(1\text{Mary} \mid \text{January}) * \Pr(\text{January})}{\sum_{j=1}^{n} \Pr(1\text{Mary} \mid H_j) * \Pr(H_j)}. \tag{6.4}$$

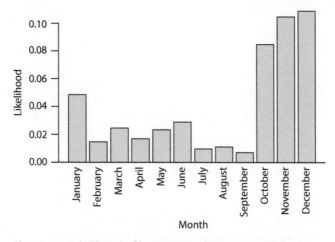

Figure 6.4 Likelihood of being named Mary by month.

The numerator focuses on the January hypothesis, which multiplies the likelihood of observing 1 Mary given the January hypothesis, Pr(1 Mary | January), multiplied by the prior probablity of the January hypothesis, Pr(January).

From the likelihood distribution in Figure 6.4, the likelihood of being named Mary in January is 0.048 (57 Marys/1180 female births in January). We multiply this by the prior probability of the January hypothesis, which is 0.06. The product of these two terms is 0.0029.

The denominator is a different beast. Here, we need to compute the likelihood of observing 1 Mary under **each** hypothesis, multiply the result by its corresponding prior, and, finally, sum the results. Here, a table specifying all 12 hypotheses can help (see Table 6.2).

Table 6.2

Month	Prior	Likelihood	Product	Denominator	Posterior
January	0.06	0.0483	0.0029	0.0342	0.0848
February	0.20	0.0145	0.0029	0.0342	0.0848
March	0.06	0.0245	0.0015	0.0342	0.0439
April	0.06	0.0168	0.0010	0.0342	0.0292
May	0.20	0.0232	0.0046	0.0342	0.1345
June	0.06	0.0287	0.0017	0.0342	0.0497
July	0.06	0.0096	0.0006	0.0342	0.0175
August	0.06	0.0110	0.0007	0.0342	0.0205
September	0.06	0.0070	0.0004	0.0342	0.0117
October	0.06	0.0847	0.0051	0.0342	0.1491
November	0.06	0.1047	0.0063	0.0342	0.1842
December	0.06	0.1091	0.0065	0.0342	0.1901
	1.00	0.4821	0.0342		1.0000

A few things to notice:

- Notice that for each month, the prior from Figure 6.2 is given, the likelihood from Figure 6.4 is given, and the product of the two is given.

- The sums are provided in the bottom row of the table. The sum of the priors must be 1.0, but this is not true for the likelihoods.
- The sum of the products (likelihood * prior) is 0.0342. This is the denominator of Bayes' Theorem for this problem, and it is fixed across the 12 hypotheses. What changes is the numerator, and this depends on which hypothesis you are analyzing.
- The posterior probability for each month is then computed as the product divided by the denominator. Note that the sum of the posterior probabilities across the hypotheses is 1.0.
- Notice the prior probability was dominated by the February and May hypotheses. The data show that most babies named "Mary" were born in October, November, or December; their likelihoods were higher. After observing the data, the posterior probabilities for February and May were reduced, while the posterior probabilities for October, November, and December were increased relative to the prior.

 ?? Is the divisor of Bayes' Theorem always a constant?

Answer: Yes. The divisor is also called a **normalizing constant**. Here, we could calculate it easily. But summing up the denominator is often a great challenge in Bayesian statistics. As we'll see later, in many problems the divisor is an intractable calculation...a topic we will visit in future sections of this book.

Now, let's look at Bobbie and Reggie's prior and posterior distribution side by side (see Figure 6.5).

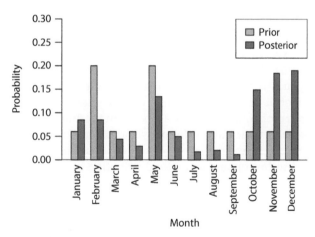

Figure 6.5 Informative prior distribution and posterior distribution.

Bobbie and Reggie started with an "informative" prior based on the "When You Were Born" books. They then collected data to calculate the likelihood of being named Mary in the year 1900. They then used Bayes' Theorem to confront each hypothesis by the data. That is, they calculated the posterior probability for each hypothesis in light of the data.

Adding in "Mary data" along with the "When You Were Born" priors won't get Bobbie out of his pickle, but you can see that posterior distribution is quite different than the prior distribution. At this point, Bobbie could use the posterior distribution as a prior distribution if other sources of information became available.

?? **What if the non-informative prior were used instead of the "When You Were Born" prior?**

Answer: Let's take a look. Here are the calculations again, this time with equal priors in column 2 (see Table 6.3).

Table 6.3

Month	Prior	Likelihood	Product	Denominator	Posterior
January	0.083	0.0483	0.0040	0.0400	0.1000
February	0.083	0.0145	0.0012	0.0400	0.0300
March	0.083	0.0245	0.0020	0.0400	0.0500
April	0.083	0.0168	0.0014	0.0400	0.0350
May	0.083	0.0232	0.0019	0.0400	0.0475
June	0.083	0.0287	0.0024	0.0400	0.0600
July	0.083	0.0096	0.0008	0.0400	0.0200
August	0.083	0.0110	0.0009	0.0400	0.0225
September	0.083	0.0070	0.0006	0.0400	0.0150
October	0.083	0.0847	0.0070	0.0400	0.1750
November	0.083	0.1047	0.0087	0.0400	0.2175
December	0.083	0.1091	0.0091	0.0400	0.2275
	1.000	**0.4821**	**0.0400**		**1.0000**

Once again, notice that the sum of the prior probabilities across hypotheses is 1.00, as is the sum of the posterior probabilities across hypotheses. For this example with a non-informative prior, we noted earlier that the likelihood drives the results. You can see this in action here for the January hypothesis (as an example). If you were only making use of the data, you would say that the probability that Mary was born in January is 0.0483/0.4821 = 0.1000. Here, the numerator is 0.0483, which is the probability that "our" Mary was born in January. The denominator is the sum of all 12 likelihoods. Thus, even if we dropped the prior altogether, we could get the posterior probability = 0.1000 as shown in the final column. This illustrates what we mean when we said "with a non-informative prior, the likelihood of the data drives the results." However, with a Bayesian analysis, a prior distribution is required!

A graph of the prior distribution and posterior distribution looks like the one shown in Figure 6.6.

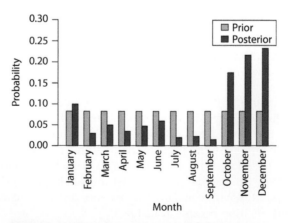

Figure 6.6 Non-informative prior distribution and posterior distribution.

Now, let's compare the two posterior distributions, where one used an informative prior and one used a non-informative prior. The chart in Figure 6.7 shows the **posterior distributions** for the "non-informative" (flat) versus "informative" (When You Were Born) priors that we've looked at thus far, with the non-informative results shown in blue:

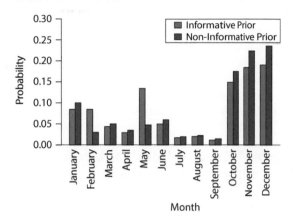

Figure 6.7 Posterior distribution with informative vs. uninformative priors.

 ?? So, the choice of the prior really affects the results?

Answer: Yes, this is why the selection of priors is important. If you have some information at your disposal (assuming it is credible), you should use it. However, you should be able to justify it.

The chart in Figure 6.7 is an example of a **prior sensitivity analysis** (Berger et al., 2000). We are comparing two posterior distributions and assessing the effect of each prior distribution's influence on the posterior, which in turn affects our conclusions. A prior sensitivity analysis is just one tool in the Robust Bayesian Analysis toolkit.

The word robust is defined as "strong and healthy; vigorous; sturdy in construction; strong and effective in all or most situations and conditions." According to Wikipedia, "Robust Bayesian Analysis, also called Bayesian sensitivity analysis, investigates the robustness of answers from a Bayesian analysis to uncertainty about the precise details of the analysis" [article accessed August 15, 2017].

 ?? Are there other times when the prior drives the results?

Answer: Well, think about it. What if 1000 girls were born each month, and 15 of them were always named Mary. We have observed a person who is named Mary. How likely is it that she was born in January? February? The answer is that the likelihoods are identical for each hypothesis. In this case, the posterior distribution will be identical to the prior distribution!

> In Bayesian analysis and scientific deduction, a primary goal of the analyst is to collect data that will discriminate the hypotheses.

?? What is so tricky about setting the prior?

Answer: The tricky part comes into play when you really don't have any information to set the prior and are trying to be as objective as possible. That is, you want to level the playing field so that all hypotheses have equal weight. The idea behind these approaches is to select a prior that has absolutely no bearing on the posterior; that is, you would draw similar conclusions from a statistical analysis that is not Bayesian. We've demonstrated such an approach when you have discrete hypotheses such as the birthday problem.

However, setting a purely uninformative prior is no small task. In the words of Zhu and Lu (2004), "flat priors are not necessarily non-informative, and non-informative priors are not necessarily flat." We are getting a bit ahead of ourselves though. For now, Hobbs and Hooten (2015) present friendly and informative thoughts on setting priors.

?? I've heard the terms "objective" and "subjective" with reference to Bayesian analysis. What do these mean?

Answer: Subjectivity generally is associated with a **belief**. In a sense, all priors are subjective because the analyst must select one and, in doing so, exercises subjectivity. The term **objective prior** normally refers to a prior that can be justified by rationality and consistency. The general goal in using an objective prior is that different analysts with the same information should arrive at the same conclusions.

This is well beyond our scope here, but we encourage you to dig deeper into this topic, as it is fodder for a rich discussion.

?? What really happened to Bobbie and Mary?

Answer: Eventually, Bobbie remembered that he took Mary to a play for her birthday. After a bit of detective work, he determined Mary's birthday was May 8th.
And they lived happily ever after.

?? Isn't that nice?

Answer: Yes, it really is! We have one more problem in Bayesian inference in this section, and we think you'll enjoy it. See you soon.

CHAPTER 7

The Portrait Problem: Bayesian Inference with Joint Likelihood

Ready for some fun? Let's take a look at Figure 7.1.

Rev. T. Bayes
Improver of the Columnar Method developed by Barrett.

Figure 7.1 Is this a portrait of Thomas Bayes?

Although you see this portrait frequently associated with Thomas Bayes, there is quite a bit of dispute about whether it is really him (*the* Thomas Bayes). In this chapter, we'll use a Bayesian inference approach to address the problem. In the spirit of fun, the data used for our example will be **completely fabricated**.

A key concept in this chapter is that you can combine **multiple sources of data** in a Bayesian inference framework. By the end of this chapter, you will be familiar with the following terms:

- Joint likelihood
- Frock coat
- Parson's wig.

Let's return to the portrait. The editors of the IMS Bulletin, a news source from the Institute of Mathematical Statistics, posted the picture of "Thomas Bayes" shown above and offered this challenge to its readers:

Bayesian Statistics for Beginners: A Step-by-Step Approach. Therese M. Donovan and Ruth M. Mickey,
Oxford University Press (2019). © Ruth M. Mickey 2019.
DOI: 10.1093/oso/9780198841296.001.0001

- **Who is this gentleman?**
- **When and where was he born?**

The person submitting the most plausible answer was to receive a prize!

The image is taken from a 1936 book called History of Life Insurance by Terence O'Donnell (American Conservation Co., Chicago). What an exciting book this must be! The image appears with the caption "Rev. T. Bayes: Improver of the Columnar Method developed by Barrett." It's not clear who added the caption or where the image came from.

 ?? Who won the bet?

Answer: According to the website, "The most plausible answer received in the Bulletin Editorial office is from Professor David R. Bellhouse, University of Western Ontario, London, Ontario, Canada. Professor Bellhouse wrote:

"The picture in The IMS Bulletin is supposedly of Thomas Bayes, who died in 1761 aged 59 and so was born in 1701 or 1702. I have added "supposedly" since I believe that the picture is of doubtful authenticity. There are three clues in the picture which lead me to this conclusion . . . For the purpose of comparison, consider the pictures [see Figure 7.2] of three other Nonconformist ministers: on the left Joshua Bayes, Thomas'es (sic) father (d. 1746); in the middle Richard Price (this portrait is dated 1776), who read Bayes' paper before the Royal Society; and on the right Philip Doddridge (1702–1751), who was a friend of Bayes' brother-in-law, Thomas Cotton."

Figure 7.2 (Left) Joshua Bayes; (middle) Richard Price; (right) Phillip Doddridge.

Bellhouse continues: *"The first thing to note in this picture is the apparent absence of a wig, or if a wig is present, it is definitely the wrong style for the period. It is likely that Bayes would have worn a wig similar to Doddridge's, which was going out of fashion in the 1740s, or a wig similar to Price's, which was coming into style at the same time. The second thing to note is that Bayes appears to be wearing a clerical gown like his father or a larger frock coat with a high collar. On viewing the other two pictures, we can see that the gown is not in style for Bayes' generation and the frock coat with a large collar is definitely anachronistic. Finally, Price is wearing a stock or wide collar on his shirt which appears around his neck in the picture; this was fashionable from about 1730 to 1770. Since Doddridge, one generation younger, appears without any stock or shirt collar, it is questionable whether Bayes would have worn a stock. However, the nineteenth century-looking clerical collar in this picture is again anachronistic. For reference, I have used C. Willett Cunnington and P. Cunnington, Handbook of English Costume in the Eighteenth Century, pub. Faber & Faber, London, 1964."*

So, Professor Bellhouse made his arguments based on fashion, which included the use of wigs. In case you were wondering, a frock coat is a man's double-breasted, long-skirted coat, now worn primarily on formal occasions. And, in case you were wondering, a Nonconformist is a Protestant Christian who did not "conform" to England's Act of Uniformity of 1662. Incidentally, Dr. Bellhouse wrote a biography of Thomas Bayes to celebrate the tercentenary of his birth, which you can download here.

 ?? ## Why did men wear wigs in the 1700's?

Answer from wiki.answers.com: Baldness, bugs, and image.

Answer from boston1775.blogspot.com: In the 1700's, a gentleman's white wig not only told other gentlemen that he was one of them; it could also signify what kind of gentleman he was. There were general styles worn by businessmen and planters, but there were also particular styles linked to professions. Doctors, for example, wore a "physick's wig," which Karin Calvert says "had a woolly, teased appearance known as a natty bob." Another specialized subset was the "parson's wig," suitable for ministers and characterized by its "rows of neat curls." Figure 7.3 shows some examples and variations on the style.

Figure 7.3 "Wigs".

 ?? ## So, how can we determine the probability that the man in the photo is Thomas Bayes?

Answer: We are asked to determine the probability that the man pictured is actually Thomas Bayes. Why not use a Bayesian inference approach to tackle this problem?

Remember that Bayes' Theorem can be expressed as:

$$\Pr(H_i \mid \text{data}) = \frac{\Pr(\text{data} \mid H_i) * \Pr(H_i)}{\sum_{j=1}^{n} \Pr(\text{data} \mid H_j) * P(H_j)}. \tag{7.1}$$

In words, the posterior probability for hypothesis i is the likelihood of observing the data under hypothesis i multiplied by the prior probability of hypothesis i, divided by the sum of likelihood times the prior across all hypotheses from $j = 1$ to n.

Where do we begin? We asked our colleague Anne Clark, who studies medieval saints at the University of Vermont, "How do you determine the authenticity of things that happened so long ago?" She replied, "You look for clues within the objects you have in hand."

If we had the portrait in hand, we could do things like analyze the paint.

 ??   Paint?

Answer: In our portrait, "Thomas" might be roughly 35–45 years old. If Thomas Bayes was born in 1701 or 1702, the portrait would have been painted in the mid-1700's, or mid-18th century (just prior to the American Revolution). The white paint that was used in this period is called "lead white."

 ?? And how can lead white help us with dating the Thomas Bayes' portrait?

Answer: Lead white is a paint that includes, well, lead! Wikipedia notes that white lead dating is a technique that is often used to pinpoint the age of an object up to 1,600 years old.

 ??   Great! Can we get started?

Answer: Umm, no. Presumably the actual portrait is long gone, and we're stuck with the printed image in O'Donnell's book. But, we can get started anyway . . . there are many other clues in the printed image that we can use. Now, let's get started.

We'll use the same steps for Bayesian analysis that we have in previous chapters:

1. Identify your hypotheses.
2. Express your belief that each hypothesis is true in terms of prior probabilities.
3. Gather the data.
4. Determine the **likelihood** of the observed data, assuming each hypothesis is true.
5. Use Bayes' Theorem to compute the posterior probabilities for each hypothesis.

 ??   Step 1. What are the hypotheses?

Answer: For this problem, we'll consider just two hypotheses: the portrait is of Thomas Bayes, and the portrait is not of Thomas Bayes.

 ??   Step 2. What are the prior probabilities that each hypothesis is true?

Answer: Your answer here! Is your answer informative or non-informative?

We need a prior distribution for our hypotheses, and for this chapter we will set the prior to 0.5 for the "Thomas Bayes Hypothesis," and 0.5 for the "Not Thomas Bayes Hypothesis," which are non-informative priors.

Step 3. What are the data?

Answer: Do you recall the author problem in Chapter 5, where we used a Bayesian inference approach to determine the probability that an unsigned Federalist Paper was penned by Alexander Hamilton or James Madison? The **data** in that problem were the words in the paper itself, and we focused on the frequency of the word **upon** in the unsigned document. For this problem, we'll consider the following sources of data that are contained with this portrait of "Thomas Bayes":

- Wigs! (or the lack thereof). We know that most ministers in the 1700's wore wigs. But not all of them! If all ministers wore wigs, there would be no chance that this portrait was of Thomas Bayes. But maybe Thomas was a non-conforming Nonconformist who refused to wear a wig.
- Similarity of "Thomas Bayes" to Joshua Bayes. How much do you look like either of your birth parents? In some cases, there are striking similarities, but not so much in other cases. We don't have the luxury of DNA testing, but *suppose* we can measure characteristics such as eyebrow shape, nose length, forehead length, and so on of Joshua Bayes (from a known portrait of Thomas Bayes' father) and "Thomas Bayes," and come up with a similarity index. Suppose this index is an integer that ranges between 0 and 100, where 0 indicates absolutely no similarity in any of the measured features, and 100 is an identical clone. Thus, the more similar that Thomas is to Joshua, the more evidence we have that the man is Thomas Bayes.

So, we will be investigating two lines of evidence that we will assume are **independent** of each other (the results of one line of reasoning will not affect the results of a second line of reasoning).

OK, then. What are the observed data with respect to wigs?

Answer: The data here are either 0 or 1, where 0 indicates a wig is absent, and a 1 indicates a wig is present. **The man is clearly not wearing a wig, so the single observation for this line of reasoning is 0**.

And what are the observed data with respect to similarity?

The score is an overall similarity between the two men, based on things like eyebrow shape, nose width, mouth shape, and any other facial feature that we can measure from the images. Perhaps if "Thomas" were wearing a wig, it might be easier to focus on the facial features only. Let's give him one (see Figure 7.4)!

Figure 7.4 (Left) "Thomas" with a wig; (right) Joshua Bayes.

At first glance, the eyebrows and nose have some resemblance, but the eyes and mouth are off. Perhaps "Thomas" inherited these traits from his mother!

For this problem, let's assume the similarity index = 55.

 ?? **So what is our final dataset for step 3?**

Answer: For this problem, we're sticking with the following data that we can observe from the portrait itself:

- Wigs = 0
- Similarity = 55.

 ?? **Step 4. What is the likelihood of the observed data under each hypothesis?**

Answer: This is a critical step.

We now have two sources of data, and we assume that each piece of information is **independent** of the other. We can now calculate the **likelihood** of observing each piece of data under each hypothesis (the portrait is of Thomas Bayes vs. the portrait is not of Thomas Bayes).

Once we have the two likelihood calculations, we can compute the joint likelihood for each hypothesis. You might recall that **joint probability** refers to the probability that multiple events will occur together; here we are interested in the probability of observing no wig **and** a similarity score of 55.

Let's work through each piece of information separately for each hypothesis and then compute the joint likelihood by multiplying the two independent likelihood results together.

 ?? **Should we start with wigs?**

Answer: Good idea. We can't directly calculate the likelihood that the man in the portrait is Thomas Bayes, but we can calculate the likelihood that a middle-aged man who sits for a portrait in the 1750's is a minister who does not wear a wig.

Suppose we were able to collect this information on the population of interest (middle-aged men who sit for a portrait in the 1750's; see Table 7.1).

Table 7.1

Person	Minister	Wig
1	0	1
2	1	0
3	0	1
4	0	1
5	1	0
6	1	0
7	1	1
8	1	0
9	1	1
10	0	1

Here, we show just the first 10 records of a 100 record dataset. We will refer to this as **dataset 1**. The first column identifies a person, the second identifies if the person is a minister or not (where 1 = minister), and the third column identifies whether the person was wearing a wig (where 1 = wig).

Table 7.2

	Wig	No Wig	Sum
Ministers	8	2	10
Non-ministers	15	75	90
Sum	**23**	**77**	**100**

We can then summarize the results of all 100 of our records as a two-way table (see Table 7.2).

Notice that this table has four "quadrants," so to speak. Our actual data are stored in the upper left quadrant, shaded the darkest blue. The upper right quadrant (shaded lighter) sums the number of ministers (10) and non-ministers (90) in our dataset. The lower left quadrant (shaded lighter) sums the number of men that wore a wig while sitting for a portrait (23) and those that did not (77). The lower right quadrant (no shading) gives the grand total, 100.

Now let's plot the results in the same Venn diagram (see Figure 7.5).

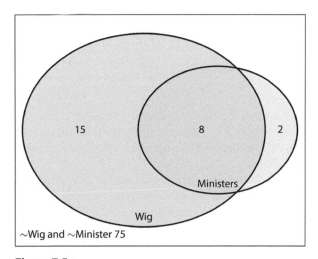

Figure 7.5

The box (which is not to scale) holds all 100 individuals in our dataset. Here, we can see that 23% of the men wear wigs while sitting for a portrait. We can also see that 10% of all of the men are ministers. And only 2% of the men sampled are ministers who do not wear wigs.

Now, let's ask how likely it is to observe our data, no wig, under the Thomas Bayes hypothesis. The answer is 0.02 (2/100 = 0.02), which is the joint probability that a man that sits for a portrait in the mid 1700's is a minister and does not wear a wig.

OK, now let's ask how likely it is to observe our data, no wig, under the Not Thomas Bayes hypothesis. This is trickier because the man in the portrait could be just about anyone. The answer here is 0.77, which is the marginal probability that a man who sits for a portrait in the mid 1700's will not wear a wig. That is the best we can do.

 ?? **What about similarity between "Thomas Bayes" and Joshua Bayes?**

Answer: Remember that we calculated a similarity score between "Thomas Bayes" and Joshua Bayes, with a result of 55. **Now we move to the question, what is the likelihood of observing a similarity score of 55 under each hypothesis?** Suppose we are able to collect similarity scores between 1000 known fathers and sons and also between 1000 pairs of unrelated men in England. The first 10 records of our dataset might look something like that shown in Table 7.3.

Table 7.3

Pair	Related	Similarity
1	0	31
2	0	80
3	0	29
4	0	71
5	1	60
6	1	61
7	1	26
8	1	39
9	1	29
10	1	75

The first column identifies a pair, the second identifies if the pair of males have a father–son relationship (where 1 = Yes), and the third column identifies their similarity score. We will refer to this as **dataset 2**; note that it is separate from our wig dataset, or **dataset 1**.

It might be useful to graph the data, where you can see that unrelated men tend to have smaller relatedness scores, but that there are some scores (such as our observed score of 55 in black) that can belong to either group (see Figure 7.6).

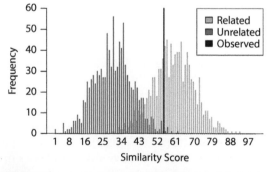

Figure 7.6 Similarity scores of related and unrelated men in the mid-1700's.

?? **How do we calculate the likelihood under each hypothesis?**

Answer: Well, we don't want to know the *exact* probability of observing 55. What we really want to find is where our observed result falls in relation to the rest of a given distribution. For instance, a score of 55 falls in the right-hand tail of the unrelated distribution (left), but it falls in the left-hand portion of the related distribution (right). To make these comparable, we need to quantify the area of each distribution that is to the *right* of our observed similarity score of 55. Thus, 55 is our minimum score of interest.

 ?? **OK, then, what is the likelihood of observing a similarity score of 55 or greater under each hypothesis?**

Answer: Your answer here!

- For the Thomas Bayes hypothesis, the two men ("Thomas" and Joshua) would be related. If we look at the green distribution above, we see that roughly 69% of the similarity scores between related men were greater than 55 (and 31% were below 55). **Thus, the likelihood of observing a similarity score of at least 55 under the Thomas Bayes hypothesis is 0.69**.
- For the Not Thomas Bayes hypothesis, the two men ("Thomas" and Joshua) would be unrelated. If we look at the orange distribution above, we see that roughly 1% of the similarity scores between unrelated men were greater than the observed value of 55 (and 99% were below 55). The likelihood of observing a similarity score of 55 or more given they are unrelated is thus 1%.

 ?? **So how do we combine both results into one likelihood for each hypothesis?**

Answer: If our two data sources are independent, we can simply multiply the two likelihood components together:

For the Thomas Bayes hypothesis, that would be

- evidence from wigs = 0.02 (the joint probability that a man that sits for a portrait in the mid-1700's is a minister who does not wear a wig)
- evidence from similarity score = 0.69 (69% of known father–son similarity scores are at least 55).

Thus, the likelihood of observing the data under the Thomas Bayes hypothesis is the product of these terms:

$$0.02 * 0.69 = 0.0138. \tag{7.2}$$

For the Not Thomas Bayes hypothesis, that would be

- evidence from wigs = 0.77 (the marginal probability that a man that sits for a portrait in the mid-1700's will not wear a wig)
- evidence from similarity score = 0.01 (1% of unrelated men have similarity scores of at least 55).

Thus, the likelihood of observing the data under the Not Thomas Bayes hypothesis is the product of these terms:

$$0.77 * 0.01 = 0.0077. \tag{7.3}$$

 ?? Step 5. What is the posterior probability that the portrait is of Thomas Bayes?

Answer: Just use Bayes' Theorem! Here, we have two hypotheses, so Bayes' Theorem can be written as:

$$\Pr(H_i \mid \text{data}) = \frac{\Pr(\text{data} \mid H_i) * \Pr(H_i)}{\sum_{j=1}^{2} \Pr(\text{data} \mid H_j) * P(H_j)}. \tag{7.4}$$

Let's assume the prior probabilities are 0.5 and 0.5. Then, the posterior probability that the man in the portrait is Thomas Bayes is shown in Table 7.4.

Table 7.4

	Prior	Likelihood	Prior * Likelihood	Posterior
Thomas Bayes	0.5	0.0138	0.0069	0.64
Not Thomas Bayes	0.5	0.0077	0.00385	0.36
Sum	1.0		0.01075	1.00

Let's go through this table together:

- Row 1 of the table is the Thomas Bayes hypothesis, and row 2 is the Not Thomas Bayes hypothesis.
- Column 1 provides the priors, which we set at 0.5 and 0.5. Remember that these must sum to 1.0.
- Column 2 is the joint likelihood of observing the two pieces of information we observed under each hypothesis:
  - wigs $= 0$
  - similarity score ≥ 55
- Column 3 multiplies the prior times the likelihood, which gives us the numerator of Bayes' Theorem for each hypothesis.
- The sum of column 3 is the denominator of Bayes' Theorem (the sum of the products of the priors times the likelihoods).
- Column 4 calculates the posterior for each hypothesis as the prior * likelihood column divided by the denominator. Note that the two posteriors sum to 1.0.

So, for the inputs for this fictitious problem, the posterior probability that the man in the portrait is Thomas Bayes is 0.64, and the posterior probability that it is someone else is 0.36. Two lines of evidence were used in this Bayesian analysis.

Don't forget that the data are fictitious! A graph of our prior and posterior distributions can now be shown in Figure 7.7.

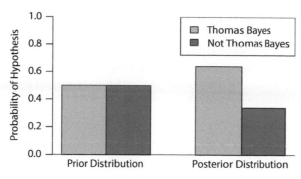

Figure 7.7

 ?? ## Are the two pieces of information really independent?

Answer: It depends. As the analyst, you would need to make an argument to justify the calculation of the joint likelihood. Here, we are claiming that our two datasets are independent and that both contribute to the likelihood calculations.

 ?? ## Can we add on more independent pieces of information?

Answer: Yes, you can. The more lines of reasoning you have, the more information you can bring to bear on the likelihood calculations and, ultimately, on the comparison of your hypotheses.

 ?? ## What if our information is not independent?

Answer: This is a lot trickier. For instance, we could ask about the use of frock coats in portraits for the same time period. If we scored the wearing of frock coats along with the use of wigs in dataset 1, we may confound the analysis because frock coats and wigs appear to go hand in hand. In addition, if you had a completely different dataset that provided similar information with respect to father–son similarity, adding it into the likelihood mix could spell trouble if they are not independent. These situations are beyond our scope here, but you might want to keep them in mind.

 ?? ## Are there any assumptions in this analysis?

Answer: Loads of them! This problem is meant to help you understand that Bayes' Theorem can easily incorporate multiple sources of data, but each piece of information that you consider must be carefully thought out. We have made multiple assumptions here, including

- wigs and similarity indices are independent, and
- a relative of Thomas Bayes is not the subject of the portrait.

You may have thought of more assumptions, and that is great.

?? **What is the main take-home point for this chapter?**

Answer: The main take-home point is that Bayesian analysis can be very, very flexible. As long as you can compute the likelihood of observing the data under each hypothesis, you are golden. However, this is not always an easy task! But, assuming it can be done, you can set up almost any kind of problem as a Bayesian inference problem.

 ?? **Looking back at the portrait, who was Barrett, developer of the columnar method?**

Answer: We were really interested in the question of whether the portrait really is of Thomas Bayes, so went ahead and purchased a copy of The History of Life Insurance by Terrence O'Donnell to explore this further. There it was—the picture—on p. 335. The picture does indeed have the caption "Rev. T. Bayes, Improver of the Columnar Method developed by Barrett." We searched high and low for any writing on Thomas Bayes, but could not find anything in the book.

George Barrett, who proposed the columnar method in the calculation of annuities (insurance), lived from 1752–1821. A Wikipedia author notes that "Barrett was the son of a farmer of Wheeler Street, a small hamlet in Surrey. At an early age, although engaged in daily labour, he made, unaided, considerable progress in mathematics, taking special interest in the class of problems connected with the duration of human life. He afterwards, during a period of twenty-five years (1786–1811), laboured assiduously at his great series of life assurance and annuity tables...to whose support he devoted a great part of his earnings." O'Donnell himself weighs in by stating, "In 1810, George Barrett presented to the Royal Society a new mode of calculating life Annuities. The old and honorable society refused to publish the contribution, no doubt considering his deductions too reactionary. However in spite of the frown of the Society, [Francis] Baily gave Barrett's findings to the world in the appendix of his very valuable work on Annuities and it immediately began to influence sound actuarial thought of the time."

Given that Thomas Bayes lived from around 1701 to 1761, it is hard to see how Bayes could have improved Barrett's method. When Bayes died, Barrett was 9 years old! "Rev. T. Bayes" may have improved Barrett's method, but it probably was not *the* Reverend Thomas Bayes we've come to know and love.

 ?? **What's next?**

Answer: This ends our section on Bayesian inference. In Section 3, we begin our journey into the world of probability distributions, which opens up many new possibilities for using Bayes' Theorem.

SECTION 3

Probability Functions

Overview

Welcome to Section 3! Now that you've had a taste of what Bayesian inference is about, it's time to take Bayes' Theorem to the next level. A major use of Bayes' Theorem involves parameter estimation, and, as such, it is critical that you have a firm understanding of parameters and variables, and how they are related in a probabilistic way.

This section consists of only two chapters.

- Chapter 8 introduces probability mass functions (pmf). This chapter introduces the idea of a random variable and presents general concepts associated with probability distributions for discrete random variables. The binomial and Bernoulli distributions are used as examples of these probability mass functions (pmf's). The pmf's can be used to specify prior distributions, likelihoods, and/or posterior distributions in Bayesian inference.
- Chapter 9 introduces probability density functions (pdf). The focus is on general concepts associated with probability density functions (pdf's), which are distributions associated with continuous random variables. The continuous uniform and normal distributions are highlighted as examples of pdf's. These and other pdf's can be used to specify prior distributions, likelihoods, and/or posterior distributions in Bayesian inference.

A solid understanding of both pmf's and pdf's is absolutely essential if you are to become a Bayesian analyst. In Bayesian analyses, we hypothesize about the values of unknown parameters. In doing so, we regard the unknown parameter as a random variable generated from some probabilistic distribution. In particular, we use probability functions to specify:

- prior distributions
- likelihood functions
- posterior distributions.

These chapters pave the way to Sections 4 and 5. Take your time with this material as it is essential for your Bayesian journey.

CHAPTER 8

Probability Mass Functions

Hopefully you have a sense now of Bayes' Theorem and how Bayesian inference is used to update our belief in alternative hypotheses. One of the primary uses of Bayesian inference is to estimate **parameters**. To do so, we need to first build a good understanding of probability distributions.

This chapter will present general concepts associated with probability distributions. We'll use the binomial and Bernoulli distributions as examples of **probability mass functions (pmf's)**. In Chapter 9, we'll use the continuous uniform and normal distributions as examples of **probability density functions (pdf's)**. Although the two chapters will contain only a few examples, the general concepts apply to other probability distributions.

These are fairly long chapters, and it may take a few readings for the material to sink in if this is new material for you . . . so make sure to get up and stretch every now and then!

By the end of this chapter, you will be able to define and use the following concepts:

- Function
- Random variable
- Probability distribution
- Parameter
- Probability mass function (**pmf**)
- Binomial **pmf**
- Bernoulli **pmf**
- Likelihood
- Likelihood profile

Since this chapter is about functions, it makes sense to start by asking the following question:

 ?? What is a function?

Answer: In math, a **function** relates an input to an output, and the classic way of writing a function is

$$f(x) = \ldots \tag{8.1}$$

For instance, consider the function shown in Figure 8.1.

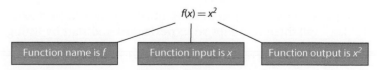

Figure 8.1

Bayesian Statistics for Beginners: A Step-by-Step Approach. Therese M. Donovan and Ruth M. Mickey, Oxford University Press (2019). © Ruth M. Mickey 2019.
DOI: 10.1093/oso/9780198841296.001.0001

The function name is f. You can name your function anything you want, but f is a commonly used name. The **inputs** go within the parentheses. Here, we have a single input called x. The **output** of the function is a set of instructions that tell us what to do with the input. In this example, we input a value for x; the instructions tell us that we should square x to give us the output.

In other words, $f(x)$ **maps** the input x to the output. We can visualize this as shown in Figure 8.2.

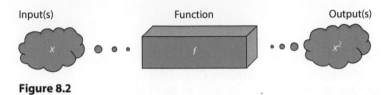

Figure 8.2

For example, if you would like to convert Fahrenheit to Celsius, you can use the function:

$$f(x) = (x{-}32)/1.8. \tag{8.2}$$

If you input 80 degrees Fahrenheit to this function, the output would be:

$$f(80) = (80{-}32)/1.8 = 26.7. \tag{8.3}$$

We are now positioned to introduce the term "variable." Dictionary.com defines variable as "capable of being varied or changed." This is certainly true of temperatures on Planet Earth: temperatures vary over time and space. Conventionally, variables are given a name, such as T or C, where T represents temperature in degrees Fahrenheit, and C represents temperature in Celsius. While the capital letters provide the **name** of the variable, **specific values** are often indicated with lower-case letters (e.g., $t = 80$, and $c = 26.7$).

The Oxford Dictionary of Statistics (Upton and Cook, 2014) states that a variable is "the characteristic measured or observed when an experiment is carried out or an observation is made. Since a non-numerical observation can always be coded numerically, a variable is usually taken to be numerical." In other words, capital T or C is the characteristic measured, and its value is usually numeric.

 ?? ## What is a random variable?

Answer: The Oxford Dictionary of Statistics (Upton and Cook, 2014) tells us, "When the value of a variable is subject to random variation, or when it is the value of a randomly chosen member of a population, it is described as a random variable—though the adjective 'random' may be omitted."

 ?? ## Can you show an example?

Answer: Yes. Suppose you flip a fair coin three times, and record each head as "H" and each tail as "T." Let's call three coin flips an **experiment** and start by listing all possible outcomes:

- HHH
- THH

- HTH
- HHT
- TTH
- THT
- HTT
- TTT

With two possible results (H or T), and three flips, there are $2^3 = 8$ possible outcomes. Count them! These outcomes make up our sample space. If this doesn't ring a bell, see Chapter 1 for a quick review.

Clearly, what you observe after 3 flips is determined at random. In this particular case, the outcome of an experiment is not a number. It's something like "HHT." So, let's define a random variable called Y, where Y is the number of heads. **Y can assume the values of 0, 1, 2, or 3. We will let lower-case y represent a *particular* value from a coin toss experiment (see Table 8.1).**

Table 8.1

	HHH	THH	HTH	HHT	TTH	THT	HTT	TTT
				Experiment Results				
$y\rightarrow$	3	2	2	2	1	1	1	0

If we were to flip a coin three times, we don't know with certainty whether its value, y, will be 0, 1, 2, or 3. But we know that is has to be one of these values. That is, our random variable is **discrete**—and it can take on a finite number of values. We can get an idea of what the value *could* be if we assigned probabilities to each. Incidentally, discrete random variables can also assume a countably infinite number of values, such as $0, 1, 2, \ldots$, to infinity and beyond.

 ?? Is a random variable a function?

Answer: Yes! A random variable is a function with inputs that are outcomes of an experiment (like "HHT") and outputs that are numerical (like "2"). In many cases, the inputs are already numerical, so the inputs are the same as the outputs.

 ?? Where do we go from here?

Answer: We're often interested in knowing the probability of observing particular outcomes. This brings us to the all-important topic of probability theory.

> Probability theory is a branch of mathematics concerned with the analysis of random phenomena. The outcome of a random event cannot be determined before it occurs, but it may be any one of several possible outcomes. The actual outcome is considered to be determined by chance (source: https://www.britannica.com/topic/probability-theory; article accessed August 17, 2017).

?? **What is the probability of observing $y = 3$ heads?**

Answer: If each of the 8 possible outcomes listed in Table 8.1 are equally likely (the coin is fair), then we can count how many times the event $y = 3$ appears out of 8 total. It appears once, so:

$$\Pr(Y = y = 3) = 1/8 = 0.125. \tag{8.4}$$

The notation Pr indicates probability. This is the probability that a random variable named Y will take on the value $y = 3$. The answer is 0.125.

 ?? **How do we move from the probability of a given value of Y to the probability distribution for all possible values of Y?**

The next step moves us from the probability of a **specific value** of Y such as $y = 3$ to **all possible values of** Y, which can be seen in the header of Table 8.2. The possible values of Y are 3, 2, 1, and 0. The number of times each of these values appears out of the 8 events is given in row 1, and the probability is given in row 2. **Notice that the sum of the probabilities of observing all possible values of Y is 1.0.**

Table 8.2

	y			
	0	1	2	3
Frequency	1	3	3	1
$\Pr(Y = y)$	0.125	0.375	0.375	0.125

 ?? **Is this an example of a probability distribution?**

Answer: Yes, it is. From the Oxford Dictionary of Statistics (Upton and Cook, 2014): "Probability distribution: A description of the possible values of a random variable, and of the probabilities of occurrence of these values."

We divided our sample space into subsets by defining the variable Y, which is the number of heads that can possibly appear with three flips of a coin, and assigned a probability to each.

 ?? **Is this also a probability mass function?**

Answer: Ah, yes . . . the title of this chapter. It is!

From The Oxford Dictionary of Statistics: "For a discrete random variable X, the probability mass function (or p.m.f.) of X is the function p such that $p(x_i) = Pr(X = x_i)$, for all i." In this definition, they use X instead of Y, and they use p instead of f for the function name.

From Wikipedia: "In probability theory and statistics, a **p**robability **m**ass **f**unction (**pmf**) is a function that gives the probability that a discrete random variable is exactly equal to some value." This definition would apply to all possible values, and the probabilities across all values must sum to 1.

Sounds like a match, doesn't it?

??   ## What if we had flipped the coin 10 times?

Answer: OK then! Each experiment now contains 10 flips. We'd do precisely the same thing. We'd write out all possible outcomes, determine the associated values of Y, and then assign to each a probability. Y is still the number of heads, and Y can assume values $0, 1, 2, \ldots, 10$. For instance, we could end up with 10 heads, which would be HHHHHHHHHH. Or we could end up with 0 heads (10 tails), which would be TTTTTTTTTT. Or we could end up with 5 heads and 5 tails, and every combination in between. With two outcomes per flip (H or T), our sample space would consist of $2^{10} = 1024$ possible outcomes!

Remember, our variable Y can take on values of $y = 0, 1, 2, \ldots, 10$. Here are a few examples:

- HHHHHHHHHH $\rightarrow$ 10
- TTTTTTTTTT $\rightarrow$ 0
- HTHTHTHTHT $\rightarrow$ 5
- HHHHHTTTTT $\rightarrow$ 5

Go ahead and list all of these outcomes and assign a probability to each. There are only 1024 calculations, so it shouldn't take too long.

 ??   ## Really?

Answer: Ha ha! Gotcha! Can you see what a chore that would be?

We need a **function** that will let us easily compute, for example, the probability of observing $y = 5$, or observing 5 heads in 10 coin flips. The **binomial probability mass function** can aid us with this task.

 ??   ## OK, what does "binomial" mean?

- **bi** means "two"
- **nomial** means "name"

So, the word "binomial" means "two names." In probability theory, it generally means "two outcomes." All binomial problems are composed of trials that have only two possible outcomes (e.g., success or failure, live or die, heads or tails, present or absent). Traditionally, we label one outcome a "success," and the other a "failure." For this problem, let's call heads a "success," which means that tails is a "failure." It doesn't matter which outcome you label a "success" and which you label a "failure," as long as you are clear about your choice!

 ??   ## When do we use binomial probability?

Answer: The binomial probability function is widely used for problems where there are a fixed number of independent tests or trials (designated n) and where each trial can have only one of two outcomes. Statistics textbooks worldwide use coin flipping as a way to demonstrate the binomial distribution.

Let's return to our 3 coin flip example, where the coin is fair. Remember that we created a variable called Y, and it can take on values $y = 0, 1, 2, 3$. Often, you'll see this written $y = 0, 1,$

$2, \ldots, n$, where n is the highest possible value that the random variable can take. The actual number of heads observed is subject to chance. Earlier, we calculated the probability of observing 0, 1, 2, 3 heads in 3 coin flips by hand (see Table 8.3).

Table 8.3

	y			
	0	1	2	3
Frequency	1	3	3	1
$\Pr(Y = y)$	0.125	0.375	0.375	0.125

Now we'll use the binomial probability mass function instead.

 ?? What does the binomial probability mass function look like?

The binomial function is written below (Equation 8.5). Notice the f notation, which denotes that it is a function named f. The inputs to the function are provided within parentheses:

$$f(y; n, p) = \binom{n}{y} p^y (1-p)^{(n-y)} \qquad y = 0, 1, \ldots, n. \tag{8.5}$$

Instead of $f(x)$, we have $f(y; n, p)$. The semicolon is read "given." This is read as the probability of y successes, given n trials and probability p of success. So this function requires three inputs: y, n, and p.

> You may also see this written as $f(y \mid n, p)$, where the vertical bar means "given." In this book, we use the semicolon notation to avoid any confusion with the conditional probability terms in Bayes' Theorem.

Two of the inputs are called **parameters**: n and p. We will discuss what a **parameter** is shortly.

- $n =$ the total number of trials (in our case, $n = 3$ coin flips or 3 trials)
- $p =$ the probability of success (in our case, the probability of flipping a heads, which is 0.5 each time if the coin is fair)

The third input, y, is the observed number of successes in the experiment. The lower-case y represents the value of a random variable, and y can take on the discrete values $0, 1, 2, \ldots, n$.

Given these inputs, the instructions for creating the output is provided on the right side of the equals sign. For example, we can use the binomial function to calculate the probability of observing 2 heads in 3 flips when the coin is fair as:

$$f(y; n, p) = \binom{n}{y} p^y (1-p)^{(n-y)} \qquad y = 0, 1, \ldots, n. \tag{8.6}$$

$$f(2; 3, 0.5) = \binom{3}{2} 0.5^2 (1-0.5)^{(3-2)}. \tag{8.7}$$

The semicolon here means "given." The entire left side of the equation can be read "the probability of observing 2 heads, *given* 3 coin flips and a probability of a heads equal to 0.5," or in general terms, "the probability of 2 successes in 3 trials given the probability of success on each trial is equal to 0.5." This is equivalent to writing:

$$\Pr(Y = 2) = f(y = 2; n = 3, p = 0.5) = \binom{3}{2}0.5^2(1-0.5)^{(3-2)}. \qquad (8.8)$$

This might look weird to you, but mathematicians use this notation because, to them, it is super-clear and concise. The left term $Pr(Y = 2)$ asks what is the probability that our random variable named Y has a value of 2. This is equivalent to the binomial **pmf** where $y = 2$, $n = 3$, and $p = 0.5$. We use the **pmf** to calculate the probability.

Let's now break the right-hand side of the binomial probability function (the output) into pieces:

- The term p^y gives p (the probability of success, or heads) raised to the number of times the success (heads) occurred (y). This term calculates the probability of observing y independent successes together in one experiment, just as we did at the beginning of the chapter.
- The term $(1 - p)^{n-y}$ gives the probability of a failure (or tails) raised to the number of times the failures (tails) occurred, which is ($n - y$). This term calculates the probability of observing y independent failures together in one experiment, just as we did at the beginning of the chapter.

To calculate the probability of flipping 2 heads out of 3 trials, we need to observe a success y times (2 heads), and we need to observe a failure $n - y$ times (1 tail). So far, our result would be:

$$0.5^2 * (1-0.5)^1 = 0.125. \qquad (8.9)$$

But if you flip a fair coin 3 times, as we've seen, there is more than one way you could end up with 2 heads and 1 tail. For instance, the sequence could be:

- HHT
- THH
- HTH.

In our problem, we are asked to compute the probability of getting 2 heads in 3 flips, given that the coin is fair. We don't care about the actual sequence of the outcomes, so we need to account for **all** of the various possible ways we can end up with 2 heads in 3 flips. The portion of the binomial probability function in brackets $\binom{n}{y}$ is called the binomial coefficient and accounts for ALL the possible ways (combinations) in which two heads and one tail could be obtained.

You can compute the binomial coefficient by hand:

$$\binom{n}{y} = \frac{n!}{y!(n-y)!} \qquad (8.10)$$

$$\frac{3*2*1}{2*1*(1)} = \frac{6}{2} = 3. \qquad (8.11)$$

Thus, there are 3 ways of getting 2 heads in 3 flips, which we confirmed earlier. We multiply 3 by 0.125 to get our final answer: 0.375. The binomial probability of getting exactly two heads out of 3 coin flips, given that the coin is fair, is:

$$f(y; n, p) = \binom{n}{y}p^y(1-p)^{(n-y)} \qquad y = 0, 1, \ldots, n \qquad (8.12)$$

$$f(y = 2; n = 3, p = 0.5) = 3 * 0.5^2 * (1-0.5)^1 = 0.375. \tag{8.13}$$

This is the same answer we got when we calculated the probability by hand (Table 8.3).

Similarly, we could use the function to compute the probability of observing, say, 0 heads and 3 tails. That would be:

$$f(y; n, p) = \binom{n}{y} p^y (1-p)^{(n-y)} \qquad y = 0, 1, \ldots, n \tag{8.14}$$

$$f(0; 3, 0.5) = \binom{3}{0} * 0.5^0 * (1-0.5)^3 \tag{8.15}$$

$$f(0; 3, 0.5) = \frac{3!}{0!(3-0)!} * 0.5^0 * (1-0.5)^3 \tag{8.16}$$

$$f(0; 3, 0.5) = 1 * 1 * 0.125 = 0.125. \tag{8.17}$$

Don't forget that zero factorial (0!) is equal to 1. Notice that in the first example we added the names of the inputs (i.e., y, n, and p), while in the second example we did not. You'll see this both ways in the literature. And now let's double-check these results with the table of results we showed previously (see Table 8.4).

Table 8.4

	y			
	0	1	2	3
Frequency	1	3	3	1
Pr($Y = y$)	0.125	0.375	0.375	0.125

Great! We hope you can see that a **pmf** such as the binomial **pmf** is a super-handy tool (function) that returns probability very quickly. The proof of this function is credited to Jakob Bernoulli (Bernoulli, 1744), a Swiss mathematician (see Figure 8.3). His tomb bears the inscription "Jakob Bernoulli, the incomparable mathematician."

Figure 8.3 Jakob Bernoulli.

?? What notation should I use to describe a binomial process like coin flipping?

If we have an experiment that has independent trials and a constant probability of success, we can indicate that a random variable Y arises from a binomial process as:

$$Y \sim \text{Binomial}(n, p). \tag{8.18}$$

If we flip a coin three times and the coin is fair, we can say:

$$Y \sim \text{Binomial}(3, 0.5). \tag{8.19}$$

?? What is a binomial distribution?

Answer: The binomial distribution is a display of **all** possible outcomes of y, given the n and p. In other words, we can use the binomial function to compute the probability of observing 0, 1, 2, 3 heads out of 3 flips, given that the coin is fair (probability of heads = probability of tails = 0.5; see Table 8.5).

Table 8.5

Successes	Probability
0	0.125
1	0.375
2	0.375
3	0.125

The sum of the probabilities that make up the binomial distribution must be 1.00. We can then graph the results as in Figure 8.4.

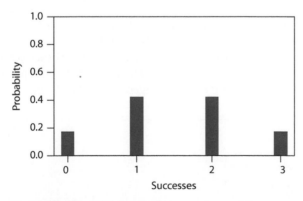

Figure 8.4 Binomial distribution: $n = 3, p = 0.5$.

This is a binomial distribution whose parameters are $n = 3$ and $p = 0.5$. The full range of possible outcomes is given on the x-axis, and the binomial probability is given on the y-axis.

Can you find the probability of observing 2 heads out of 3 coin flips, given that the coin is fair?

?? **How about the probability of observing 2.5 heads out of 3 coin flips, given that the coin is fair?**

Answer: This is an impossible result! You cannot observe 2.5 heads.

 ?? **What is a parameter?**

Answer: The Oxford Dictionary of Statistics (Upton and Cook, 2014) defines it as "a constant appearing as part of the description of a probability function...The shape of the distribution depends on the value(s) given to the parameter(s)."

The **parameters** of a probability distribution define how the distribution looks: change the parameters, and you change the shape and location of the distribution. Some authors describe parameters as the "knobs" or "controls" of a function.

Let's return to our 10 coin flip example. Figure 8.5 shows the binomial distribution when $n = 10$ and $p = 0.5$ (note that p, the probability of success, is still 0.5, but n, the number of trials, has changed from 3 to 10).

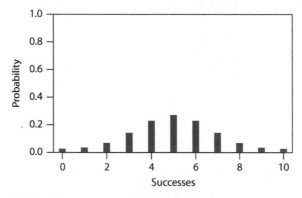

Figure 8.5 Binomial distribution: $n = 10$, $p = 0.5$.

In this example, the full range of possible values for our variable is 0 to 10. Did you notice that the binomial distribution is symmetric when p is 0.5?

Symmetry is not always the case. To illustrate, Figure 8.6 shows yet another binomial distribution, this time when $n = 100$ and $p = 0.05$ (note that p, the probability of success, was changed from 0.5 to 0.05, and n was changed from 10 to 100).

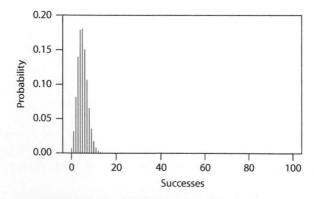

Figure 8.6 Binomial distribution: $n = 100$, $p = 0.05$.

See how different they look? All three graphs are binomial probability distributions, but they have different parameters (*n* and *p*). The x-axis should extend to the maximum number of possible successes. The y-axis often ranges between 0 and 1 but is sometimes truncated, as in Figure 8.6. Changing the y-axis does not change the distribution at all: it just magnifies or shrinks how the distribution is displayed.

No matter what range you use for the y-axis, the sum of the bars will still be equal to 1.0.

When it comes to probability mass functions, the values of the **parameters** identify *which* binomial distribution you are talking about; they determine the shape of the distribution (the distribution of probabilities along the y-axis) and its location along the x-axis. Does the "knob" or "control" image work for you?

?? What are the assumptions of the binomial probability mass function?

Answer: There are three of them.

1. The trials are independent.
2. There are two possible outcomes (success and failure) on each trial.
3. The probability of success is constant across trials.

The idea that *p* is constant is pretty straightforward: if we flip a coin 10 times, and $p = 0.5$, this means that *p* is held at 0.5 for each and every flip. The concept of independence relates to the outcomes of two or more flips. If we flip a coin and it turns up heads, this outcome has absolutely no bearing on what the result of the next flip will be.

?? Are there other probability mass functions besides the binomial?

Answer: You bet. A few that come to mind are:

- Negative binomial
- Bernoulli distribution
- Poisson distribution
- Discrete uniform distribution
- Geometric distribution
- Hypergeometric distribution

We will touch on the Bernoulli function in this chapter and discuss a few others in future chapters.

?? What do all of these functions have in common?

Answer: In each, there is a discrete random variable, say *Y*. For possible values the random variable can assume, *f*(*y*) is a probability, written $Pr(Y = y)$, and thus *f*(*y*) must assume values between 0 and 1. The values of $f(y)$ must sum to 1 for all values that Y can assume. This is the definition of a probability mass function.

 ?? ## All right then ... what is the Bernoulli distribution?

Answer:

From The Oxford Dictionary of Statistics (Upton and Cook, 2014): The distribution of a discrete random variable taking two values, usually 0 and 1.

From Wikipedia: In probability theory and statistics, the **Bernoulli distribution**, named after Swiss scientist Jakob Bernoulli, is the probability distribution of a random variable which takes the value 1 with success probability of p and the value 0 with failure probability of $q = 1 - p$.

In short, a Bernoulli distribution is a special case of a binomial distribution in which the number of trials is $n = 1$. If we have just one coin flip ($n = 1$) and the coin is fair ($p = 0.5$), then we can use the binomial **pmf** to calculate the probability that the result is heads ($y = 1$) as:

$$f(y; n, p) = \binom{n}{y} p^y (1-p)^{(n-y)} \qquad y = 0, 1, \ldots, n \tag{8.20}$$

$$f(1; 1, 0.5) = 1 * 0.5^1 * (1-0.5)^0 = 0.5. \tag{8.21}$$

If the coin was not fair, such that the probability of getting a heads is 0.4, then we would have:

$$f(y; n, p) = \binom{n}{y} p^y (1-p)^{(n-y)} \tag{8.22}$$

$$f(1; 1, 0.4) = 1 * 0.4^1 * (1-0.4)^0 = 0.4. \tag{8.23}$$

Some things are worth pointing out here. First of all, with 1 trial, the binomial coefficient is 1.

Let's run through the probability of observing a success:

- With 1 trial, if we have a success, the term p^y is the same as p^1, which is just p.
- With 1 trial, if we have a success, the term $(1 - p)^0$ is 1. Remember that anything raised to the power of 0 is 1.0.
- Multiply 1 (the binomial coefficient) by p and then by 1 and you end up with p.

Now let's run through the probability of observing a failure:

- With 1 trial, if we have a failure, the term p^y is the same as p^0, which is 1.
- With 1 trial, if we have a failure, the term $(1 - p)^1$ is $(1 - p)$.
- Multiply 1 (the binomial coefficient) by 1 and then by $(1 - p)$, and you end up with $(1 - p)$.

Thus, with a single trial, the probability of getting a success is the same thing as p, and the probability of getting a failure is the same thing as $(1 - p)$.

Because of these properties, the Bernoulli distribution is often written as:

$$f(y; 1, p) = p^y (1-p)^{(1-y)}. \tag{8.24}$$

The Bernoulli distribution provides the probabilities for all possible outcomes. With only 1 trial, there are only two possible results, and a graph of the distribution looks like the one shown in Figure 8.7.

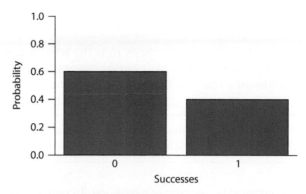

Figure 8.7 Bernoulli distribution: $n = 1$, $p = 0.4$.

We mention the Bernoulli distribution because we will be using it later in this chapter and in future chapters. Incidentally, you can think of a binomial distribution as a series of independent Bernoulli trials.

Likelihood

Now, let's gracefully roll into the concept of likelihood when dealing with probability mass functions. Likelihood is a key concept in Bayesian inference... remember that Bayes' Theorem can be expressed as shown in Figure 8.8.

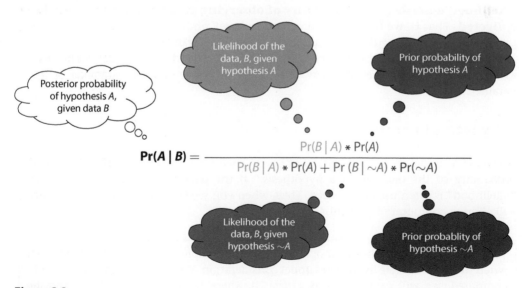

$$\text{Pr}(A \mid B) = \frac{\text{Pr}(B \mid A) * \text{Pr}(A)}{\text{Pr}(B \mid A) * \text{Pr}(A) + \text{Pr}(B \mid \sim A) * \text{Pr}(\sim A)}$$

Figure 8.8

If we let a given hypothesis stand in for A, and our data stand in for B, then this equation can be written as:

$$\Pr(H_i \mid \text{data}) = \frac{\Pr(\text{data} \mid H_i) * \Pr(H_i)}{\sum_{j=1}^{n} \Pr(\text{data} \mid H_j) * \Pr(H_j)}. \tag{8.25}$$

In words, the posterior probability for hypothesis H_i is the **likelihood** of observing the data under hypothesis i (green) multiplied by the prior probability of hypothesis i, divided by the sum of the **likelihood** $*$ the prior across all $j = 1$ to n hypotheses.

> Notice that the likelihoods are conditional on each hypothesis! In Bayesian analyses, the likelihood is interpreted as the probability of observing the data, given the hypothesis.

In short, you must understand the concept of likelihood! In previous chapters, we tried to give you a **general** feel for likelihood . . . now it is time to dig deeper.

 ?? **OK, what exactly is likelihood?**

Answer: As noted in Chapter 5, Dictionary.com provides three definitions of likelihood:

1. The state of being likely or probable; probability.
2. A probability or chance of something: There is a strong likelihood of his being elected.
3. Archaic: indication of a favorable end; promise.

Here, **likelihood** is another word for probability. As we'll see in our next chapter, likelihood can refer probability density as well. We'll save that tidbit for Chapter 9.

There is a subtle difference, though, in how the term is used in statistical analysis, where **likelihood describes the probability of observing data that have already been collected**. That is, we have data (outcomes) in hand, and we look retrospectively at the probability of collecting those data under a given set of parameters.

Indeed, the Cambridge Dictionary of Statistics (Everitt, 1998) formally defines likelihood as "the probability of a set of observations given the value of some parameter or set of parameters."

 ?? **Why is this important?**

Answer: As we've seen, a component of Bayes' Theorem is the $\Pr(\text{data} \mid H)$, which is "the probability of the data, given a hypothesis." If the data are in hand, then the term "likelihood" is often used in its stead. Thus, when you see $\Pr(\text{data} \mid H)$ in Bayes' Theorem, a likelihood computation is required.

> Throughout this book, the likelihood term in Bayes' Theorem is expressed as $\Pr(\text{data} \mid H)$ when there are discrete hypotheses under consideration. When the likelihood is actually computed, we will express this as $\mathcal{L}(\text{data}; H)$, where $\mathcal{L}$ symbolizes "likelihood" and the semicolon means "given." We do this to distinctly identify the likelihood term in Bayes' Theorem from the actual calculation.

As a side note to our readers who previously have studied maximum likelihood methods, read the box below:

> You may have seen likelihood expressed as the likelihood of the parameters given the data. For example, the likelihood of the parameter, θ, given the data, y, can be written $\mathcal{L}(\theta; y)$ or $\mathcal{L}(\theta \mid y)$. This is fairly conventional, but can be confusing when related to Bayes' Theorem. Regardless of the notation used, you must enter values for θ and y to calculate a result.

 ?? **Are there any other key points to bear in mind regarding likelihood?**

Answer: Yes. We've mentioned in previous chapters that the likelihood computations do not need to sum to 1.00. This is in sharp contrast to the prior and posterior distributions in a Bayesian analysis, which must sum to 1.00. Don't ever forget this!

 ?? **Can we quickly confirm that the likelihoods do not need to sum to 1.0 here?**

Answer: Of course. Suppose you are given 2 heads out of 3 coin flips, but you don't know p. We can plug in many alternative values (hypotheses) for p and use the binomial pmf to compute the likelihood for each and every combination (see Table 8.6).

Table 8.6

p	Likelihood
0	0
0.1	0.027
0.2	0.096
0.3	0.189
0.4	0.288
0.5	0.375
0.6	0.432
0.7	0.441
0.8	0.384
0.9	0.243
1	0

Here, the left column represents different hypothesized values for p (the probability of success). The right column represents the likelihood of observing 2 heads out of 3 flips given a particular value for p.

A few things worth pointing out:

- p can assume a value in the range of 0 to 1, so there are an infinite number of possibilities we could examine; we just recorded 11 cases.

- When the probability of success is 0, the likelihood of observing 2 heads out of 3 flips is 0. It's impossible to observe ANY heads if the probability of flipping a head is 0!
- When the probability of success is 1.0, the likelihood of observing 2 heads out of 3 flips is 0. If the coin MUST land heads, then it is impossible that you observe any tails!
- The sum of the second column does not equal 1.00.

Let's graph our results across a **full spectrum** of p alternatives (see Figure 8.9).

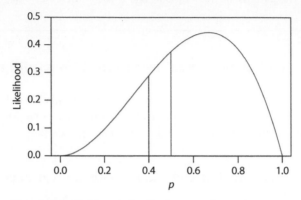

Figure 8.9 Likelihood distribution: $n = 3, y = 2$.

This is a **likelihood profile** of the binomial function when $n = 3$ and $y = 2$. On the x-axis is the unknown parameter, p, which ranges from 0 to 1. On the y-axis is the likelihood value. The graph shows the range of likelihood values possible, given the data $y = 2, n = 3$. Let's look more closely:

- Find the likelihood of observing $y = 2$ if $p = 0.4$ in Table 8.6 and match it to Figure 8.9.
- Find the likelihood of observing $y = 2$ if $p = 0.5$ in Table 8.6 and match it to Figure 8.9.
- The gray dotted lines will help you, and we'll revisit these soon.

Some other things are worth pointing out:

- We can plug in ANY value for p between 0 and 1 and calculate the likelihood of the data given the parameters.
- That means there are an *infinite* number of possible values plotted along the x-axis. For instance, we could compute the likelihood that p is 0.3, 0.301 0.302, and so on.
- The likelihood profile is often drawn as curved lines rather with bars if the parameter depicted along the x-axis is continuous.
- And the area under the likelihood curve is not equal to 1.0.

 ?? **How would this be used in a Bayesian inference problem?**

Answer: OK! Let's try to put together everything we've learned so far!

Let's go through a quick example where we are considering just 2 hypotheses. Remember that Bayes' Theorem can be expressed as shown below, with the likelihood portion shown in green:

$$\Pr(H_i \mid \text{data}) = \frac{\Pr(\text{data} \mid H_i) * \Pr(H_i)}{\sum_{j=1}^{n} \Pr(\text{data} \mid H_j) * \Pr(H_j)}. \tag{8.26}$$

In fact, although this is **not** conventional, we could write Bayes' Theorem as:

$$\Pr(H_i \mid \text{data}) = \frac{\mathcal{L}(\text{data} \mid H_i) * \Pr(H_i)}{\sum_{j=1}^{n} \mathcal{L}(\text{data} \mid H_j) * \Pr(H_j)}. \tag{8.27}$$

Suppose we have **two hypotheses** for our coin in terms of fairness. A friend gives you the coin and tells you that the coin is either fair ($p = 0.5$) or that it is weighted such that the probability of observing heads is only 0.4. **These are the only two options.**

We can now write out the posterior probability for hypothesis 1 with the expanded denominator of Bayes' Theorem as:

$$\Pr(H_1 \mid \text{data}) = \frac{\Pr(\text{data} \mid H_1) * \Pr(H_1)}{\Pr(\text{data} \mid H_1) * \Pr(H_1) + \Pr(\text{data} \mid H_2) * \Pr(H_2)}. \tag{8.28}$$

Let's review the steps of Bayesian analysis:

Step 1. What are the hypotheses?

Answer:

- H_1 is the fair hypothesis: the coin is fair so that the probability of heads is 0.5 ($p = 0.5$).
- H_2 is the unfair hypothesis: the coin is weighted so that the probability of heads is 0.4 ($p = 0.4$).

There are two discrete hypotheses (and we'll assume these are the only two possibilities).

Step 2. What were the prior probabilities for each hypothesis?

- Let's set the prior probability for each hypothesis $= 0.5$. In other words, we give equal weights to the hypothesis that the coin is fair and to the hypothesis that the coin is weighted. In this case, we have no *a priori* reason to think that one hypothesis is more likely to be true than the other.

So our prior distribution looks like Figure 8.10:

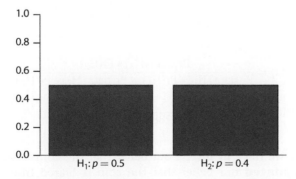

Figure 8.10 Prior distribution for a coin's probability of heads.

Thus, we are using a **Bernoulli distribution** to set our priors for the two alternative hypotheses ($p = 0.5$ and $p = 0.4$), and the probability associated with each hypothesis is 0.5. In contrast to the Hamilton and Madison hypotheses in Chapter 5, here we have alternative hypotheses for a **parameter**.

Step 3. Collect data.

- Let's assume you tossed a coin 3 times, and ended up with 2 heads.

Step 4. Compute the likelihood of the data under each hypothesis.

- For the fair hypothesis ($p = 0.5$):

$$\Pr(\text{data} \mid H_1) = \mathcal{L}(\text{data} \mid H_1) = \binom{n}{y} p^y (1-p)^{(n-y)} \tag{8.29}$$

$$\Pr(\text{data} \mid H_1) = \mathcal{L}(\text{data}; H_1) = \binom{3}{2} 0.5^2 (1-0.5)^{(3-2)} = 3 * 0.5^2 * 0.5^1 = 0.375. \tag{8.30}$$

- For the unfair hypothesis ($p = 0.4$):

$$\Pr(\text{data} \mid H_2) = \mathcal{L}(\text{data}; H_2) = \binom{n}{y} p^y (1-p)^{(n-y)} \tag{8.31}$$

$$\Pr(\text{data} \mid H_2) = \mathcal{L}(\text{data}; H_2) = \binom{3}{2} 0.4^2 (1-0.4)^{(3-2)} = 3 * 0.4^2 * 0.6^1 = 0.288. \tag{8.32}$$

You can cross-check these results in the likelihood profile in Figure 8.9.

Step 5. Use Bayes' Theorem to update the priors to posteriors.

Now it is a matter of plugging in the numbers.

- For the fair coin hypothesis ($p = 0.5$):

$$\Pr(H_1 \mid \text{data}) = \frac{\Pr(\text{data} \mid H_1) * \Pr(H_1)}{\Pr(\text{data} \mid H_1) * \Pr(H_1) + \Pr(\text{data} \mid H_2) * \Pr(H_2)} \tag{8.33}$$

$$\Pr(H_1 \mid \text{data}) = \frac{0.375 * 0.5}{0.375 * 0.5 + 0.288 * 0.5} = 0.566. \tag{8.34}$$

- For the unfair coin hypothesis ($p = 0.4$):

$$\Pr(H_2 \mid \text{data}) = \frac{\Pr(\text{data} \mid H_2) * \Pr(H_2)}{\Pr(\text{data} \mid H_1) * \Pr(H_1) + \Pr(\text{data} \mid H_2) * \Pr(H_2)} \tag{8.35}$$

$$\Pr(H_2 \mid \text{data}) = \frac{0.288 * 0.5}{0.375 * 0.5 + 0.288 * 0.5} = 0.434. \tag{8.36}$$

Thus, after 3 coin flips, we have **updated** our belief that the coin is fair from 0.5 to 0.566, and we have **updated** our belief that the coin is biased from 0.5 to 0.434 (see Figure 8.11).

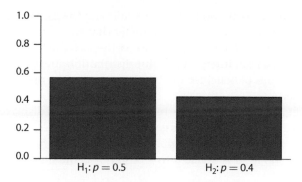

Figure 8.11 Posterior probability distribution.

We hope this makes perfect sense! We are given a coin, and told that it is either fair or weighted. These are our two hypotheses, and we gave them equal weight for the prior distribution. We then collected data: 2 heads in 3 coin flips. We calculated the likelihood of observing the data under each hypothesis. We then used Bayes' Theorem to calculate the posterior probability of each hypothesis. After the analysis, we still have two hypotheses, but the probability that H_1 is correct is now higher than the probability that H_2 is correct.

Thus, our posterior distribution for the two alternative hypotheses for p is another Bernoulli distribution.

 ?? Can we depict this problem graphically?

Answer: Yes. Let's take a step back to get a "big picture" view of this process (see Figure 8.12).

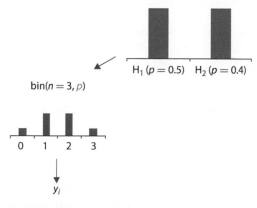

Figure 8.12

This sort of diagram was popularized by John Kruschke in his book, Doing Bayesian Data Analysis (Kruschke, 2015) and is intended to communicate the structure of the **prior** and **likelihood**. At the bottom of this diagram, we have our observed data, y_i. The data were generated from a binomial distribution with $n = 3$ rolls and the parameter, p, shown in red. The parameter, p, in turn, is the unknown parameter that we are trying to estimate. Here, we use a Bernoulli distribution (shown in blue) to set the "weights" on the **two** alternative hypotheses for p. Thus, you can interpret the blue distribution above as a prior distribution which provides our weights of belief for the two hypotheses. The drawing of this distribution can be generalized.

But there is another interpretation for the blue prior distribution displayed: the unknown parameter, p, is a **random variable** that is generated from a Bernoulli distribution. We are trying to characterize *which* Bernoulli distribution produced it. Through Bayes' Theorem, the prior Bernoulli distribution is updated to a posterior Bernoulli distribution in light of new data.

With either interpretation, the goal is to make probabilistic statements about our belief in p and the distribution that characterizes it, and to update those beliefs or knowledge in light of new data.

 ?? Can we compare this problem with the authorship problem?

Answer: For both problems, we had two hypotheses. For the author problem, our hypotheses were Madison vs. Hamilton, and for this chapter's problem the hypotheses were about the value of a parameter: $p = 0.5$ vs. $p = 0.4$.

In the author problem, we informally calculated the likelihood of the data (use of the word **upon**) under each hypothesis based on previous analysis of each man's writing. In contrast, in this chapter we used a **probability mass function** to calculate the likelihood of the data under each hypothesis. The use of probability mass functions to compute the likelihood of the data is common in Bayesian analysis (though not required, as we've seen).

 ?? What if we considered all possible hypotheses for p between 0 and 1 instead of just two specific hypotheses?

Answer: In that case, the prior distribution of p will be a **continuous distribution** instead of a discrete distribution. Chapter 9 focuses on probability density functions, and we provide the background needed to address this situation.

 ?? Can we summarize the main points of this chapter?

Answer: Of course. In this chapter:
- We introduced functions as a recipe that take in inputs and generate outputs based on a set of instructions.
- We learned about the binomial probability mass function, whose inputs are n (the number of trials), p (the probability of success in each trial), and y (the number of observed successes across the n trials):

$$f(y; n, p) = \binom{n}{y} p^y (1-p)^{(n-y)}. \tag{8.37}$$

- In this function, n and p are parameters, whereas y is a randomly observed discrete value. The binomial distribution maps the probability associated with all possible values of the random variable, Y.

- We stressed that the shape of a probability distribution is controlled by its parameters: change the parameters, and the shape and location of the distribution changes.
- We then introduced the term "likelihood," and stressed that likelihood is another word for probability. We use the term "likelihood" when the data are in hand.
- We repeatedly stressed that any Bayesian analysis will involve a likelihood calculation because the term Pr(data | hypothesis) is the likelihood of observing the data, assuming the hypothesis is true.
- Finally, we wrapped up this chapter with a simple Bayesian inference problem in which we had two hypotheses, collected binomial data, and then updated the priors to posteriors using Bayes' Theorem. **There are several other probability mass functions that you may use in Bayesian analysis. We focused on the binomial pmf, but the underlying concepts presented here should apply to new functions. Each probability function has unique inputs and provides probability as an output. The sum of the probabilities across all outcomes must be 1**.

 ?? OK, what's next?

Answer: Chapter 9 will tackle **probability density functions**, another critical tool in your Bayesian toolbox.

CHAPTER 9

Probability Density Functions

In this chapter, we'll continue to build our understanding of probability distributions. We will focus on general concepts associated with probability density functions, which are functions associated with random variables that are **continuous** in nature. For this chapter, we'll be focusing on the continuous uniform and normal distributions. Although we focus on only these distributions, the general concepts in this chapter will apply to other continuous probability distributions. This is a fairly long chapter, and it may take a few readings for the material to sink in, so make sure to get up and stretch every now and then. By the end of this chapter, you should be able to define and use the following terms for a continuous random variable:

- Random variable
- Probability distribution
- Parameter
- Probability density
- Likelihood
- Likelihood profile.

We'll start with the same question that we started with in Chapter 8:

 ??

What is a function?

Answer: In math, a **function** relates an input to an output, and the classic way of writing a function is

$$f(x) = \ldots \tag{9.1}$$

Here, the function name is f, the inputs are denoted by the letter x, and the dots represent instructions that will generate the output. For instance, consider the function shown in Figure 9.1.

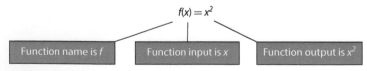

Figure 9.1

The function name is f. The inputs go between the parentheses. Here, we have a single input called x. The output of the function is a set of instructions that tell us what to do with the input. Here, we input a value for x; the instructions tell us that we should square x to give us the output.

Bayesian Statistics for Beginners: A Step-by-Step Approach. Therese M. Donovan and Ruth M. Mickey, Oxford University Press (2019). © Ruth M. Mickey 2019.
DOI: 10.1093/oso/9780198841296.001.0001

In other words, $f(x)$ **maps** the input x to the output. We can visualize this as shown in Figure 9.2.

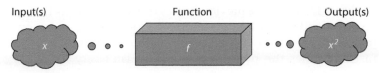

Figure 9.2

In Chapter 8, we introduced probability mass functions. We noted that for a discrete random variable named Y, the probability mass function, $f(y) = Pr(Y = y)$, provides the probability for each value of y.

And we introduced the **binomial probability mass function** as a function that lets you evaluate, for example, the probability of observing 2 heads in 3 flips when the coin is fair as:

$$f(y; n, p) = \binom{n}{y} p^y (1-p)^{(n-y)} \tag{9.2}$$

$$f(2; 3, 0.5) = \binom{3}{2} 0.5^2 (1 - 0.5)^{(3-2)} = 0.375. \tag{9.3}$$

The function's name is f, and the inputs are the observed random variable, y (the number of successes, which in this case is the 2 heads), n (the total trials, which is three flips), and p (the probability of success, which is 0.5). Don't forget that a variable is considered random when its value is subject to random variation.

In this example, the discrete random variable named Y could be 0, 1, 2, or 3. We just used the binomial **pmf** to calculate the probability of observing a random variable $y = 2$, given three tosses and a fair coin.

Probability mass functions deal with random variables, such as the number of heads observed, that are discrete (or distinct) in nature. In this chapter, however, we focus on random variables that are **continuous** in nature.

 ?? **Can you give me an example of a continuous random variable?**

Answer: Of course. Let's start with the following example from Wikipedia (accessed August 17, 2017). Suppose a species of bacterium typically lives 4 to 6 hours. In this example, we can say that capital X represents a random variable that stands for the lifespan of a bacterium. This is an example of a **continuous random variable**, because the lifespan can assume values over some interval (4 to 6 hours, such as 4.1, 4.2, 5.9999, etc.). We can define the possible outcomes of X as:

$$4 \leq X \leq 6. \tag{9.4}$$

 ?? **What is the probability that a bacterium lives *exactly* 5 hours?**

Answer: The answer is actually 0. A lot of bacteria live for approximately 5 hours, but there is a negligible chance that any given bacterium dies at **exactly** 5.0000000000...hours. We can always get more precise!

Instead we might ask: What is the probability that the bacterium dies **between** 5 hours and 5.01 hours? Let's say the answer is 0.02 (i.e., 2%). Next, what is the probability that the bacterium dies between 5 hours and 5.001 hours? The answer is probably around 0.002, because this is 1/10th of the previous interval. The probability that the bacterium dies between 5 hours and 5.0001 hours is probably about 0.0002, and so on. Notice that as the time increment decreases, the probability also decreases. When the time increment is 0 (an exact time), the probability is 0. This isn't very helpful if our goal is probability!

 ?? So, what can we do?

Answer: For cases where a random variable is **continuous**, you need to consider probability in the context of a **probability density function**.

 ?? Can we see an example of a probability density function?

Answer: Sure! How about this one:

$$f(x) = 0.5. \tag{9.5}$$

Here, the function name is f, and it has one argument called x. We can name the function and argument anything we'd like. Regardless of what it is named, the function returns the probability density, and it is always 0.5 (with some constraints that we'll discuss below).

 ?? I see ... and what distribution would result from this pdf?

Answer: A uniform distribution where values range between 4 and 6. Technically, the probability density function is written:

$$f(x) = 0.5, \quad 4 \leq x \leq 6. \tag{9.6}$$

Notice the use of the comma to separate the function, $f(x) = 0.5$, from its constraints, $4 \leq x \leq 6$. In other words, x must fall within this interval.

The random variable named X can be drawn from this distribution. We can write:

$$X \sim U(4, 6). \tag{9.7}$$

The symbol $\sim$ means "is distributed as." So we read this equation "the random variable named X is distributed as a uniform distribution with a minimum value of 4 and a maximum value of 6."

The uniform distribution is usually indicated by a capital U, and this distribution has two parameters, the minimum (sometimes called a) and maximum (sometimes called b). Thus, the uniform distribution is uniquely described as $U(a, b)$. If you change a or b, you have a different uniform distribution.

Now let's take a look at this uniform distribution. The lower-case x is the actual value of X (see Figure 9.3). The blue line in this figure represents the **p**robability **d**ensity **f**unction $f(x) = 0.5$. Notice that the y-axis is labeled "Density." Notice also that we don't use the word "probability" here. If all values of x have the exact same density, it makes sense that the **pdf**

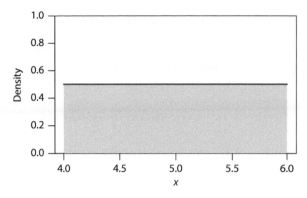

Figure 9.3

of a uniform distribution is a **constant**, $f(x) = 0.5$, right? It's just a matter of figuring out what this constant is.

 ?? **Why is the density 0.5 in this example?**

Answer: Because just as the bars in the binomial distribution or any other pmf must sum to 1.00, the **area under the density f(x) must equal 1.00.** Let's explore this idea a little more...

The uniform distribution is sometimes called a **rectangular distribution** because it looks like a rectangle. Wild! And rectangles have **area**, which is calculated by multiplying the **length** of the rectangle along the x-axis by the **height** of the rectangle along the y-axis. The length of our rectangle is calculated by subtracting the minimum, 4, from the maximum, 6, which is 2.0. **If the area of our uniform distribution represents probability, then we know that the area of this rectangle must be 1.0.** So, we now know:

- area = 1.0
- length = 2.0

And now solve for the height, which is called **density**:

$$\text{area} = \text{length} * \text{height} \tag{9.8}$$

$$1.0 = 2.0 * \text{height} \tag{9.9}$$

$$\text{height} = \frac{1.0}{2.0} = 0.50. \tag{9.10}$$

The height of the distribution at any given point is called its **density** (Figure 9.3, blue), whereas the total area under the distribution must be 1.00 (Figure 9.3, gray). The function that yields the density is called a **probability density function**.

Now, to repeat the most crucial point: the area under the pdf is equal to 1.0. Thus, the probability that X is between 4 and 6 is 1.0, or:

$$\Pr(4 \le X \le 6) = 1.00. \tag{9.11}$$

?? **Can we formally define a uniform pdf?**

Answer: You read our minds! As we mentioned, the uniform **pdf** has two parameters, called a and b, where a represents the lower bound of the distribution, and b represents the upper bound of the distribution:

$$f(x; a, b) = \frac{1}{b - a}. \tag{9.12}$$

This is the uniform probability density function (**pdf**). For our example, where $a = 4$ and $b = 6$, the density is calculated as:

$$f(x; a, b) = \frac{1}{6 - 4} = \frac{1}{2} = 0.5. \tag{9.13}$$

Let's not forget our constraints! The formal definition of the uniform pdf is written:

$$f(x; a, b) = \begin{cases} \dfrac{1}{b - a} & \text{for } a \le x \le b \\ 0 & \text{for } x < a \text{ or } x > b. \end{cases} \tag{9.14}$$

In our example:

$$f(x; a, b) = \begin{cases} 0.5 & \text{for } 4 \le x \le 6 \\ 0 & \text{for } x < 4 \; x > 6. \end{cases} \tag{9.15}$$

In words, the probability density is 0.5 for all x between 4 and 6 hours; it is 0 for any values of x that are less than 4 hours or greater than 6 hours.

 ?? **What is the probability that x is between 4.5 and 5.5 hours for our uniform distribution?**

Answer: OK, now we need to move from **probability density** to **probability**. Our uniform distribution has $a = 4$ and $b = 6$. The probability that a random variable named X is between 4.5 and 5.5 hours could be depicted as shown in Figure 9.4.

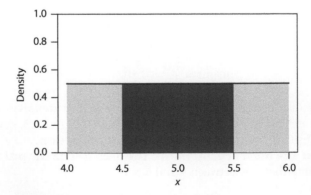

Figure 9.4

Intuitively, if the full grey box above represents an area of 1.00, then the red box represents 1/2 of the total area, so the probability of drawing x between 4.5 and 5.5 hours is 0.5. Here, we are talking about **probability** (what proportion of the full gray rectangle consists of the red rectangle) and not **density**. However, we will use the density to calculate the result. The area of the red rectangle is:

- length = 5.5 − 4.5 = 1
- height (density) = 0.5
- area = length ∗ height = 0.5.

The probability that a bacterium has a lifespan between 4.5 and 5.5 hours is 0.5, that is:

$$\Pr(4.5 \leq X \leq 5.5) = 0.5. \tag{9.16}$$

What is the probability that x is exactly 5 hours?

Answer: Once again, the answer is zero! Assuming that 5 means *exactly* 5 (and not a smitch more or less), then we are dealing with a line, not a rectangle. Lines don't have area, so the answer technically is 0. When you ask what is the probability that $x = 5$, you might really mean what is the probability of 5, give or take a tiny amount.

Are there other examples of continuous probability density functions?

Answer: Yes, there are many to choose from.
Let's return to the Wikipedia example posed in the beginning of the chapter. Suppose a species of bacterium typically lives 4 to 6 hours. We just explored this question earlier with a uniform distribution. **But now let's assume that most bacteria live 5 hours on average, with fewer living to 4 or 6.** Let's us explore this answer using a **normal distribution**.

What exactly is the normal distribution?

Answer: The Online Statistics Education book (Lane, 2011) tells us: "The normal distribution is the most important and most widely used distribution in statistics. It is sometimes called the 'bell curve,' although the tonal qualities of such a bell would be less than pleasing." The normal (or Gaussian) distribution is a very common **continuous probability distribution**.

And what does "Gaussian" refer to?

Answer: The normal distribution is also called the Gaussian distribution, named for the German mathematician Carl Friedrich Gauss, who "had an exceptional influence in many fields of mathematics and science and is ranked as one of history's most influential mathematicians" (see Figure 9.5). Gauss' works were collected and published posthumously in 1863.

Figure 9.5 Carl Friedrich Gauss.

 ?? What does a normal (Gaussian) distribution look like?

An example of a normal distribution is shown in Figure 9.6.

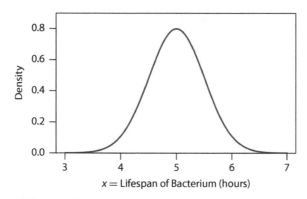

Figure 9.6 Normal distribution.

Here we are interested in the lifespan of a bacterium, which is given on the x-axis. The y-axis is labeled Density, as it was in our uniform **pdf** example. The normal distribution is a bell-shaped curve. The middle of the curve is centered on the **mean**, or average, which is represented by the Greek letter, mu (μ). Here, $\mu = 5$. The spread of the curve is controlled by the **standard deviation**, which is represented by the Greek letter, sigma (σ). Here, $\sigma = 0.5$.

Keep these points in mind:

- First, the peak of the normal distribution is centered on μ. It is a **symmetrical** curve, so half of the distribution is to the right of the mean, and half is to the left. In other words, the mean of a normal distribution is the same thing as its median (the value at which half of the random variables are greater than the median, and half are less).
- Second, the distribution is shown with a smooth curve rather than discrete bars. Consequently, the values of x underneath the curve comprise a continuous range of values rather than being discrete values.
- Third, the values along the x-axis can be negative! We don't show it here, but the normal distribution can be used for any range of positive and negative values.
- Fourth, the spread of the distribution is controlled by the standard deviation, σ. The higher the σ, the more spread there is.

 ?? So, the normal distribution has two parameters?

Answer: Yes. Change the mean (μ) or standard deviation (σ), and you change the location and/or spread of the curve. Remember the parameters are like the "control knobs." The parameter, μ is called the **location parameter**: it controls where the distribution is centered over the x-axis. In contrast, the parameter σ is called the **scale parameter**: it controls the shape or spread of the distribution.

Sometimes people will denote the parameters of a normal pdf as μ and σ^2 instead, where σ^2 is just $\sigma * \sigma$. This is called the variance of the distribution. Figure 9.7 shows some examples with different control settings.

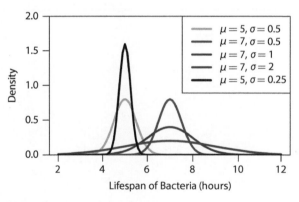

Figure 9.7 Normal distributions.

Let's study these distributions more closely.

- The red, purple, and blue distributions all have the same mean, $\mu = 7$. What differs among them is the standard deviation, σ. The red distribution has the lowest standard deviation (0.5), while the blue distribution has the highest standard deviation (2.0).
- The green and black distributions have the same mean, 5.0. What differs between them is the standard deviation. The green distribution has a standard deviation of 0.5, while the black distribution has a standard deviation of 0.25.
- The green and the red distributions have the same standard deviation (0.5). What differs between them is their mean: the blue distribution has a mean of 5, while the red distribution has a mean of 7.

?? How were these distributions generated?

Answer: These graphs were generated by plugging in a range of hours (from 0 to 15 in increments of 0.01) into the **normal probability density function** and then graphing all of the results. The y-axis is labeled density because the normal probability density function returns a probability density as opposed to actual probability (which is constrained between 0 and 1). The area under all of the bells is the same: namely, 1.

?? What exactly is the normal (Gaussian) pdf?

Answer: The normal probability density function is a **pdf**, just like the uniform **pdf** we examined previously.

The function is as follows:

$$f(x; \mu, \sigma) = \frac{1}{\sqrt{2\pi}\sigma} e^{-(x-\mu)^2/(2\sigma^2)} \qquad -\infty \leq x \leq \infty. \tag{9.17}$$

This function is named f and it has just three inputs. Two of the inputs are parameters: the mean (which is μ) and the standard deviation (which is σ). The third input, x, represents an **outcome**, such as bacterial lifespan in hours.

Look for this function on the Deutsche Mark (see Figure 9.8)!

Figure 9.8 Deutsche Mark.

?? Can you give me an example of how to use the normal pdf?

Answer: Suppose the average lifespan of a bacterium is 5 hours. We can use the normal **pdf** to get the **probability density** of observing, say, $x = 4.5$ hours, when the mean lifespan is $\mu = 5$ hours and the standard deviation $\sigma = 0.5$ hours:

$$f(x; \mu, \sigma) = \frac{1}{\sqrt{2\pi}\sigma} e^{-(x-\mu)^2/(2\sigma^2)} \tag{9.18}$$

$$f(4.5; 5, 0.5) = \frac{1}{\sqrt{2\pi} * 0.5} e^{-(4.5-5)^2/(2*0.5^2)} = 0.4839414. \qquad (9.19)$$

Remember that this is **NOT** probability ... it is probability density!

?? So, we plug in multiple values for x and generate the distribution?

Answer: Yes! Figure 9.9 shows the Gaussian distribution with $\mu = 5$ and $\sigma = 0.5$ (blue). Look for the density when $x = 4.5$ as calculated above. Also plotted is the Gaussian distribution with $\mu = 5$ and $\sigma = 0.25$ (black).

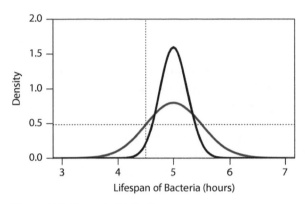

Figure 9.9 Normal distributions.

A critical thing to notice here is that both curves are generated by the normal pdf, but they have different parameters. As with the continuous uniform distribution, the area under each curve is 1.0. That is, the area is the same under the two curves. Since the black curve has a small standard deviation compared to the blue curve, the density is squished upward in order to accommodate the narrow spread. Notice that the density can be greater than 1.0.

?? How do we go from probability density to probability with a normal pdf?

Answer: Remember how we calculated the **density** for the continuous uniform distribution? For the bacterium lifespan problem, we said that lifespan is distributed as a uniform **pdf** between 4 and 6 hours, $X \sim U(4, 6)$. We knew the area of the rectangle was 1.0, we knew the length of the rectangle (2.0), and we simply solved to get the height of the rectangle, which is the density. We then calculated the probability that a bacterium would live **between** 4.5 and 5.5 hours by determining the proportion of U(4,6) where $4.5 \leq x \leq 5.5$, which is a smaller rectangle.

We can use the same sort of reasoning with the normal distribution, only we'll need more rectangles. Let's have a look, focusing on the blue distribution in Figure 9.9, where $\mu = 5$, $\sigma = 0.5$. Here it is again in Figure 9.10, with some rectangles overlaid on top.

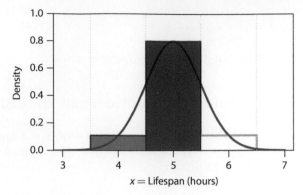

Figure 9.10 $\mu = 5$, $\sigma = 0.5$.

Here, the x-axis has been divided into 5 rectangles, where each rectangle has a length of 1 hour and a height that is the normal pdf density (two of these rectangles are in the tails and are barely visible). This is a very rough approximation of a normal distribution!

See if you can find the rectangle where the lifespan is **between** 3.5 hours and 4.5 hours. Now, this particular rectangle (blue) has a height, or density, of 0.1079819. The area of this rectangle is length (1.00) times height (0.1079819) = 0.1079819.

 ?? **What is the probability that a bacterium has a lifespan between 4.5 and 5.5 hours?**

Answer: Now look for the rectangle where lifespan is **between** 4.5 and 5.5 hours. This particular rectangle (red) has a density of 0.7978846. The area of this rectangle is length (1.0) ∗ height (0.7978846) = 0.7978846.

This is our first attempt to answer the question, "What is the probability that a bacterium has a lifespan between 4.5 and 5.5 hours?" **Hang on to this answer, as we will revisit this same question in a few minutes.**

 ?? **Is the total area equal to 1.0?**

Answer: If Gauss did his homework correctly, we can calculate the area of each rectangle in the graph and then add the areas together. We should get an answer close to 1.0. Let's try it. We have 5 rectangles, each with a length of 1 and height that is given by its density. And the answer is . . .

$$1.0143837. \tag{9.20}$$

Not bad at all! It's not precisely 1.0 because our 5-rectangle distribution only **approximates** the normal **pdf** shown in blue, where the area under the curve is exactly 1.0.

Perhaps we can improve our result by creating 41 rectangles instead of five. Let's have a look (see Figure 9.11).

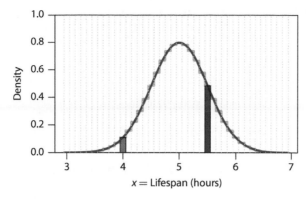

Figure 9.11 $\mu = 5$, $\sigma = 0.5$.

- Now, the x-axis has been divided into 41 slivers, where each sliver has a length of 0.1 hours.
- See if you can find the rectangle where lifespan is **between** 3.95 hours and 4.05 hours. Now, this particular rectangle (blue) has a density of 0.1079819. The area of this rectangle is $0.1 * 0.1079819 = 0.0107982$.
- Now look for the rectangle where lifespan is **between** 5.55 and 5.65 hours. This particular rectangle (red) has a density of 0.3883721. The area of this rectangle is $0.1 * 0.3883721 = 0.0388372$.

To get the area under the curve, we can repeat this little exercise for each of the 41 slivers and then add up the results. And the answer is...

$$0.9999599. \tag{9.21}$$

Notice that this answer is closer to 1.0 than our previous example, where we had only 5 rectangles.

 ?? How would one express this mathematically?

Answer: If we have 41 **discrete** rectangles, then the sum of the rectangles can be expressed as:

$$\text{Area} = \sum_{j=1}^{41} \text{length} * \text{density}. \tag{9.22}$$

The general idea is that we can more closely approximate the normal **pdf** as we create thinner and thinner rectangles.

 ?? With 41 slivers, what is the probability that a bacterium has a lifespan between 4.5 and 5.5 hours?

Answer: This is the same question we answered previously with our 5-rectangle approximation. Here are the steps: Find the rectangles centered on 4.5 and 5.5. Take half of the area of each (because half the area is outside the boundary of interest). Then, add in the area of all rectangles that fall between these two anchors. The sum is the probability, and the answer is:

$$0.6810742. \tag{9.23}$$

Remember that the answer from our 5-rectangle approximation was:

$$0.7978846. \tag{9.24}$$

Hang on to this result, and we'll revisit it one more time in a minute or two!

 ?? **What is the total area if our rectangles became really, really skinny?**

Answer: When the rectangles **approach** a width of 0, the sum symbol, $\sum$ is replaced by the integral symbol, $\int$, and the area under the curve is exactly 1.000. A general way to write this is:

$$\int_{-\infty}^{\infty} f(x)dx = 1.0. \tag{9.25}$$

In this general version:

- $f(x)$ is a **pdf**, and you can plug in any **pdf** you'd like. We've introduced the continuous uniform pdf and normal pdf so far.
- We want to get the area under the curve for a specified range of values of x. The range of values is tucked next to the integral symbol, with the lower end of x provided on the bottom of the integral symbol (here, negative infinity) and the upper end of x provided at the top of the integral symbol (here, positive infinity). That covers all possibilities!
- The integral symbol means "get the total area under the curve" for the specified range of x. Since we cover all possibilities for x, the total area under the curve is 1.0.
- To the right of the **pdf** is the symbol dx, which is the width of the rectangles (approaching 0), and not d times x. If dx is the width of the rectangles, and $f(x)$ is the height of each rectangle, then $f(x)dx$ is the area of a very small rectangle.
- To the right of the equal sign is 1.0; the area under the curve for all possible values of x between negative and positive infinity is 1.0.

Wolfram defines an integral as "a mathematical object that can be interpreted as an area or a generalization of area. Integrals, together with derivatives, are the fundamental objects of calculus. Other words for integral include antiderivative and primitive."

 ?? **Do all probability density functions have an area under the curve = 1.0?**

Answer: Yes. That is a characteristic that defines all probability density functions. Click here to see the derivation for the normal **pdf**!

 ?? **What would the integral look like for the normal pdf?**

Answer: Just replace $f(x)$ above with the normal **pdf**, which is:

$$f(x; \mu, \sigma) = \frac{1}{\sqrt{2\pi}\sigma} e^{-(x-\mu)^2/(2\sigma^2)}. \tag{9.26}$$

Then, we have:

$$\int_{-\infty}^{\infty} f(x; \mu, \sigma)dx = 1.0 \tag{9.27}$$

$$\int_{-\infty}^{\infty} \frac{1}{\sqrt{2\pi}\sigma} e^{-(x-\mu)^2/(2\sigma^2)}dx = 1.0. \tag{9.28}$$

 ?? OK, one more time! What is the probability that a bacterium has a lifespan between 4.5 and 5.5 hours?

Answer: Before, we divided the distribution into thin rectangles, found the area of the rectangles in question, and added them up. Now we are considering a subset of *x*'s, but **we make use of the density function to find the exact area**.

Let's step through this carefully:

$$\Pr(4.5 \leq X \leq 5.5) = \int_{4.5}^{5.5} \frac{1}{\sqrt{2\pi}\sigma} e^{-(x-\mu)^2/(2\sigma^2)}dx. \tag{9.29}$$

Now, we need to evaluate this to get our answer. The process of finding this answer is called **integration**. And the result is a **probability**.

 ?? How does one go about integrating?

Answer: Well, we have the **pdf**, and we know the shape of the curve. Now we need a tool to help us find the area under a specific part of the curve. This area is probability. Computing the Gaussian integral is a bit beyond the scope of this book, and you really don't need to solve it by hand for our purposes.

Fortunately, the normal distribution is such a commonly used distribution that we can easily find the areas that we want from calculators or computers, which often use the "super-skinny rectangle" approach to approximate the answer. Technically, this is called Riemann's (*ree-mahn*) approximation. We can use virtually any calculator or math software to generate the probability that a bacterium's lifespan is between 4.5 and 5.5 hours, given a normal **pdf** with $\mu = 5$ and $\sigma = 0.5$. The calculation gives 0.6822689. Our 41 rectangles approach suggested that the probability was 0.6810742, which is not a bad approximation.

> For many problems, however, integration is a complex mathematical calculation, or worse, is completely intractable!

 ?? Are there refresher courses that review this material?

Answer: Yes, there are several excellent sources out there. Forget about memorizing anything...the only thing you need for this book is an intuition for how things work. Here are some of our favorites:

- Khan Academy
- Better Explained
- How to Enjoy Calculus by Eli Pine

??   **What other probability density functions are there?**

Answer: Loads of them! Here's a look at a few that you may have come across:

- normal
- log-normal
- beta
- gamma
- exponential
- Weibull
- Cauchy

We will explore a few of these in future chapters, but encourage you to check these out on Wikipedia. Who knows? Some may be very relevant to your work!

Likelihood

Now let's gracefully switch to the topic of likelihood. Likelihood is a key concept in Bayesian inference, so we need to spend time with it.

 ??   **OK, what exactly is likelihood?**

Answer: You may recall that Dictionary.com provides three definitions of likelihood:

1. The state of being likely or probable; probability.
2. A probability or chance of something: There is a strong likelihood of his being elected.
3. Archaic: indication of a favorable end; promise.

So, **likelihood** is another word for probability. However, because our data are continuous in nature, likelihood here can be taken to mean probability density. Likelihood involves the collection of data (variables), and we look retrospectively at the probability or probability density of collecting those data under a given set of parameters.

We noted in Chapter 8 that the discrete version of Bayes' Theorem can be written as shown below, with the likelihood portion shown in red, the priors in blue, and the posterior in purple:

$$\Pr(H_i \mid data) = \frac{\Pr(data \mid H_i) * \Pr(H_i)}{\sum_{j=1}^{n} \Pr(data \mid H_i) * \Pr(H_j)}. \tag{9.30}$$

Now, let's consider Bayes' Theorem when the hypotheses for a parameter are infinite. Here, Bayes' Theorem takes a new form. Suppose we are trying to estimate a single parameter called θ. Bayes' Theorem in this case is specified as:

$$P(\theta \mid data) = \frac{P(data \mid \theta) * P(\theta)}{\int P(data \mid \theta) * P(\theta) d\theta}. \tag{9.31}$$

This is the generic version of Bayes' Theorem when the posterior distribution for a single parameter, θ, given the observed data, is represented by a **pdf**.

Notice the notation $P(\theta \mid \text{data})$ for the posterior distribution of θ. Here, P indicates probability density. Both the prior ($P(\theta)$) and the posterior $P(\theta \mid \text{data})$ distributions are **pdf**s. In contrast, with the discrete version of Bayes' Theorem, the prior and posterior distributions are **pmf**s, denoted with Pr.

The likelihood in Bayes' Theorem may be written $P(\text{data} \mid \theta)$ or $\Pr(\text{data} \mid H_i)$—which version you use depends on the problem you are solving.

Throughout this book, we will express computations for the likelihood term of Bayes' Theorem as $\mathcal{L}(\text{data}; \theta)$ where $\mathcal{L}$ symbolizes "likelihood" and the semicolon means "given." We do this to differentiate the likelihood term in Bayes' Theorem from the actual computation.

As a side note to our readers who previously have studied maximum likelihood methods, read the box below:

You may have seen likelihood expressed as the likelihood of the parameters given the data, or $\mathcal{L}(\theta; \text{data})$, which is conventional. Regardless of the notation used, you must enter values for the parameters and observed data to calculate a result.

Earlier in this chapter, we used the normal **pdf** to get the **probability density** of observing x, given the mean lifespan is μ and the standard deviation is σ:

$$f(x; \mu, \sigma) = \frac{1}{\sqrt{2\pi}\sigma} e^{-(x-\mu)^2/(2\sigma^2)}. \tag{9.32}$$

This is used *before* we observe the data. When we speak of likelihood, in contrast, we have the variables (data) in hand, and one or more of the parameters are unknown. To compute the likelihood of the data, we first need to make an assumption about how the data were generated. Here, we assume that the data are generated from a normal distribution. Let's let X be our random variable, and X is the lifetime of a bacterium in hours. We can write:

$$X \sim N(\mu, \sigma). \tag{9.33}$$

Suppose we hypothesize that $\mu = 5.0$, and assume that σ is known to be 0.5. And further suppose that we draw a random bacterium that lives $x = 4.5$ hours. We can ask, "What is the likelihood that $x = 4.5$ given that μ is 5.0 and $\sigma = 0.5$?" We will use the normal **pdf** to answer this question, which is

$$\mathcal{L}(x = 4.5; \mu = 5.0, \sigma = 0.5) = \frac{1}{\sqrt{2\pi}\sigma} e^{-(x-\mu)^2/(2\sigma^2)}. \tag{9.34}$$

Now, let's plug in the values that we know to get our answer:

$$\mathcal{L}(x = 4.5; \mu = 5.0, \sigma = 0.5) = \frac{1}{\sqrt{2\pi}0.5} e^{-(4.5-5.0)^2/(2*0.5^2)} = 0.4839414. \tag{9.35}$$

This is a probability density. Look for this result in blue in Figure 9.12.

Suppose now that we hypothesize that $\mu = 4.1$ (with σ known to be 0.5). Now let's calculate the likelihood:

$$\mathcal{L}(x = 4.5; \mu = 4.1, \sigma = 0.5) = \frac{1}{\sqrt{2\pi}0.5} e^{-(4.5-4.1)^2/(2*0.5^2)} = 0.5793831. \tag{9.36}$$

Look for this result in green in Figure 9.12.

As you can see, we have an infinite number of possibilities for μ. Nevertheless, we can generate the **likelihood profile** by repeatedly plugging in values and mapping the shape of the surface. Here it is in Figure 9.12.

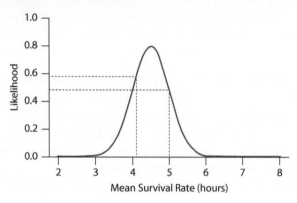

Figure 9.12 Likelihood profile for μ.

A few things are worth pointing out:

- First, notice that the y-axis is labeled "Likelihood" instead of "Density" to indicate that we are plotting the density associated with alternative values for an unknown *parameter*. The x-axis is the unknown parameter, μ.
- Second, the function that can be used to draw this curve is called a likelihood function. Here, we are given data in the form of observed random variables. We observed $x = 4.5$. Assuming we know $\sigma = 0.5$, we use the normal **pdf** to ask how likely alternative values of μ are.
- Third, the area under a likelihood function does NOT need to integrate to 1.0! We saw this in Chapter 8 as well.

 ?? ## What if you don't know that σ is 0.5?

Answer: We generate the likelihood surface for **two parameters**. In other words, we play the 'what if' game for both μ and σ.

To illustrate, suppose we selected three bacteria at random that lived 3.6, 4.7, and 5.8 years, respectively. These are our observed data. We are trying to estimate μ and σ from the distribution that generated these data. We choose a suite of combinations of μ and σ. Then, for **each** datapoint, we calculate the likelihood of observing the data.

We use the normal **pdf** as our baseline:

$$f(x; \mu, \sigma) = \frac{1}{\sqrt{2\pi}\sigma} e^{-(x-\mu)^2/(2\sigma^2)}. \tag{9.37}$$

For instance, if $\mu = 5$ and $\sigma = 1.0$, then the likelihood of observing $x = 3.6$ is calculated as:

$$\mathcal{L}(x = 3.6; \mu = 5.0, \sigma = 1.0) = \frac{1}{\sqrt{2\pi}(1.0)}e^{-(3.6-5)^2/(2(1.0)^2)} \qquad (9.38)$$

$$\mathcal{L}(x = 3.6; \mu = 5.0, \sigma = 1.0) = 0.1497275. \qquad (9.39)$$

For $x = 4.7$, we have:

$$\mathcal{L}(x = 4.7; \mu = 5.0, \sigma = 1.0) = \frac{1}{\sqrt{2\pi}(1.0)}e^{-(4.7-5)^2/(2(1.0)^2)} = 0.3813878. \qquad (9.40)$$

For $x = 5.8$, we have:

$$\mathcal{L}(x = 4.8; \mu = 5.0, \sigma = 1.0) = \frac{1}{\sqrt{2\pi}(1.0)}e^{-(5.8-5)^2/(2(1.0)^2)} = 0.2896916. \qquad (9.41)$$

If we assume that all three datapoints were independent, we can compute the likelihood of observing the full dataset (which is 3 observations) given $\mu = 5$ and $\sigma = 1$ as the product of the independent likelihoods:

$$\mathcal{L}(\text{data}; \mu = 5.0, \sigma = 1.0) = 0.1497275 * 0.3813878 * 0.2896916 = 0.01654262. \qquad (9.42)$$

If we played this game across different combinations of μ and σ, we could create a **likelihood surface** that looks like the one in Figure 9.13.

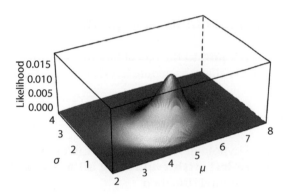

Figure 9.13

This is a **likelihood surface** for the unknown parameters, μ and σ. See if you can find the likelihood value when $\mu = 5$ and $\sigma = 1$. It should be 0.01654262 if our calculations were correct.

Our observed dataset consists of 3 values of lifespan: 3.6, 4.7, and 5.8. The mean of this dataset is 4.7 and the standard deviation is 1.1. You should see that the **most likely** parameter estimates in the graph above correspond to these values.

 ?? How can this be used in a Bayesian inference problem?

Answer: Ah, yes! Let's not lose sight of the big picture!

Suppose we are trying to estimate a single parameter called θ. Bayes' Theorem in this case is specified as:

$$P(\theta \,|\, \text{data}) = \frac{P(\text{data} \,|\, \theta) * P(\theta)}{\displaystyle\int P(\text{data} \,|\, \theta) * P(\theta) d\theta}. \tag{9.43}$$

This is the generic version of Bayes' Theorem when the posterior distribution for a single parameter, given the observed data, is represented by a **pdf**.

- The posterior distribution is designated $P(\theta \,|\, \text{data})$, **where P is probability density (*not* probability)**. Pay attention to this notation! This is the left side of the equation, and must integrate to 1.
- On the right side of the equation, the numerator multiplies the prior probability **density** of θ, which is written $P(\theta)$, by the likelihood of observing the data under a given hypothesis for θ, which is written $P(\text{data} \,|\, \theta)$. THIS is the likelihood we've been talking about, and it can be a continuous or discrete function. Remember that the likelihood does not have to integrate or sum to 1!
- In the denominator, we see the same terms, but this time we also see a few more symbols. The symbol $\int$ means "integrate", which roughly means "sum up all the pieces" for each tiny change in θ, which is written $d\theta$. In other words, the denominator accounts for the prior density * likelihood for all possible hypotheses for θ, and sums them.

Note that this version deals with a single parameter called θ. But you can use Bayes' Theorem to estimate multiple parameters. We'll touch on this topic more in future chapters.

 ?? Can you estimate the probability of a specific hypothesis for theta?

Answer: Nope! $P(\theta \,|\, \text{data})$ is a **pdf**, so the probability that θ assumes a specific value is 0. Remember that the probability that a bacterium lives for exactly 5.0000000 hours is 0!

 ?? So, there are no specific hypotheses?

Answer: There are infinite hypotheses, and since you can't estimate the probability of a specific hypothesis, **you are left with evaluating all of them!** In other words, you must estimate the entire posterior distribution.

 ?? Can we see a Bayesian inference problem with infinite hypotheses?

Answer: OK! Let's try to put together everything we've learned so far!

Let's go through a quick example where we are considering hypotheses for an unknown mean, which is a parameter from the Gaussian distribution. How about the average lifespan of a bacterium? For this example, we'll assume σ is known to be 0.5.

Here, Bayes' Theorem can be expressed as:

$$P(\mu \,|\, \text{data}) = \frac{P(\text{data} \,|\, \mu) * P(\mu)}{\displaystyle\int P(\text{data} \,|\, \mu) * P(\mu) d\mu}. \tag{9.44}$$

We have an infinite number of hypotheses for μ, and we want to update our beliefs after collecting some data. Let's review the steps of Bayesian analysis.

Step 1. What are the hypotheses?

Answer: There are an infinite number of hypotheses for μ. You could have some bounds though.

Step 2. What were the prior probabilities for each hypothesis?

Since there are an infinite number of hypotheses, it makes sense that we use a **pdf** to represent the prior distribution. Here, we have some choices. If we think that μ can range between 4 and 6, we can use a uniform distribution to indicate that all hypotheses have the same "weight" before considering the data (density $= 0.5$; see Figure 9.14).

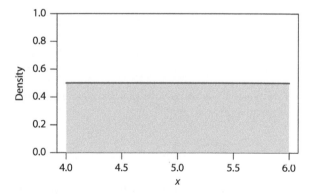

Figure 9.14

This represents our prior distribution. This is called a **proper prior** because the density function integrates to 1.

Step 3. Collect data.

Suppose we draw a random bacterium that lives 4.7 years.

Step 4. Compute the likelihood of the data under each hypothesis.

Now we play the what if game, and use the normal **pdf** to compute the likelihood of observing $x = 4.7$ under different values of μ. Suppose, for the sake of example, that we know that $\sigma = 0.5$. We start by writing the normal **pdf**:

$$f(x; \mu, \sigma) = \frac{1}{\sqrt{2\pi}\sigma} e^{-(x-\mu)^2/(2\sigma^2)}. \tag{9.45}$$

But instead of calling the function f, let's call it $\mathcal{L}$. And, let's fully specify our model. Here is the likelihood value for just one value of μ:

$$\mathcal{L}(x = 4.7; \sigma = 0.5, \mu = 4.0) = \frac{1}{\sqrt{2\pi}(0.5)} e^{-(4.7-4.0)^2/(2(0.5)^2)} = 0.2994549 \tag{9.46}$$

and so on . . . look for this value in Figure 9.15.

As you can see, we have an infinite number of possibilities for μ. Nevertheless, we can use the normal **pdf** repeatedly to generate the **likelihood profile**. Here it is in Figure 9.15.

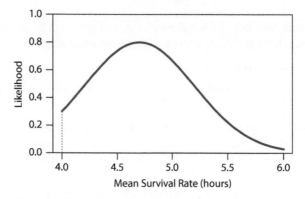

Figure 9.15 Likelihood profile for the unknown mean.

Here, we have the likelihood of observing the data, $x = 4.7$, under each hypothesized value for μ. Given just a single datapoint, the most likely survival rate is 4.7. You can see from the shape of this curve that several other hypotheses have decent height, too. The hypothesis with the lowest likelihood is $\mu = 6.0$.

Step 5. Use Bayes' Theorem to update the priors to posteriors.
Here, Bayes' Theorem can be expressed as:

$$P(\mu \mid \text{data}) = \frac{P(\text{data} \mid \mu) * P(\mu)}{\int P(\text{data} \mid \mu) * P(\mu) d\mu}. \tag{9.47}$$

Remember, our prior distribution is the uniform **pdf**. The placeholder for that function is shown in blue:

$$P(\mu \mid \text{data}) = \frac{P(\text{data} \mid \mu) * P(\mu)}{\int P(\text{data} \mid \mu) * P(\mu) d\mu}. \tag{9.48}$$

And our likelihood function is generated using the Gaussian **pdf**. The placeholder for that function is shown in red:

$$P(\mu \mid \text{data}) = \frac{P(\text{data} \mid \mu) * P(\mu)}{\int P(\text{data} \mid \mu) * P(\mu) d\mu}. \tag{9.49}$$

The posterior distribution is another **pdf** of some sort. The placeholder for that function is shown in purple:

$$P(\mu \mid data) = \frac{P(data \mid \mu) * P(\mu)}{\int P(data \mid \mu) * P(\mu)d\mu}. \tag{9.50}$$

Now it's just a matter of solving.

?? Can we depict this problem graphically?

Answer: Yes. Let's focus on the big picture and create a diagram for this problem (Kruschke plot) as we did in Chapter 8 (see Figure 9.16). Remember, this diagram is intended to communicate the structure of the prior and likelihood.

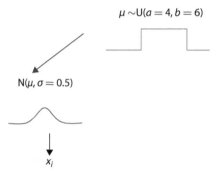

Figure 9.16

At the bottom of this diagram, we have our observed data, x_i. The data were generated from a normal distribution with an unknown parameter, μ, and $\sigma = 0.5$. The parameter, μ, in turn, is the unknown parameter that we are trying to estimate. Here, we use a uniform distribution with parameters $a = 4$ and $b = 6$ to set the "weights" on the alternative hypotheses for μ. Thus, you can interpret the blue distribution above as a prior distribution that provides our weights of belief, or current knowledge, regarding μ. The drawing of this distribution can be generalized.

This interpretation falls in line with that given by Sander Greenland (2006), who writes, "It is often said (incorrectly) that parameters are treated as fixed by the frequentist but as random by the Bayesians. For frequentists and Bayesians alike, the value of a parameter may have been fixed from the start or may have been generated from a physically random mechanism. In either case, both suppose it has taken on some fixed value that we would like to know. The Bayesian uses formal probability models to express personal uncertainty about that value. The 'randomness' in these models represents personal uncertainty about the parameter's value; it is not a property of the parameter (although we should hope it accurately reflects properties of the mechanisms that produced the parameter)."

S. Greenland. "Bayesian perspectives for epidemiological research: I. Foundations and basic methods." *International Journal of Epidemiology* 35.3 (2006): 765–74.

But there is another interpretation for the blue prior distribution displayed: the unknown parameter, μ, is a **random variable** that arises from a uniform distribution. In this example, μ assumes the values between $\mu = 4$ and $\mu = 6$. The Bayesian machinery will update the prior distribution to a posterior distribution in light of new data.

 ?? If we couldn't integrate the normal distribution, how on earth are we going to integrate the denominator of Bayes' Theorem?

Answer: Sometimes it is intractable! For those cases, we need a special "tool" that can help us estimate the posterior distribution without a mathematical, closed-form solution. That tool is called Markov Chain Monte Carlo (MCMC), and we will introduce this topic later in the book.

There are a few cases, however, where you can use a particular **pdf** as a prior distribution, collect data of a specific flavor, and then derive the posterior **pdf**. In these special cases, the **pdf** of the prior and posterior are the same probability density function, but their *parameters* may differ. The prior distribution is called a conjugate prior (Raiffa and Schlaeffer, 1961), and the effect of the data can then be interpreted in terms of changes in parameter values (Upton and Cook, 2014).

Here are some examples:

- beta pdf prior + binomial data → beta pdf posterior
- gamma pdf prior + Poisson data → gamma pdf posterior
- normal pdf prior + normal data → normal pdf posterior
- Dirichlet pdf prior + multinomial data → Dirichlet pdf posterior

We'll dive into these in Section 4. We covered a lot of ground. Now make sure to stretch and relax!

SECTION 4

Bayesian Conjugates

Overview

Welcome to Section 4! As we've mentioned, a major use of Bayesian inference involves parameter estimation. In this section, we will learn about three Bayesian conjugates that can aid in parameter estimation. As you'll see, these are special cases where a Bayesian prior probability density function for an unknown parameter of interest can be updated to the posterior pdf when data that are collected are a specific flavor. In these special cases, an analytical solution exists that makes this update possible, and avoids the integration required in the denominator of Bayes' Theorem. In the process, we'll learn two new probability density functions (beta and gamma), and a new probability mass function (Poisson).

- Chapter 10 introduces the beta-binomial conjugate. In the White House Problem, we use a beta distribution to set the priors for all hypotheses of p, the probability that a famous person could get into the White House without invitation. We then collected binomial data to determine the number of times a famous person could gain entry out of a fixed number of attempts, and update the prior distribution to the posterior distribution in light of this new information. In short, a beta prior distribution for the unknown parameter + binomial data → beta posterior distribution for the unknown parameter, p.
- Chapter 11 introduces the gamma-Poisson conjugate. In the Shark Attack Problem, we use a gamma distribution as the prior distribution of λ, the mean number of shark attacks in a given year. We then collect Poisson data to determine the number of shark attacks in a given year, and update the prior distribution to the posterior distribution in light of this new information. In short, a gamma prior distribution + Poisson data → gamma posterior distribution.
- Chapter 12 introduces the normal-normal conjugate. In the Maple Syrup Problem, we use Bayesian methods to estimate the two parameters that identify a normal distribution, μ and σ. To use the conjugate solution, we assume σ is known and focus our attention on μ. We use a normal distribution as the prior distribution of μ, the mean number of millions of gallons of maple syrup produced in Vermont in a year. We then determined the amount of syrup produced in multiple years and assumed that the amount followed a normal distribution with known σ, and updated the prior distribution to the posterior

distribution in light of this new information. In short, a normal prior distribution + normally distributed data → normal posterior distribution.

There are many other conjugate solutions, but we've highlighted three commonly used solutions. In each chapter, we set the scene with a fun problem and then use a conjugate solution to update the prior distribution for an unknown parameter to the posterior distribution. We will revisit each of these chapters in Section 5, where we will show you how to estimate the posterior distribution with a different method, MCMC.

CHAPTER 10

The White House Problem:
The Beta-Binomial Conjugate

We hope you now have a very solid understanding of probability distributions. In this chapter (and the next two that follow), we show you how to use Bayes' Theorem to estimate the parameters of a probability distribution. Indeed, a very common use of Bayesian inference involves **parameter estimation**, where the analysis uses probability density functions (**pdf's**).

By the end of this chapter, you will have a thorough knowledge of the following:

* Beta distribution
* Binomial data
* Hyperparameters
* Conjugate prior
* Credible intervals

The simplest way to show you how Bayes' Theorem is used to estimate parameters is to dive right into a new problem. Let's run through an example of parameter estimation with Bayesian inference.

This one was taken from The Mike Wise Radio Show (accessed August 18, 2017).

NBA star Shaquille O'Neal and a friend debated whether or not he could get into the White House without an appointment. The wager: 1000 push-ups.

Here's a picture of Shaq in case you haven't met him (see Figure 10.1); he's 7′ 1″, weighs 325 pounds, and has earned tons of accolades in the NBA. On top of that, he's a rapper and has a Ph.D. At the time of the bet, Barack Obama was President of the United States. Shaq knows that President Obama is a huge basketball fan and coach and is sure he can get in to meet with the President.

For this chapter, we're going to broaden this problem a bit and ask, "What is the probability that any famous person (like Shaq) can drop by the White House without an appointment?"

 What do YOU think Shaq's probability of getting into the White House is?

Answer: Your answer here!

Bayesian Statistics for Beginners: A Step-by-Step Approach. Therese M. Donovan and Ruth M. Mickey, Oxford University Press (2019). © Ruth M. Mickey 2019.
DOI: 10.1093/oso/9780198841296.001.0001

Figure 10.1 Shaquille O'Neal.

 ?? **What probability function would be appropriate for Shaq's bet?**

Answer: Well, let's see. Shaq is going to attempt to get into the White House, so that represents a **trial**. He will either succeed or fail. Shaq thinks his probability of success is quite high. The outcome of this trial determines who wins the bet. Ring any bells?

This is a binomial problem. You might recall that the binomial probability mass function is:

$$f(y; n, p) = \binom{n}{y} p^y (1-p)^{(n-y)} \qquad y = 0, 1, \ldots, n \qquad (10.1)$$

The number of trials is denoted n. The number of observed successes is denoted as y. The probability of success is denoted as p.

If we *knew* that Shaq's probability of success is say, 0.7, and that Shaq had only **one** chance, we could use the binomial probability mass function (pmf) to show the probability of each outcome. The probability that Shaq would succeed is:

$$f(1; 1, 0.7) = \binom{1}{1} 0.7^1 (1 - 0.7)^{(1-1)} = 0.7. \qquad (10.2)$$

The probability that Shaq would fail is:

$$f(0; 1, 0.7) = \binom{1}{0} 0.7^0 (1 - 0.7)^{(1-0)} = 0.3 \qquad (10.3)$$

We can graph this probability distribution as shown in Figure 10.2.

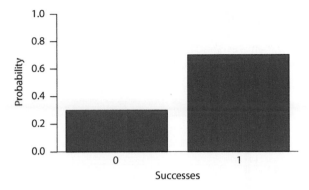

Figure 10.2 Binomial distribution: $n = 1$, $p = 0.7$.

With one trial, the probability of getting in is 0.7, and the probability of not getting in is 0.3. The sum of the probabilities equals 1.0. Incidentally, a binomial distribution with just 1 trial is also called a Bernoulli distribution. Remember?

If the bet let Shaq try **three times**, he could compute the probability of observing 0, 1, 2, 3 successes out of three trials, given $p = 0.7$:

$$f(y; 3, 0.7) = \binom{3}{y} 0.7^y (1 - 0.7)^{(3-y)}. \tag{10.4}$$

The binomial distribution for $n = 3$ and $p = 0.7$ looks like the one shown in Figure 10.3.

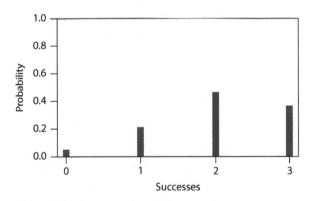

Figure 10.3 Binomial distribution: $n = 3$, $p = 0.7$.

With 3 attempts, and p (the probability of success) being 0.7, Shaq would have a small chance of never getting in (0 successes out of 3 trials; a probability of 0.027).

?? Are these trials independent?

Answer: You might recall that a key assumption of the binomial distribution is that the trials are independent. If Shaq makes all three attempts, it could be considered an example of non-independence. You might argue, however, that if the White House guards change so that the same guards don't confront Shaq, the trials would be independent. However, different guards might have different standards for uninvited guests! Let's "play on," assuming that the trials are independent.

?? **But what about Shaq's friend?**

Shaq's friend, however, believes Shaq's chances are much lower than 0.7. If p is, say 0.1, then Shaq's chances would look much different. Figure 10.4 shows the binomial distribution with $n = 3$ trials and $p = 0.1$.

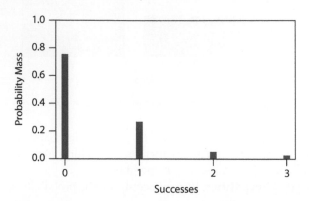

Figure 10.4 Binomial distribution: $n = 3$, $p = 0.1$.

In this case, Shaq would have a large chance of failure (in fact, a probability of 0.729).

Here's the kicker: We don't know what p (the probability of success) is! Obviously, there is some dispute over what value it is, or there would not be a bet! So, here we are confronted with a parameter estimation problem.

You are trying to estimate p, the probability that Shaquille O'Neal or some other famous person can get into the White House without an invitation.

Our goal here is to use a Bayesian inference approach to estimate the probability that a celebrity can get into the White House without an invitation. We'll use the same steps that we have in previous chapters:

1. Identify your hypotheses—these would be the alternative hypotheses for p, ranging from 0 to 1.00.
2. Express your belief that each hypothesis is true in terms of prior densities.
3. Gather the data—Shaq makes his attempt, and will either fail or succeed.
4. Determine the **likelihood** of the observed data, assuming each hypothesis is true.
5. Use Bayes' Theorem to compute the posterior densities for each value of p (i.e., the posterior distribution).

Now, let's go through the steps one at a time.

 ?? **Step 1. What are the hypotheses for p?**

Answer: We know that p, the probability of success in the binomial distribution, can take on any value between 0 and 1.0. Shaq *could* say that the prior probability of p is close to 1.0; his friend could disagree and suggest that the prior probability is closer to 0.01. These are just two hypotheses for p. However, there's nothing to stop us from considering the full range of hypotheses between 0 and 1, which is **infinite** ($p = 0.01$, p = 0.011, $p = 0.0111$, etc.).

?? Step 2. What are the prior densities for these hypotheses?

Answer: We need to assign a prior for each hypothesized value of p. Here, we will use the **beta distribution** to set prior probabilities for each and every hypothesis for p. An example of a beta probability distribution is shown in Figure 10.5.

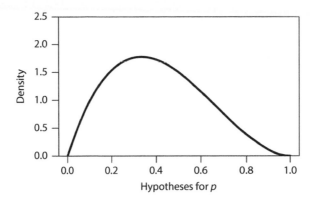

Figure 10.5

The x-axis for the beta distribution ranges between 0 and 1. That is to say, random variables that are beta distributed must be between 0 and 1. The y-axis gives the **probability density**, which we learned about in the last chapter. Remember that density is **not** probability, per se, because the area under the curve must be 1. But you can think of the densities associated with each value of p as the "weight" for each hypothesis. The distribution above would put greater weight for p's between, say, 0.2 and 0.6, than other values.

?? What is the beta distribution?

Answer: The Oxford Dictionary of Statistics tells us that the beta distribution is "a distribution often used as a prior distribution for a proportion." Wikipedia has this to say: "In probability theory and statistics, the beta distribution is a family of continuous probability distributions defined on the interval (0, 1) parameterized by two positive shape parameters, typically denoted by alpha (α) and beta (β). The beta distribution can be suited to the statistical modelling of proportions in applications where values of proportions equal to 0 or 1 do not occur."

So, the beta distribution is a statistical distribution whose x-axis spans the interval from 0 to 1. The beta distribution has two parameters that control its shape and position, and these parameters are named alpha (α) and beta (β). The values for α and β must be positive (0 and negative entries won't work). The beta probability distribution shown in Figure 10.5 has parameters $\alpha = 2$ and $\beta = 3$.

Observations drawn from this distribution can take on values between 0 and 1. This can be denoted generally as:

$$X \sim \text{beta}(\alpha, \beta) \tag{10.5}$$

Here, X is a continuous random variable and is distributed as a beta distribution with parameters α and β.

The beta probability density function looks like this:

$$f(x; \alpha, \beta) = \frac{1}{B(\alpha, \beta)} x^{\alpha-1}(1-x)^{\beta-1}, \qquad 0 < x < 1. \tag{10.6}$$

Can you find the two parameters, α and β here? The beta function, B, is a normalization constant to ensure that the area under the curve is 1. Don't confuse the beta function with the beta pdf!

 ?? ## Can you show other examples of beta distributions?

Answer: Of course. Figure 10.6 shows some other examples.

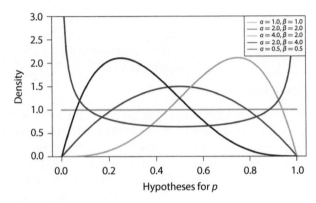

Figure 10.6

Since Shaq and his friends have totally different ideas of what p should be, they might settle on using a beta distribution with $\alpha = 0.5$ and $\beta = 0.5$ as the prior distribution for p, which gives the U-shaped result in blue.

 ?? ## I don't have a good feeling for what the α and β parameters do in terms of controlling the shape and location of the distribution. Can you help?

Answer: Generally speaking, the bigger α is relative to β, the more we shift the **weight** of the curve to the right, and the bigger β is relative to α, the more we shift the weight of the curve to the left.

But there is another way to get a handle on α and β. The mean of a beta distribution (also called its first moment) can be calculated as:

$$\mu = \frac{\alpha}{\alpha + \beta}. \tag{10.7}$$

And the variance of a beta distribution can be calculated as:

$$\sigma^2 = \frac{\alpha * \beta}{(\alpha + \beta)^2 * (\alpha + \beta + 1)}. \tag{10.8}$$

?? Can I see an example of how this would be used?

Answer: Sure. Suppose **YOU** think that the average probability of Shaq getting into the White House is $p = 0.10$ but think that neighboring values also are likely and set the standard deviation, σ, to 0.05. Remember that the standard deviation is the square root of variance, or σ^2. Let's rearrange, plug in these values, and solve:

$$\beta = \mu - 1 + \frac{\mu * (1-\mu)^2}{\sigma^2} = 0.1 - 1 + \frac{0.1 * (1-0.1)^2}{0.05^2} = 31.5 \tag{10.9}$$

$$\alpha = \frac{\beta * \mu}{1 - \mu} = \frac{31.5 * 0.1}{1 - 0.1} = 3.5. \tag{10.10}$$

You can derive this on your own at http://www.wolframalpha.com/. The resulting beta distribution looks like the one shown in Figure 10.7.

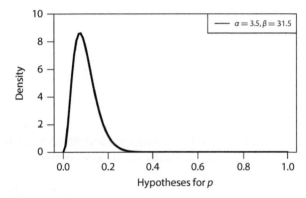

Figure 10.7

This little shortcut—called **moment matching**—can help you **parameterize** the beta distribution using means and standard deviations, which are probably more familiar to you.

?? What prior distribution did Shaq and his friend settle on?

Answer: Let's assume they go with a beta distribution with α and β set to 0.5. They think p is either really high or really low. We can designate these as $\alpha_0 = 0.5$ and $\beta_0 = 0.5$, where the "naught" subscripts alert us that these are the parameters of a prior distribution. Technically, these are known as **hyperparameters**.

?? Hyperparameters?

Answer: In Bayesian statistics, a hyperparameter is a parameter of a prior or posterior distribution. This term is used to distinguish the parameters of the prior or posterior distribution (see Figure 10.8) from the unknown parameter of interest. Our focus here is on the parameters of the prior distribution for the unknown parameter of interest, p.

Shaq and his friend had to select a prior distribution, and we can imagine them arguing over which prior distribution to use. The prior distribution in Figure 10.8 may have resulted after hours of heated discussion. Of course, Shaq and his friend may have opted to do a bit of research and make inquiries to the FBI regarding other uninvited but attempted visits by famous people!

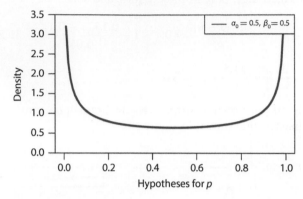

Figure 10.8 Prior distribution.

 ?? Step 3. Now what?

Answer: Collect data! That is step 3. **Let's assume that Shaq makes 1 attempt and fails to get in.** In the binomial function terms, the number of trials $n = 1$, and the number of successes $y = 0$.

Let's create a diagram that illustrates the process by which the data were generated (see Figure 10.9).

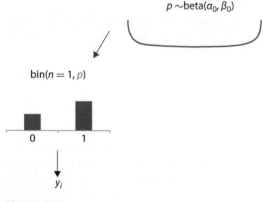

Figure 10.9

At the bottom of this diagram, we have our observed data, y_i. If Shaq makes one attempt, the data arise from a binomial distribution with parameters $n = 1$ and p. You may recall that a binomial distribution with $n = 1$ is also called a Bernoulli distribution. The parameter, p, in turn, is the unknown parameter that we are trying to estimate. Here, we use a beta distribution to set the "weights" on each and every hypothesis for the unknown parameter, p.

?? Step 4. And then?

Answer: Step 4 is to determine the likelihood of the observed data, assuming each hypothesis is true.

Here is where your previous study of binomial probability mass function comes into play. Because the data, 1 failure out of 1 attempt, are in hand, we will call the result "likelihood."

$$\mathcal{L}(y; n, p) = \binom{n}{y} p^y (1 - p)^{(n-y)}. \tag{10.11}$$

The number of trials is denoted n. The number of observed successes is denoted as y. The probability of success is denoted as p.

Now, for each hypothesized value of p, let's compute the binomial likelihood of observing 0 successes out of 1 trial.

Just to give you an example, under the $p = 0.3$ hypothesis, the likelihood of observing 0 successes out of 1 trial, given $p = 0.3$ is computed as:

$$\mathcal{L}(y = 0; n = 1, p = 0.3) = \binom{1}{0} 0.3^0 (1 - 0.3)^{(1-0)} = 0.695. \tag{10.12}$$

 ?? Do you see a problem here?

Answer: Your answer here!

You might have noted something critical: Because p is a continuous variable between 0 and 1, we don't have 1 hypothesis for p—we have an infinite number of hypotheses!

 ?? Now what?

Answer: Well, we need to set this Bayesian inference problem up for a **continuous random variable**. Suppose we are trying to estimate a single parameter called θ. If θ is continuous, you have an infinite number of hypotheses for it. Bayes' Theorem in this case is specified as:

$$P(\theta \mid \text{data}) = \frac{P(\text{data} \mid \theta) * P(\theta)}{\int P(\text{data} \mid \theta) * P(\theta) d\theta}. \tag{10.13}$$

This is the generic version of Bayes' Theorem when the posterior distribution for a single parameter, given the observed data, is represented by a pdf. This is designated $P(\theta \mid \text{data})$, **where P is probability density (*not* probability)**. Pay attention to this notation! This is the left side of the equation. On the right side of the equation, the numerator multiplies the prior probability **density** of θ, which is written $P(\theta)$, by the likelihood of observing the data under a given hypothesis for θ, which is written $P(\text{data} \mid \theta)$. Technically, the likelihood could be a **pmf** or a **pdf**, depending on the problem. For example, in this chapter's illustration, the likelihood is a pmf. In the denominator, we see the same terms, but this time we also see a few more symbols. The symbol $\int$ means "integrate," which roughly means "sum up all the pieces" for each tiny change in θ, which is written $d\theta$. In other words, the denominator accounts for the prior density * likelihood for all possible hypotheses for theta and then sums them.

For the Shaq problem, we replace θ with p, so Bayes' Theorem looks like this (where the priors are colored blue, the likelihoods red, and the posterior purple):

$$P(p \mid \text{data}) = \frac{P(\text{data} \mid p) * P(p)}{\int_0^1 P(\text{data} \mid p) * P(p)dp}. \tag{10.14}$$

But here's the kicker: The integration of the denominator is often tedious, and sometimes impossible!

?? How do we make headway?

Answer: Start integrating!

?? Really?

Answer: Just kidding! For this particular problem, thankfully, no. There is an analytical shortcut that makes updating possible in a snap. Here it is:

- $\alpha_{\text{posterior}} = \alpha_0 + y$.
- $\beta_{\text{posterior}} = \beta_0 + n - y$.

For the White House Problem, our prior distribution is a beta distribution with $\alpha_0 = 0.5$ and $\beta_0 = 0.5$. Shaq made an attempt, so $n = 1$. He failed to get into the White house, so $y = 0$. We can now use this shortcut to calculate the parameters of the posterior distribution:

- posterior $\alpha = \alpha_0 + y = 0.5 + 0 = 0.5$.
- posterior $\beta = \beta_0 + n - y = 0.5 + 1 - 0 = 1.5$.

Incidentally, the parameters of the posterior distribution are also called **hyperparameters**; posterior hyperparameters, to be exact.

Now we can look at the prior and posterior distributions for p (see Figure 10.10).

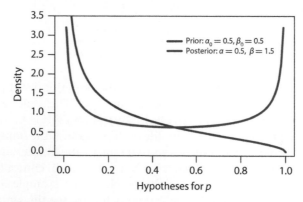

Figure 10.10

We started off with a prior distribution shown in blue, where Shaq and his friend believed they had a very high or very low probability of gaining entry to the White House without an invitation. Then we collected binomial data: one failure out of one trial. We then used the shortcut to generate the posterior distribution for all hypotheses of p.

As a result of Shaq's failed attempt, we now have **new** knowledge about the support for each and every hypothesis of p. Notice how the posterior really shifted towards the lower end of the p spectrum.

?? Would the posterior be different if Shaq used a different prior distribution?

Answer: Yes! Can we talk about priors for a minute?

Let's suppose that Shaq and his friend were concerned about using a non-informative (vague) prior distribution. They may have been tempted to set equal weight on all values of p, which would look like the distribution in blue below, where $\alpha = 1$ and $\beta = 1$. With this prior distribution, and with 0 successes out of 1 attempt to get into the White House without an invitation, the posterior hyperparameters are:

- $\alpha_{posterior} = \alpha_0 + y = 1 + 0 = 1$.
- $\beta_{posterior} = \beta_0 + n - y = 1 + 1 - 0 = 2$.

This prior distribution is shown in Figure 10.11 as solid blue, and the posterior distribution is shown as solid purple. The Figure 10.10 result is shown as a dashed line for comparison. These differences highlight that the choice of prior affects the posterior.

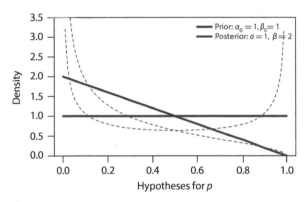

Figure 10.11

?? Is the flat prior really non-informative?

Answer: No. Here is a strange twist: the U-shaped prior that was actually used (α and $\beta = 0.5$) is **less informative** (more vague) than the "flat prior" above, where α and $\beta = 1$. The conjugate solutions show that as we make these parameters tiny, they have less influence on the resultant posterior distributions compared to the data. In other words, the data play a large role in determining the parameters of the beta posterior.

Thus, a non-informative prior for a beta distribution will be one in which α and β are tiny. In the words of Zhu and Lu, "flat priors are not necessarily non-informative, and non-informative priors are not necessarily flat." Once again, we refer you to Hobbs and Hooten

(2015) for a thoughtful discussion on setting priors. These authors prefer the term "vague prior" instead of non-informative prior because it's difficult or impossible to have a prior that is truly non-informative when dealing with a parameter that is continuous.

?? What if Shaq makes a second attempt?

Answer: Let's assume the first prior distribution mentioned was used (α_0 and $\beta_0 = 0.5$), and then updated after 1 failed attempt ($\alpha_0 = 0.5$ and $\beta_0 = 1.5$). We now set a prior distribution based on our updated knowledge, and then collect more data. **Suppose Shaq fails again.** Now, the parameters for our **next** posterior distribution will be as follows:

- $\alpha_{posterior} = \alpha_0 + y = 0.5 + 0 = 0.5$.
- $\beta_{posterior} = \beta_0 + n - y = 1.5 + 1 - 0 = 2.5$.

Now we can look at the prior and posterior distributions for p (see Figure 10.12).

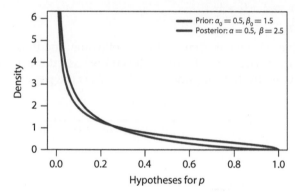

Figure 10.12

Here you can see the major benefit of Bayesian inference: as we collect more data, we update our beliefs. We start with a prior, collect data and then use Bayes' Theorem to generate posteriors. These posteriors then become the priors for the next round of data collection. If we track our beliefs, we track our **learning**.

?? How is this shortcut possible?

Answer: The answer is that the beta distribution is a conjugate distribution that can be updated with binomial data. This is why we named this chapter "The beta-binomial conjugate."

These shortcuts were introduced by Howard Raiffa (Figure 10.13, left) and Robert Schlaifer (Figure 10.13, right) in their work on Bayesian decision theory. Their classic 1961 book is titled Applied Statistical Decision Theory.

The conjugate shortcuts are conveniently provided on a Wikipedia page on Bayesian conjugates. A visual overview of that page's beta-binomial conjugate can be depicted as shown in Figure 10.14.

Figure 10.13 (Left) Howard Raiffa (HBS Archives Photograph Collection: Faculty and Staff, Baker Library, Harvard Business School (olvwork376291)). (Right) Robert Schlaifer (HBS Archives Photograph Collection: Faculty and Staff, Baker Library, Harvard Business School (olvwork383065)).

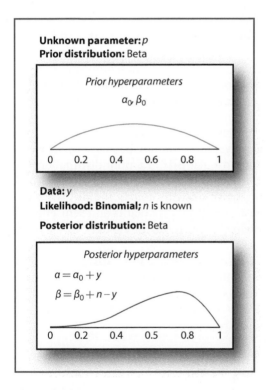

Figure 10.14

Here, each attempt was considered an experiment, where Shaq made a single attempt and either succeeded or failed. Before making any attempt, we set a prior distribution on p. He then failed to get into the White House. We then used this new information to get an updated posterior distribution. This in turn became our prior distribution for Shaq's second attempt. He failed again, and this information was used to update to yet another posterior distribution.

 ?? Could he have just tried twice before we updated the prior?

Answer: Yes. We would end up in the same place if we started with a prior distribution and considered 2 attempts (both failures) before updating. Let's confirm this:

- The prior:
  - $\alpha_0 = 0.5$; $\beta_0 = 0.5$
- The data:
  - $n = 2$
  - $y = 0$
- The posterior:
  - $\alpha = 0.5 + 0 = 0.5$
  - $\beta = 0.5 + 2 - 0 = 2.5$

This is the same answer we got before.

 ?? What exactly does the word "conjugate" mean?

Answer: First, Dictionary.com defines the word "conjugate" in several ways. In grammar it means "to recite or display all or some subsets of the inflected forms of a verb." For example, the forms of "to be" are "I am, you are, he is, we are, you are, they are." As a noun or adjective, the word conjugate means "joined together," especially in pairs. Thus, conjugates have a common theme that ties them together.

We mentioned the statistical definition in Chapter 9, where we indicated that there are cases where you can use a particular **pdf** as a prior distribution, collect data of a specific flavor, and then derive the posterior **pdf**. In these special cases, the **pdf** of the prior and posterior are the same probability density function, but their *parameters* may differ. The prior distribution is called a conjugate prior (Raiffa and Schlaeffer, 1961), and the effect of the data can then be interpreted in terms of changes in parameter values (Upton and Cook, 2014).

In the White House Problem, we used a **beta distribution** to set the priors for all hypotheses of p, the probability that a famous person could get into the White House without an invitation. We then collected binomial data, and, rather than computing the posterior by the use of Bayes' Theorem, we used the conjugate shortcut to generate the posterior distribution for p.

 ?? Why does the shortcut work?

Answer: Wikipedia again: "A conjugate prior is an algebraic convenience, giving a closed-form expression for the posterior; otherwise a difficult numerical integration may be necessary." In other words, the shortcut solution that provides the same answer you'd get if you worked through the integration yourself. The shortcut allows us to avoid the integral in Bayes' Theorem (although the result would be the same).

?? Can you show me the proof?

Answer: Gladly. The proof is in Appendix 1.

 ?? Can the beta distribution be used as a conjugate prior for data other than binomial data?

Answer: Yes. Wikipedia once more: "In Bayesian inference, the beta distribution is the conjugate prior probability distribution for the Bernoulli, binomial, negative binomial and geometric distributions. The beta distribution is a suitable model for the random behavior of percentages and proportions." These distributions all depend on a parameter, p that can assume values between 0 and 1.

 ?? Will we have more practice with conjugate priors?

Answer: Yes! Chapters 11 and 12 will give us more practice:

- gamma prior + Poisson data → gamma posterior
- normal prior + normal data → normal posterior

Some problems don't have the luxury of a conjugate prior. For those problems, we need another method for avoiding the integral in the denominator of Bayes' Theorem. That method, called the MCMC (for Markov Chain Monte Carlo) method, is described in Section 5.

 ?? OK. I'll wait. Suppose I do an analysis and have a new posterior distribution. How should I present my results?

Answer: A picture may be worth a thousand words. If you can display the posterior distribution, do so.

In addition, often people summarize the posterior distribution with some simple statistics that are familiar to most people: the mean, median, mode (all measures of middleness), plus the quantiles.

For well-known probability distributions such as the beta distribution, you can simply look up the expected values that characterize the distribution on Wolfram Mathworld:

- Mean $= \dfrac{\alpha}{\alpha + \beta}$.
- Mode $= \dfrac{\alpha - 1}{\alpha + \beta - 2}$ for α and $\beta > 1$.
- Variance $= \dfrac{\alpha * \beta}{(\alpha + \beta)^2 * (\alpha + \beta + 1)}$.

Let's do this for the Shaq problem. Let's calculate these for our new posterior, which has $\alpha = 0.5$ and $\beta = 2.5$:

- Mean $= \dfrac{0.5}{0.5 + 2.5} = 0.1667$.
- Mode cannot be calculated because $\alpha < 1$. Snap!
- Variance $= \dfrac{0.5 * 2.5}{(0.5 + 2.5)^2 * (0.5 + 2.5 + 1)} = 0.0617$.

?? How should I describe my confidence in the hypothesized values for *p*?

Answer: Report the Bayesian "credible intervals." Bayesian credible intervals are common-ly reported in Bayesian inference problems that involve parameter estimation. According to the Oxford Dictionary of Social Research Methods (Elliot et al., 2016), a credible interval represents "an interval in the domain of the posterior or predictive distribution. For a 95 percent credible interval, the value of interest lies with a 95 percent probability in the interval. In other words, given the data and the model, there is a 95 percent chance the true value lies in that interval."

A Wikipedia article describes three methods for choosing a credible interval (see also Kruschke, 2015):

- Choosing the narrowest interval, which for a unimodal distribution will involve choos-ing those values of highest probability density including the mode. This is sometimes called the highest posterior density interval.
- Choosing the interval where the probability of being below the interval is as likely as being above it. This interval will include the median. This is sometimes called the equal-tailed interval.
- Assuming that the mean exists, choosing the interval for which the mean is the central point.

Let's use our most up-to-date estimation of *p* (given Shaq's two failed attempts) and go with option 2 above, that is, look for the 90% credible interval. Our posterior distribution is a beta distribution with $\alpha = 0.5$ and $\beta = 2.5$. We need to find the area under the curve where 5% of the distribution is in the upper tail, and 5% is in the lower tail. We can use any computer software program to find these values, which are $p = 0.00087$ for the lower end, and $p = 0.57$ for the upper end (as shown by the dashed green lines in Figure 10.15).

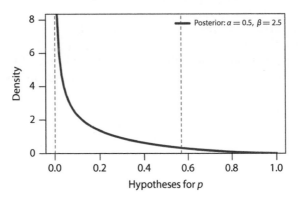

Figure 10.15

Interpretation: The probability that *p* is between 0.000867 and 0.57 is 0.90. The word "probability" is correct here because we've calculated an area under the pdf.

 ?? Can we summarize this chapter?

Answer: Indeed.

- We began by mentioning that Shaquille O'Neal made a bet with his friend that he could get into the White House without an invitation.
- We broadened this problem to ask, "What is the probability that any famous person can get into the White House without an invitation?"
- We introduced the **beta distribution** as a prior **pdf** that can be used to set hypotheses for p, the probability of success, noting that the beta distribution returns a probability density, not a probability. We selected hyperparameters for the beta distribution that reflected both Shaq and his friend's opinions about his probability of entry.
- We then collected data that is binomial in flavor. Shaq made one attempt and did not get in.
- We then calculated the likelihood of observing the data under some hypothesized values for p.
- But then we noted that we have an infinite number of hypotheses for p, so that the denominator of Bayes' Theorem now has an integral. This stopped us in our tracks.
- We relied on a shortcut—a conjugate solution—which allowed us to update our prior beta distribution to a posterior beta distribution using the data (one trial, one failed attempt).
- We then let Shaq make a second attempt. This time, we used our newly updated posterior as the prior distribution and then updated it again, given the outcome of the second attempt. We discussed this as a way to formalize "learning."
- Finally, we discussed ways of presenting your results, including calculation of credible intervals.

 ?? ## What really happened to Shaq?

Answer: Sadly, he was denied entrance. Shaq recounted the story to a Washington Post sportswriter this way (accessed August 17, 2017):

> "I went to the gate," [Shaq] said. "They were nice. They said, 'Shaq, we can't do it.' I said, 'I understand. Understood.'"

The sportswriter said that "Shaq left 1600 Pennsylvania Avenue in defeat Sunday, 1,000 push-ups in arrears. He said he was popping them off in bursts of 20 or 30 all day, and that he had already logged about 100 by the time we saw him in the production room Monday evening."

 ?? ## Will we see Shaq again?

We will revisit this chapter again in Section 5, where we will show you a different method for estimating the posterior distribution. For now, though, Chapter 11 introduces another conjugate called the gamma-Poisson conjugate. The chapter features blood and gore. Beware!

CHAPTER 11

The Shark Attack Problem: The Gamma-Poisson Conjugate

We interrupt this book to bring you the following news flash from the National Post:

Shark attacks attributed to random Poisson burst

Great headline, eh? This article appeared in the National Post on September 7, 2001. An unusually large number of shark attacks off the coast of Florida in 2001 spurred a flurry of excitement. What is the cause? Will I be next? In the story, Dr. David Kelton of Penn State University explained the attacks this way:

"Just because you see events happening in a rash like this does not imply that there's some physical driver causing them to happen. It is characteristic of random processes that they exhibit this bursty behavior."

Comforted now? What random process was he referring to? The headline refers to a **Poisson process**. You might have guessed this probability distribution is used to estimate

Bayesian Statistics for Beginners: A Step-by-Step Approach. Therese M. Donovan and Ruth M. Mickey, Oxford University Press (2019). © Ruth M. Mickey 2019.
DOI: 10.1093/oso/9780198841296.001.0001

the number of times an event will occur. You'd be right! In this chapter, we'll take a look at the problem of shark attacks and put it into a Bayesian framework.

By the end of this chapter, you should be able to define and use the following:

- The Poisson probability mass function
- The gamma probability density function
- The gamma-Poisson conjugate

As usual, we'll start with a question.

?? So, what exactly is the Poisson distribution?

Answer: Because shark attacks are a public health problem, let's see what Oxford Dictionary of Public Heath (Last, 2007) has to say: "A distribution function to describe rare events, or the sampling distribution of isolated counts in a continuum of space or time, such as cases of uncommon diseases."

In the Shark Attack Problem, Dr. Kelton explains: "There is something in probability theory called a Poisson process, giving amazingly good descriptions of such random independent events occurring through time, like customer arrivals at a fast-food store, cosmic rays striking a planet, accidents in a factory, airplane mishaps, and maybe shark attacks." We need to consider pmf's because we have discrete numbers; that is, the number of attacks is not continuous.

> The Poisson distribution is a mathematical rule that assigns probabilities to the number of occurrences observed.

?? To what does "Poisson" refer?

The Poisson distribution is named for the French mathematician Simeon Poisson (see Figure 11.1), who derived this distribution in 1837. This is the same year that Queen Victoria of England took the throne. Coincidentally, "poisson" means "fish" in French, which is fitting for the Shark Attack Problem!

Let's assume that the average rate is **2.1** attacks per year, based on the number of attacks reported annually at Florida beaches in the past 10 years (see Table 11.1).

Knowing this, you can use the Poisson probability mass function to determine the Poisson probability that there will be, say, 10 attacks next year.

?? OK, what exactly is the function?

The Poisson probability mass function is written:

$$\Pr(X = x; \lambda) = \frac{\lambda^x e^{-\lambda}}{x!} \qquad x = 0, 1, 2, \ldots \tag{11.1}$$

where λ is the mean number of occurrences in a given period of time, x is the number of occurrences we are interested in, and **e** is the natural logarithm constant (approximately 2.718), also called Euler's number. The left side of the equation can be read, "What is the probability that a random variable called X has a value of x, given λ." Notice the three dots to the right of the equation. Unlike the binomial distribution, where the number of possible

Figure 11.1 Simeon Poisson.

Table 11.1

Year	Attacks
1	1
2	0
3	2
4	1
5	1
6	3
7	2
8	3
9	4
10	4

values is fixed $(x = 0, 1, \ldots, n)$, the possible values that x can assume for the Poisson distribution $(x = 0, 1, \ldots)$ is not finite; rather it is "countably infinite."

 ?? What are the parameters of the function?

Answer: The Poisson **pmf** has just **one** parameter that controls both its shape and location. This parameter is called "lambda" and is written λ. It represents the mean number of occurrences in a fixed space or time period, such as the mean number of births in a year, the mean number of accidents in a factory in a year, or the mean number of shark attacks in a year. So, λ must be > 0 or the event will never occur. By the way, λ is also the variance of the distribution.

Let's work through an example to see the Poisson pmf in action. We have information in Table 11.1 that suggests the average rate of shark attacks in Florida is 2.1 attacks per year. What is the probability that there will be 10 attacks next year?

The probability would be calculated as:

$$\Pr(x; \lambda) = \frac{\lambda^x e^{-\lambda}}{x!} \qquad (11.2)$$

$$\Pr(10; 2.1) = \frac{2.1^{10} e^{-2.1}}{10!} = \frac{(1668 * 0.12)}{3628800} = 0.0000563. \qquad (11.3)$$

?? What is the probability of 3 attacks if $\lambda = 2.1$?

Answer: No problem! That would be:

$$\Pr(x; \lambda) = \frac{\lambda^x e^{-\lambda}}{x!} \qquad (11.4)$$

$$\Pr(3; 2.1) = \frac{2.1^3 e^{-2.1}}{3!} = \frac{(9.61 * 0.12)}{3 * 2 * 1} = 0.189. \qquad (11.5)$$

We could calculate the probability of observing exactly 0, 1, 2, 3, 4, 5, 6, 7, ... attacks as well, and display the results as the **Poisson probability distribution** (see Figure 11.2).

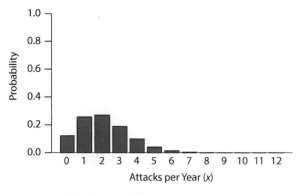

Figure 11.2 Poisson distribution with $\lambda = 2.1$.

Look for the probabilities associated with 3 and 10 attacks. Also notice that the y-axis is labeled "Probability" and that the sum of the bars must equal 1.0. Now take a look at the x-axis; it consists of non-negative integers only. The graph is depicted in bars to indicate there are only discrete outcomes; that is, that there is no way to observe, say, 2.5 shark attacks in a year. In a nutshell, the outcome x is a non-negative integer, but the average rate, λ, can be any positive number such as 2.1. Thus, the Poisson function is a probability mass function (pmf).

There is only one parameter or "knob" to twiddle with for this probability function, and that is λ. Let's have a look at the Poisson distribution when $\lambda = 1, 3$, and 7 (see Figure 11.3).

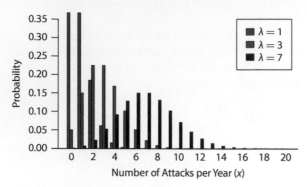

Figure 11.3 Poisson distribution.

These graphs highlight the fact that the Poisson probability distribution is governed by just one parameter, λ. In looking at these three distributions, you might have noticed a few things:

- First, the mean of the distribution is λ.
- Second, as λ gets larger, the Poisson distribution starts to look like a normal distribution.
- Third, the spread, or variance, of the distribution is the same as λ. So as λ increases, so does the spread of the Poisson distribution. See if you can confirm this visually.

 ?? What is so "bursty" about a Poisson process?

Answer: The news headline read: "Shark attacks attributed to random Poisson bursts." Let's have a second look at the Poisson distribution for $\lambda = 2.1$, shown in Figure 11.4. The number of shark attacks $x = 1, 2$, and 3 are all fairly likely outcomes, but 0, 4, and 5 are also in the running. With an average attack rate of 2.1 per year, it would be rare, but not impossible, to observe > 7 attacks.

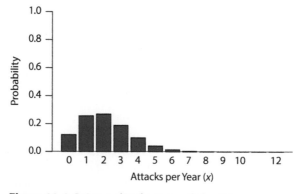

Figure 11.4 Poisson distribution with $\lambda = 2.1$.

Let's try an experiment where we randomly draw **50 observations** from the Poisson distribution in Figure 11.4 ($\lambda = 2.1$) and record our results in a table with 10 rows and 5 columns (see Table 11.2).

Table 11.2

2	4	2	2	3
1	3	1	1	0
1	2	1	1	1
3	1	3	1	3
3	4	2	1	3
2	1	3	3	3
5	2	3	1	0
5	2	4	5	1
4	5	0	4	1
7	2	3	1	3

What you are seeing are random draws from a Poisson distribution where $\lambda = 2.1$. Heading down the columns, we drew a 2, then 1, then a 1, then 3, and so on. This is an example of a random Poisson process like the one described in the shark attack article. We can say that each observation, x, arises from a Poisson distribution. That is, the random variable X is the number of shark attacks in a year, and X is distributed as a Poisson distribution:

$$X \sim \text{Poisson}(\lambda). \tag{11.6}$$

Now, take a look at the last four entries in column 1 and the first entry in column 2. Do you see a series of attacks? If you did not know that these values came from a Poisson distribution with $\lambda = 2.1$, you might conclude that shark populations are on the rise and that the rate of attacks is actually increasing. With random variables, occasionally you can string together a series of outcomes that strike you as "non-random," even if they are. Make sense?

 ?? **What does the Poisson pmf have to do with a Bayesian analysis?**

Answer: Ah, yes!

Are you ready for another Bayesian inference problem?

Suppose you are on a Bay(es) Watch patrol team, and 5 attacks occurred this year. Your supervisor doubts that the current estimate of $\lambda = 2.1$ is valid.

She has charged you with getting an up-to-date estimate of the annual rate of shark attacks. Thus, this is a parameter estimation problem, and you've decided to use a Bayesian approach.

 ??   What is your estimate of lambda?

Answer: Your answer here!

Well, you have just one new observation, which is that there were 5 attacks this year. You note that it's possible to observe 5 shark attacks under a variety of λ estimates. Look for the probability of observing 5 attacks when $\lambda = 2$, 5, or 7 (see Figure 11.5).

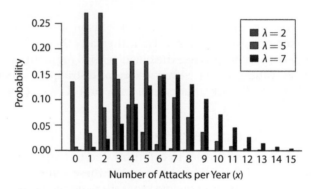

Figure 11.5 Poisson distribution.

Because there are multiple alternative Poisson distributions that can yield 5 shark attacks with a non-0 probability, it makes sense that we focus our attention on considering **all** possible values for lambda, and then confront each hypothesis with our data of 5 shark attacks. For each hypothesis for λ, we then ask: **What is the likelihood of observing 5 shark attacks under each hypothesis for λ?** That way, you are estimating a probability distribution for λ and giving credence to an infinite number of hypotheses instead of estimating just a single point.

This is a natural Bayesian problem. With a Bayesian inference problem, we need a prior distribution that represents our 'belief' in each alternative value for λ and the likelihood of the data to generate the posterior distribution. We'll get there before you know it.

 ??   Sounds good! Where do I start?

Answer: Before we leap in, let's take time to acquaint ourselves with a new probability density function. Remember that λ can go from 0 to infinity. If our hypotheses for λ are graphed along the x-axis, we need an x-axis that starts at 0 and increases to infinity. Because λ can be any real number greater than 0, the x-axis must be **continuous**. Then on the y-axis, we compute the **probability density** of each possible λ. An example of a prior distribution for the various hypotheses for λ might look like something like the one shown in Figure 11.6.

Of course, you could use a different prior distribution if you'd like.

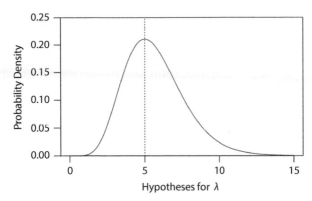

Figure 11.6 Prior probability distribution for λ.

 ?? OK. Is there some sort of probability distribution I can use to help me set the prior distribution?

Answer: Yes! In Chapter 10, we were intent on estimating the posterior distribution for p, the probability that a famous person like Shaq could get into the White House without an invitation. We used a **beta distribution** to set our priors for each and every p (p = probability of success, which ranges between 0 and 1). The data we collected constituted a binomial problem (number of successes out of number of trials). Then, we obtained a new posterior distribution for p in light of the new data.

In this problem, the beta distribution won't work for us. Why? Because λ can be greater than 1. Fortunately there is another distribution that is suitable, and it is called the **gamma distribution**.

 ?? Gamma distribution?

Answer: Wolfram and the Oxford Dictionary of Statistics (Upton and Cook, 2014) provide descriptions of the gamma distribution.

If you checked out these links, you may have noticed that there are different ways of expressing the gamma distribution, which is explained in a Wikipedia entry (accessed August 17, 2017): "In probability theory and statistics, the gamma distribution is a two-parameter family of continuous probability distributions. There are three parameterizations in common use:

- With a **shape** parameter k and a **scale** parameter θ.
- With a **shape** parameter $\alpha = k$ and an inverse scale parameter $\beta = \frac{1}{\theta}$, called a **rate** parameter.
- With a shape parameter k and a **mean** parameter $\mu = \frac{k}{\beta}$.

The second parameterization, with α and β, is more common in Bayesian statistics, where the gamma distribution is used as a **conjugate prior distribution** for various types of inverse scale (i.e., rate) parameters, such as the λ of an exponential distribution or a Poisson distribution."

?? Uh-huh. Can you simplify?

Answer: The gamma distribution is a continuous probability distribution; the x-axis is continuous, not discrete, and begins with 0 and goes to infinity. There are three ways to "create" a gamma distribution. In our Bayesian analysis, we are interested in a gamma distribution whose shape and location are controlled by two parameters: the shape parameter is called alpha, α, and the rate parameter is called beta, β.

> Note: If you dig into other references, take care to note that different software packages may use different names for the shape, rate, and scale parameters. In this book, we will stick with the names α for shape, β for rate, and θ for scale.

Let's take a look at a few examples (see Figure 11.7).

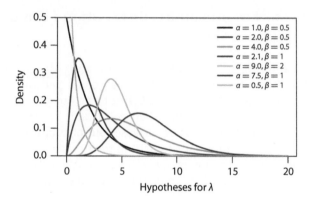

Figure 11.7 Gamma distributions.

Note that we have drawn these distributions as continuous functions (lines)—highlighting that the gamma distribution is a **probability density function**—there are an **infinite** number of hypotheses for λ, the average number of shark attacks per year.

?? Do α and β have anything to do with the α and β from the beta distribution in Chapter 10?

Answer: Nope, not a thing. The beta distribution is controlled by two parameters that are also named α and β, but these could have been named something else. It just so happens that α and β are common names for parameters in probability functions, much like Spot and Daisy are common names for dogs!

?? Why are there three ways to parameterize this function?

Answer: Can you see that all three parameterizations are related to one another? In the first form, we have a shape and scale parameter. In the second version, the rate parameter is just 1 divided by the scale parameter. In the third version, the mean is the shape parameter divided by the rate parameter!

Which of the three alternative parameterizations you use depends on the problem you are tackling. Here, we'll stick with the second version which has α as the shape parameter and β as the rate parameter, as that is the version commonly used for Poisson hypotheses for λ. After all, λ from the Poisson distribution is a rate!

?? How do we use this distribution to estimate probability?

Answer: Well, because it's a probability density function, you can't calculate probability for a specific value. Rather, you are calculating the probability density.

If you are given a particular gamma distribution, say, $\alpha = 2$ and $\beta = 1$, you can compute the density associated with, say, $x = 3$ with this function:

$$g(x; \alpha, \beta) = \frac{\beta^\alpha x^{\alpha-1} e^{-\beta x}}{\Gamma(\alpha)} \qquad 0 \leq x \leq \infty. \tag{11.7}$$

?? Looks Greek to me!

This is the probability density function for the gamma distribution. Here, the function name is g, and the inputs are x, α, and β. The random variable named X must assume values greater than or equal to 0, and α and β must be positive. The output will provide the probability density of x given α and β.

The funny symbol, Γ, in the denominator is the capital Greek letter for G (called gamma), and represents the gamma function, which is not to be confused with the gamma distribution! The gamma function is needed to ensure that the function integrates to 1.0 (that is, the total area under the curve = 1.0). Let's try an example.

The probability density of observing $x = 3$, given $\alpha = 2$ and $\beta = 1$ is:

$$g(x; \alpha, \beta) = \frac{\beta^\alpha x^{\alpha-1} e^{-\beta x}}{\Gamma(\alpha)} \tag{11.8}$$

$$g(3; 2, 1) = \frac{1^2 3^{2-1} e^{-1*3}}{\Gamma(2)}. \tag{11.9}$$

?? But what do we do about $\Gamma(2)$?

Answer: When α is a **positive integer**, $\Gamma(\alpha) = (\alpha-1)!$, which in our case ends up being $\Gamma(2) = (2-1)! = 1! = 1.0$. So we are left with:

$$g(3; 2, 1) = \frac{1^2 3^{2-1} e^{-1*3}}{\Gamma(2)} \tag{11.10}$$

$$g(3; 2, 1) = \frac{1^2 3^{2-1} e^{-1*3}}{(2-1)!} \tag{11.11}$$

$$g(3; 2, 1) = 1 * 3 * e^{-3} = 0.14936. \tag{11.12}$$

If α is not a positive integer, you can compute the value of the gamma function with an online calculator.

Here is a graph of the gamma distribution for $\alpha = 2$ and $\beta = 1$. See if you can find the probability density for $x = 3$ in Figure 11.8.

It's worth pointing out again that this is a probability density function. The y-axis gives density, and the x-axis features a continuous variable.

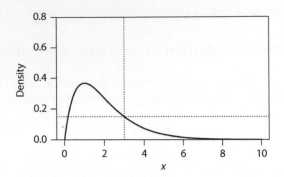

Figure 11.8 Gamma distribution: $a = 2, \beta = 1$.

 ?? **OK, so how do we use the gamma distribution for the Shark Attack Problem?**

Answer: Our goal here is to use a Bayesian inference approach to estimate the rate of shark attacks per year, or λ.

We'll use the same steps that we have in previous chapters:

1. Identify your hypotheses—these would be the alternative hypotheses for λ, ranging from 0 to infinity.
2. Express your belief that each hypothesis is true in terms of prior densities.
3. Gather the data—5 attacks occurred this year.
4. Determine the **likelihood** of the observed data, assuming each hypothesis is true.
5. Use Bayes' Theorem to compute the posterior densities for each value of λ (i.e., the posterior distribution).

Before we leap in, let's create a graphical diagram that illustrates the process by which the data were generated (see Figure 11.9).

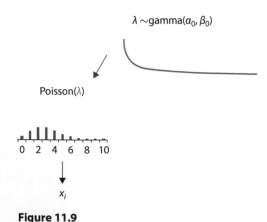

Figure 11.9

At the very bottom of this diagram, we have our observed data, x_i. The data are random observations that arise from a Poisson distribution, defined by the parameter λ. The parameter, λ, in turn, is the unknown parameter that we are trying to estimate. We use a gamma distribution to "weight" each and every hypothesis for λ, treating the unknown parameter as a random variable from a gamma distribution.

Now that we have specified the problem, let's step through these one at a time.

Step 1. What are the hypotheses for lambda?

Answer: Your answer here!

We know that λ can take on any value from 0 to infinity, so we have an infinite number of hypotheses for λ.

Step 2. What are the prior densities for these hypotheses?

Answer: We need to assign a prior for each hypothesized value of λ, and we already introduced the **gamma distribution** as a suitable prior probability distribution. We've also decided that the gamma distribution that uses the shape and rate parameterization suits our needs.

OK, what values for α and β should we use?

Answer: Your answer here!

Answer: It makes sense that we dust off the 10 years of shark attack data in Table 11.1 and use that information to inform our prior distribution.

But the mean of that dataset is 2.1. How do I convert this to the α and β parameters for a gamma distribution?

Answer: Here's where the Wolfram MathWorld, the Oxford Dictionary of Statistics, or Wikipedia pages on probability distributions can be very helpful. For instance, Figure 11.10 shows a typical Wikipedia panel featuring the gamma distribution (with our parameterization highlighted in red).

These panels provide summary information such as the mean, median, and variance, for a given gamma distribution when the shape α and rate β parameterization is used. Scan down and look at the mean:

$$\mu = E[X] = \frac{\alpha}{\beta}. \tag{11.13}$$

The mode is:

$$\frac{\alpha-1}{\beta} \tag{11.14}$$

when $\alpha \geq 1$. The variance is:

$$\frac{\alpha}{\beta^2}. \tag{11.15}$$

Parameters	• $\alpha > 0$ shape • $\beta > 0$ rate
Support	$x \in (0, \infty)$
Probability density function (pdf)	$\dfrac{\beta^\alpha}{\Gamma(\alpha)} x^{\alpha-1} e^{-\beta x}$
Mean	$E[X] = \dfrac{\alpha}{\beta}$ $E[\ln X] = \psi(\alpha) - \ln(\beta)$ (see digamma function)
Median	No simple closed form
Mode	$\dfrac{\alpha - 1}{\beta}$ for $\alpha \geq 1$
Variance	$\text{Var}[X] = \dfrac{\alpha}{\beta^2}$ $\text{Var}[\ln X] = \psi 1(\alpha)$ (see trigamma function)

Figure 11.10

Just as in the beta distribution in the last chapter, the magnitude of α and β are not nearly as important as the **relation** between them. For instance, if the mean is 2.1, you might set α to 2.1 and β to 1.0. Or you might set α to 4.2 and β to 2.0.

If the mode (the peak of the curve) is at 2.1, you might set α to 3.1 and β to 1.0. Or you might set α to 31 and β to 14.3 (see Figure 11.11).

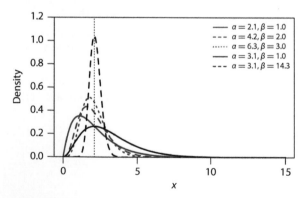

Figure 11.11 Gamma distributions.

A few things worth noticing:

• The area under all five distributions is 1.
• Notice that as α and β become larger, the distribution becomes tighter.

- α and β for all of the blue distributions were calculated assuming the mean of the distribution = 2.1. Notice that the mean is not the top of curve, though, because there is "weight" in the right tail of the distribution.
- When α and β are parameterized in terms of the mode, the peak of the curve is the mode (black distributions).

With these graphs, it's clear that even though you have prior information that the mean number of shark attacks is 2.1 per year, you still need to exercise some subjectivity with respect to which prior distribution to use. You could also consider the variance of the data in Table 11.1 in making your selection (we'll cover how to do that in Chapter 13).

 ?? Which gamma distribution should we use for this problem?

Answer: Let's go with the prior distribution of $\alpha_0 = 2.1$ and $\beta_0 = 1$ (solid blue line in Figure 11.12), which has a mean rate of 2.1. Notice that we use a "naught" subscript to designate these as the hyperparameters of the prior distribution.

 ?? Hyperparameters?

Answer: In Bayesian statistics, a hyperparameter is a parameter of a prior or posterior distribution. This term is used to distinguish the parameters of the prior or posterior distribution from the unknown parameter of interest, which is λ.

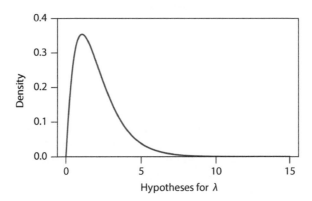

Figure 11.12 Prior distribution: $\alpha_0 = 2.1$, $\beta_0 = 1$.

 ?? Step 3. Now what?

Answer: Collect data! **We have observed 5 shark attacks.**

 ?? Step 4. And then?

Answer: Step 4 is to determine the likelihood of the observed data, assuming each hypothesis is true.

Now, for each hypothesized value of λ, let's compute the Poisson likelihood of observing five attacks in one year.

We touched on the Poisson **pmf** at the beginning of the chapter, where we showed that the Poisson pmf can be used to calculate the probability of observing 0, 1, 2, 3, . . . attacks, *given* lambda. This time, we have observed the value x, and now we are using the function to calculate the **likelihood** of observing the data under each hypothesized λ.

To give you an example, under the $\lambda = 2.1$ hypothesis, the likelihood of observing five shark attacks is computed as:

$$\mathcal{L}(X = x; \lambda) = f(x; \lambda) = \frac{\lambda^x e^{-\lambda}}{x!} \tag{11.16}$$

$$\mathcal{L}(x = 5; \lambda = 2.1) = \frac{2.1^5 e^{-2.1}}{5!} = \frac{(40.84 * 0.12)}{120} = 0.04167704. \tag{11.17}$$

A plot of the likelihoods for a range of values for λ is called a likelihood profile (see Figure 11.13).

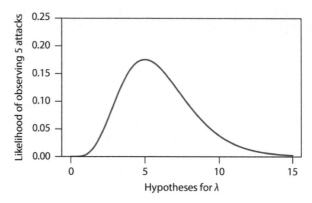

Figure 11.13

 ?? And Step 5?

Answer: The final step is to use Bayes' Theorem to compute the posterior probabilities for each value of λ. As in Chapter 10, we need to set this Bayesian inference problem up for a **continuous random variable**. Suppose we are trying to estimate a single parameter called θ. If θ is continuous, you have an infinite number of hypotheses for it. Bayes' Theorem in this case is specified as:

$$P(\theta \mid \text{data}) = \frac{P(\text{data} \mid \theta) * P(\theta)}{\int P(\text{data} \mid \theta) * P(\theta) d\theta}. \tag{11.18}$$

This is the generic version of Bayes' Theorem when the posterior distribution for a single parameter, given the observed data, is represented by a pdf. This is designated $P(\theta \mid \text{data})$, **where P is probability density**. This is the left side of the equation. On the right side of the equation, the numerator multiplies the prior probability **density** of θ, which is written $P(\theta)$, by the likelihood of observing the data under a given hypothesis for θ, which is written $P(\text{data} \mid \theta)$. Technically, the likelihood can be a pdf or a pmf. In this chapter's illustration, for instance, the likelihood is a pmf because it is a Poisson distribution. In the denominator, we see the same terms, but this time we also see a few more symbols. The symbol $\int$ means "integrate," which roughly means "sum up all the pieces" for each tiny change in θ, which is written $d\theta$. In other words, the denominator accounts for the prior density * likelihood for all possible hypotheses for theta, and sums them.

For the Shark Attack Problem, we replace θ with λ, so Bayes' Theorem looks like this, with the posterior shown in purple, the prior in blue, and the likelihood in red:

$$P(\lambda \,|\, \text{data}) = \frac{P(\text{data} \,|\, \lambda) * P(\lambda)}{\displaystyle\int_0^\infty P(\text{data}|\lambda) * P(\lambda)d\lambda}. \tag{11.19}$$

Remember that this is a density, not a probability, because the probability of a given value of λ will always be 0 for a continuous pdf (refer to Chapter 9 for a refresher). Thus, in this problem, we are interested in all possible values of λ, or the **entire posterior distribution**.

But here's the kicker: The integration of the denominator can be very difficult to calculate!

?? How do we make headway?

Answer: Well, for this particular problem, there is an analytical shortcut that makes updating the prior to the posterior distribution a snap. Integration not required! Remember that our prior distribution set $\alpha_0 = 2.1$ and $\beta_0 = 1$.

Furthermore, we have observed 5 shark attacks in one year. So here, $x = 5$. Let's let $n =$ the number of years; here $n = 1$. **Here is the shortcut:**

The updated α parameter is α_0 plus the sum of the Poisson observations. We have just one observation, which was of 5 attacks:

$$\alpha_{\text{posterior}} = \alpha_0 + \sum_{i=1}^{n} x_i. \tag{11.20}$$

So, the summation term results in the number 5, because $x_1=5$. Then we have:

$$\alpha_{\text{posterior}} = 2.1 + 5 = 7.1. \tag{11.21}$$

The updated β parameter is β_0 plus n, the number of Poisson observations:

$$\beta_{\text{posterior}} = \beta_0 + n \tag{11.22}$$

$$\beta_{\text{posterior}} = 1 + 1 = 2. \tag{11.23}$$

These parameters are referred to as the posterior hyperparameters. Now we can look at the prior and posterior distributions for λ (see Figure 11.14).

We started off with a prior distribution shown in blue. Then we collected data in the form of a single Poisson random variable: five attacks. We then used the shortcut to generate the posterior distribution for all hypotheses of λ.

As a result of our data, we now have **new** beliefs in each and every hypothesis of λ. The posterior distribution is shown in purple. Notice how the posterior shifted toward our observed data for this problem.

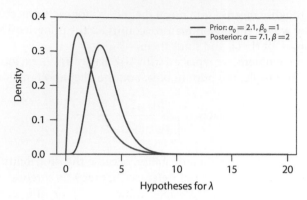

Figure 11.14 Gamma prior and posterior.

 ?? How is this shortcut possible?

Answer: The answer is that the gamma distribution is a conjugate distribution that can be updated with Poisson data (Raiffa and Schlaeffer, 1961). This is why we named this chapter "The gamma-Poisson conjugate."

We mentioned the statistical definition of a conjugate distribution in Chapter 11, where we indicated that there are cases where you can use a particular **pdf** as a prior distribution, collect data of a specific flavor, and then derive the posterior **pdf** with a closed-form solution. In these special cases, the **pdf**'s of the prior and posterior distributions are the same probability density function, but their *parameters* may differ. The prior distribution is called a conjugate prior, and the effect of the data can then be interpreted in terms of changes in parameter values (Upton and Cook, 2014).

In the Shark Attack Problem, we used a **gamma distribution** to set the priors for all hypotheses of λ, the rate of shark attacks per year. We used the information provided in Table 11.1 to help set the prior distribution. We then collected Poisson data (5 shark attacks), and then used a shortcut to generate the posterior distribution for λ.

Raiffa and Schlaeffer's (1961) conjugate solution is conveniently given in a Wikipedia page on Bayesian conjugates. A quick overview of that page's gamma-Poisson conjugate can be visualized with the graph in Figure 11.15.

Let's work our way through this figure. The top item indicates that the model parameter of interest, λ, is a rate. The prior distribution is a gamma distribution, defined by hyperparameters α_0 and β_0. The "naughts" signal that these are hyperparameters for the prior. The data collected are Poisson in flavor, so the likelihood to be used is the Poisson pmf. With this setup, the posterior hyperparameters (the conjugate shortcuts) are provided.

 ?? Can I see the proof?

Answer: Of course! Check out Appendix 2.

 ?? What if I used a different prior distribution?

Answer: You could have used a different prior distribution, and your inferences provided by the posterior distribution would probably be different. You must defend your choice for

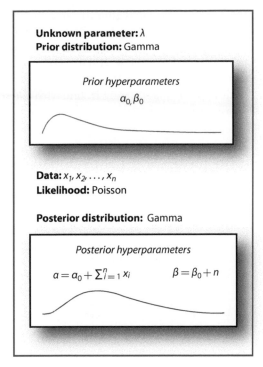

Unknown parameter: λ
Prior distribution: Gamma

Prior hyperparameters
a_0, β_0

Data: $x_1, x_2, \ldots, x_n$
Likelihood: Poisson

Posterior distribution: Gamma

Posterior hyperparameters

$a = a_0 + \sum_{i=1}^{n} x_i$ $\beta = \beta_0 + n$

Figure 11.15

the prior! Here, we had previous knowledge of the annual rate of shark attacks and thus used an informative prior. To ignore this previous information is contrary to the scientific method.

 ?? Why do I need to use a Bayesian approach? Couldn't I just use the data from Table 11.1 and add in the extra observation of 5 shark attacks?

Answer: You certainly could do that. But then you wouldn't be capitalizing on the notion of learning, and on the notion of multiple, competing hypotheses for λ. Table 11.3 shows the dataset again, this time with one extra observation (our new observation of 5 shark attacks).

Table 11.3

Year	Attacks
1	1
2	0
3	2
4	1
5	1
6	3
7	2
8	3
9	4
10	4
11	5

The mean of this dataset is 2.36. Yes, this does give you an estimate of λ, and you can ask your friendly neighborhood statistician to help you draw conclusions about the mean number of shark attacks in Florida based on your sample. However, this result fails to acknowledge support for alternative values of λ. Bayesians are interested in the probability distribution for λ, where the density associated with each alternative hypothesis for λ gives a measure of "plausibility." This approach is squarely in line with the scientific method. As we collect new information, we can use Bayesian approaches to refine how plausible each and every hypothesis for λ is.

 ?? ## OK, what happens to the posterior if I collect new data?

Answer: Try it! The prior distribution will now be the posterior distribution after observing 5 shark attacks: $\alpha = 7.1$ and $\beta = 2$.

Let's suppose you collect 5 more years of shark attack data. Let's define the variable X_1 as the number of shark attacks in year $1, \ldots,$ and X_5 as the number of shark attacks in year 5. You observe the results shown in Table 11.4.

Table 11.4

Year	Attacks
12	1
13	2
14	0
15	3
16	4

Note that we have 5 new random observations from the Poisson distribution, so $n = 5$. Also notice that the sum of the attacks, $\sum_{i=1}^{n} x_i$, is 10.

Remember that the prior distribution is now defined by $\alpha_0 = 7.1$ and $\beta_0 = 2$. Now let's incorporate the $n = 5$ years of new data (see Figure 11.16).

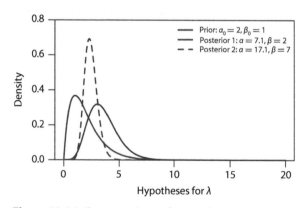

Figure 11.16 Gamma prior and posterior.

The updated α parameter is the prior α_0 plus the sum of the Poisson observations:

$$\alpha_{\text{posterior}} = \alpha_0 + \sum_{i=1}^{n} x_i \tag{11.24}$$

$$\alpha_{posterior} = 7.1 + 10 = 17.1. \tag{11.25}$$

The updated β parameter is the prior β_0 plus n, the number of Poisson observations:

$$\beta_{posterior} = \beta_0 + n \tag{11.26}$$

$$\beta_{posterior} = 2 + 5 = 7. \tag{11.27}$$

A few things are worth pointing out here:

- We started off with a prior shown in blue
- We updated this based on a single year's worth of data: 5 shark attacks. This shifted the posterior to the right, but with a single year's worth of data, the distribution is still quite broad.
- We then collected 5 more years' worth of data, during which a total of 10 attacks were observed. This shifted the posterior back to the left, and the distribution narrowed considerably.
- The mean of this new distribution is $\frac{\alpha}{\beta} = \frac{17.1}{7} = 2.44$.
- The variance is $\frac{\alpha}{\beta^2} = \frac{17.1}{7^2} = 0.35$.
- The standard deviation is 0.59.

Of course, you would want to report the Bayesian credible intervals.

 ?? ## What parameters do you use for α_0 and β_0 if you want a vague prior?

Answer: If you didn't have the information in Table 11.1 and wanted to use a vague prior, you could set α_0 and β_0 to something really small, like 0.01. Let's try our very first analysis again, using a prior with α_0 and $\beta_0 = 0.01$, and then update it after observing 5 shark attacks.
The updated alpha parameter is the prior alpha plus the sum of the Poisson observations:

$$\alpha_{posterior} = \alpha_0 + \sum_{i=1}^{n} x_i \tag{11.28}$$

$$\alpha_{posterior} = 0.01 + 5 = 5.01. \tag{11.29}$$

The updated beta parameter is the prior beta plus n, the number of Poisson observations:

$$\beta_{posterior} = \beta_0 + n \tag{11.30}$$

$$\beta_{posterior} = 0.01 + 1 = 1.01. \tag{11.31}$$

If you were estimating the mean number of attacks based on the data alone, you would estimate that λ is 5.0. Here, the mean of the posterior distribution is very close to that, and can be calculated as:

$$\frac{\alpha}{\beta} = \frac{5.01}{1.01} = 4.96. \tag{11.32}$$

It's evident that the prior is not contributing much to the posterior distribution (see Figure 11.17).

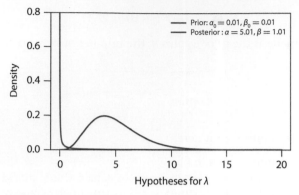

Figure 11.17 Gamma prior and posterior.

The prior certainly does not look "flat." But notice in this case, since α_0 and β_0 were so small for the prior distribution, they had almost no bearing on the shape of the posterior distribution. But, why would you use this prior distribution if the data in Table 11.1 were at your disposal?

 ?? **OK, how should I present my Bayesian results?**

Answer: If you can show the prior and posterior distributions, that would be fabulous! But you can also summarize properties of the posterior distribution such as the mean and variance by using some of the calculations presented on Wikipedia's gamma distribution page. And don't forget the credible intervals!

 ?? **Can I use a different prior distribution than the gamma distribution?**

Answer: You may! The log-normal distribution can be used as the prior distribution for setting alternative hypotheses for λ. Figure 11.18 shows a few examples of the log-normal distribution, which has two parameters.

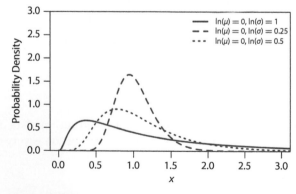

Figure 11.18

Here you can see that the distribution ranges from 0 to ∞ along the x-axis. This distribution is very useful for many analyses. However, since it is not a conjugate distribution, there is no easy way—no shortcut—to get the posterior. You could estimate the posterior distribution with an MCMC analysis, but that is a matter for another day.

 ?? **Can we summarize this chapter?**

Answer: Why not?

- We started by mentioning an article that indicated that a string of numerous shark attacks could be a random process.
- We introduced the Poisson distribution as a probability mass function which has a single parameter, λ, which is the average rate of attacks per year. We showed how you can get a burst of a high numbers of attacks even when λ is relatively low.
- We identified a new Bayesian challenge, where the goal was to estimate λ.
- We then introduced the gamma distribution as one that can be used to set hypotheses for λ, the average number of shark attacks per year, noting that the gamma distribution returns a probability density, not a probability. We used an informative prior distribution that was based on previous data.
- We then collected new data for one year. There were five shark attacks.
- We then calculated the likelihood of observing the data under some hypothesized values for λ.
- But then we noted that we have an infinite number of hypotheses for λ; the denominator of Bayes' Theorem is an integral.
- We used a shortcut, an analytical solution that allowed us to update our prior gamma distribution to a posterior gamma distribution using the data.
- We then collected five more years of data. This time, we used our newly updated posterior as the prior distribution, and then updated it again using the newly observed data. Thus, we are using Bayesian inference as a way to formalize "learning."

We will revisit this chapter again in the MCMC section of this book, where we will show you a different method for estimating the posterior distribution. For now, though, we have one more conjugate chapter in this section, and it is a "sticky" one. We hope to see you there!

CHAPTER 12

The Maple Syrup Problem: The Normal-Normal Conjugate

In this chapter, we return to the **normal** distribution and set up a Bayesian inference problem that deals with estimating parameters from the normal distribution. You might recall that the normal distribution has **two** parameters that control the location and spread of the distribution: the mean and the standard deviation. These are called mu (μ) and sigma (σ), respectively. We'll now be dealing with the **joint** distribution of the two parameters in the normal distribution, where you need to assign a prior associated with a **combination** of μ and σ.

Here's a roadmap for this chapter. Pay attention! We'll begin with a motivating example that highlights the need to estimate the two parameters of the normal distribution. We'll first tackle the problem by considering just 10 discrete hypotheses for μ and σ so that you can see Bayes' Theorem at work with a joint prior. Then we'll move to the case where the number of joint hypotheses is infinite and in which the prior distribution is a continuous, joint distribution. This complicates the analysis considerably. But, as with the previous two chapters, this chapter introduces a **conjugate** solution, which can be applied only if you assume that one of the parameters of the normal distribution is known. Specifically, we assume that σ is known. Hence the focus is on estimating the unknown parameter of the normal distribution, μ. We will revisit this chapter again in Section 5, where we show you how to solve the problem when both μ and σ are unknown with an MCMC analysis using Gibb's sampling.

By the end of this chapter, you should be able to define and use the following:

- The normal distribution pdf in terms of the mean and standard deviation (μ and σ) and in terms of the mean and precision (μ and τ)
- The normal-normal conjugate

We'll start with a story written by Brendan Borrell in 2013 for Bloomberg Business:

On the morning of July 30, 2012, an accountant named Michel Gauvreau arrived at the Global Strategic Maple Syrup Reserve, housed in a huge red brick warehouse on the side of the Trans-Canadian Highway… about two hours northeast of Montreal. Inside, baby-blue barrels of maple syrup were stacked six high in rows hundreds deep. Full, each barrel weighs about 620 pounds. With grade A syrup trading at about $32 per gallon, that adds up to $1,800 a barrel, approximately 13 times the price of crude oil.

The fiscal year was coming to a close, and the Federation of Quebec Maple Syrup Producers had hired Gauvreau's company to audit its inventory…There were around 16,000 barrels here, about one-tenth of Quebec's annual production. The gap between the rows was barely wide enough to walk through, and the rubber soles of Gauvreau's steel-tip boots stuck to the sugar-coated concrete floor.

Bayesian Statistics for Beginners: A Step-by-Step Approach. Therese M. Donovan and Ruth M. Mickey, Oxford University Press (2019). © Ruth M. Mickey 2019.
DOI: 10.1093/oso/9780198841296.001.0001

He scaled a row of barrels and was nearing the top of the stack when one of them rocked with his weight. He nearly fell. Regaining his balance, he rattled the barrel: It was light because it was empty. He soon found others that were empty. After notifying the Federation's leaders and returning with them to examine the stockpile, they unscrewed the cap on a full barrel. The liquid inside was not goopy, brown, or redolent with the wintry scent of vanilla, caramel, and childhood; it was thin, clear, and odorless. It was water... Sixty percent, or six million pounds of syrup, had vanished, worth about $18 million wholesale.

That's some story! It has even caught Hollywood's attention! This begs a few questions, though.

What exactly is maple syrup?

Answer: Yum!

Here's what Encyclopedia Britannica has to say: "Maple syrup, sweet-water sap of certain North American maple trees, chiefly the sugar maple, *Acer saccharum*, but also the black maple, *Acer nigrum*. It was utilized by the Indians of the Great Lakes and St. Lawrence River regions prior to the arrival of European settlers and is still produced solely in North America.

The sweet-water sap from which maple syrup is made is different from the circulatory sap of the growing tree. When the tree is dormant, the sap will flow from any wound in the sapwood, such as a taphole, each time a period of freezing is followed by a period of thawing. The sap contains 1 1/2 to 3 percent solids, mostly sucrose, but does not contain the colour or flavour of maple syrup, which are imparted to the sap as it is concentrated by evaporation in open pans. About 30 to 50 gallons (115 to 190 litres) of sap yield one gallon of syrup."

Why does Canada have a maple-syrup cartel?

Answer: The answer is explained in an article by Josh Sandburn from Time magazine:

It may seem bizarre that Canada has a maple-syrup cartel at all. But think of it this way: Quebec, which produces about 77% of the world's maple syrup, is the Saudi Arabia of the sweet, sticky stuff, and the FPAQ [Fédération des producteurs acéricoles du Québec] is its OPEC. The stated goal of the cartel, in this case, is keeping prices relatively stable.

The problem with maple syrup is that the natural supply of it varies dramatically from year to year. "It's highly dependent on the weather," explains Pascal Theriault, an agricultural economist at McGill University in Montreal. "The maple trees need optimal climate conditions—cold nights, temperate days—to produce the right sap," explained a recent article in the Washington Post. "That doesn't always happen, and production varies sharply each spring."

I see. And how does this apply to Bayesian inference?

Answer: Since the cartel has lost its syrup, **for this chapter we are going to assume that they want to incorporate the great state of Vermont (USA) into their game.** After all, Vermont is Quebec's neighbor with healthy sugar maple forests (see Figure 12.1). And by the way, the name **Vermont** stems from the Latin *Viridis Montis*, which means **Green Mountain**.

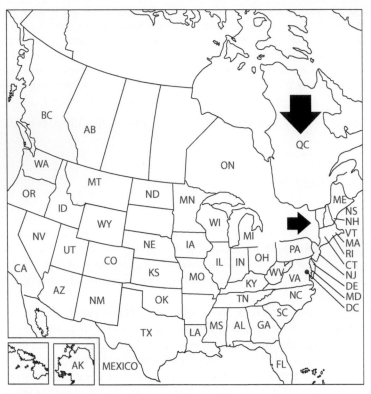

Figure 12.1

Suppose the cartel is interested in annexing Vermont's syrup production. They would need to have some idea of Vermont's average annual production of syrup, right? And because production varies from year to year, they would want to quantify the variation, right?

?? So how would we capture the essence of this with a probability distribution?

Answer: We can use the **normal** probability density function. We studied the normal pdf in detail in Chapter 9, so we'll just briefly review it now, starting with an example (see Figure 12.2).

The formula that generates the normal distribution for the random variable X (millions of gallons of maple syrup produced) is called the **normal probability density function**. Here it is, and remember that the mean is represented by the symbol μ, and the standard deviation is represented by the symbol σ:

$$f(x; \mu, \sigma) = \frac{1}{\sqrt{2\pi}\,\sigma} e^{-\frac{(x-\mu)^2}{2\sigma^2}} \quad -\infty \leq x \leq \infty \tag{12.1}$$

As an example, suppose the average number of gallons (in millions) of maple syrup produced per year is $\mu = 10.0$, with a standard deviation σ of 2.3. The probability density of observing 6.2 million gallons, say, would be:

$$f(x; \mu, \sigma) = \frac{1}{\sqrt{2\pi}\sigma} e^{-\frac{(x-\mu)^2}{2\sigma^2}} \tag{12.2}$$

$$f(6.2; 10, 2.3) = \frac{1}{\sqrt{2\pi} * 2.3} e^{-\frac{(6.2-10)^2}{2*2.3^2}} \tag{12.3}$$

$$f(6.2; 10, 2.3) = \frac{1}{5.765} e^{-\frac{14.44}{24.334}} = 0.044. \tag{12.4}$$

See if you can find this result in Figure 12.2.

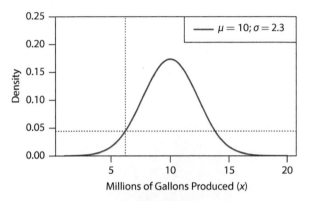

Figure 12.2 Pdf of a normal distribution.

 ?? **What is the probability density associated with 4.8 million gallons of syrup?**

Answer: No problem! That would be:

$$f(x; \mu, \sigma) = \frac{1}{\sqrt{2\pi}\,\sigma} e^{-\frac{(x-\mu)^2}{2\sigma^2}} \tag{12.5}$$

$$f(4.8; 10, 2.3) = \frac{1}{\sqrt{2\pi} * 2.3} e^{-\frac{(4.8-10)^2}{2*2.3^2}} = 0.0135 \tag{12.6}$$

Look for the probability density associated with 4.8 million gallons in the distribution shown in Figure 12.3.

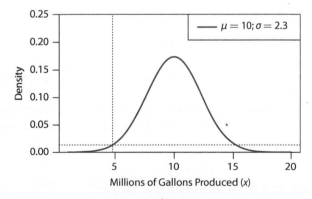

Figure 12.3 Pdf of a normal distribution.

?? **What does the normal pdf have to do with the Canadian syrup cartel?**

Answer: Your answer here!

The cartel is planning on annexing Vermont! Of course, this is purely hypothetical! They need to know how much syrup is produced in Vermont on average, as well as the standard deviation in production. In other words, they need to know *which* normal **pdf** best describes maple syrup production in Vermont. Having this new source of syrup can help stabilize the stockpile, and knowing the mean and standard deviation of Vermont's production can help them with pricing and decision making.

So, our challenge is to estimate the two parameters of the normal distribution in terms of syrup production in Vermont.

Are you ready for another Bayesian inference problem?

 ?? **OK. Where do I start?**

Our goal here is to use a Bayesian inference approach to estimate the two parameters of the normal distribution: the mean, μ, and the standard deviation, σ. This differs from that in Chapters 10 and 11, where our goal was to estimate only one parameter of a **pdf** distribution (p for the White House Problem, and λ for the Shark Attack Problem). Here, for illustrative purposes, we will consider just 10 discrete hypotheses that represent alternative combinations of μ and σ.

We'll use the same steps that we did in previous chapters:

1. Identify your hypotheses—these would be the alternative hypotheses for μ, and alternative hypotheses for σ.
2. Express your belief that each hypothesis is true in terms of prior probabilities.
3. Gather the data.
4. Determine the **likelihood** of the observed data under each hypothesis.
5. Use Bayes' Theorem to compute the posterior probabilities for each hypothesis.

Let's step through these one at a time.

 ?? **Step 1. What are the hypotheses for μ and σ?**

Answer: Your answer here!

Well, let's see. Let's assume that the following statements are reasonable:

- The mean ranges between 5 and 15 million gallons per year.
- The standard deviation ranges between 1.2 and 2.8 millions of gallons.

To begin, let's assume we give the cartel 10 dots to place on a chart that represent 10 combinations of μ and σ together. This is a small number of discrete hypotheses for μ and σ. The cartel can place the dots where they'd like, putting more dots in locations that they feel are more likely for the state of Vermont. When they are finished, the chart might look something like that in Figure 12.4.

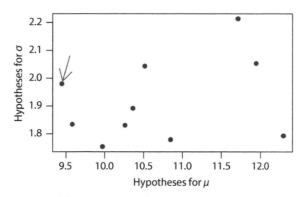

Figure 12.4 Scatterplot example.

Each dot represents the cartel's belief in a particular **combination** for the mean and the standard deviation. For instance, the arrow in the chart indicates that one cartel member believes that the mean is ~9.45 and the standard deviation is ~1.98.

These 10 dots represent 10 discrete hypotheses for μ and σ.

 ?? ## Step 2. What are the prior probabilities for each hypothesis?

Answer: A key consideration here is that we are assigning prior probabilities for each hypothesized **combination** of μ and σ. They are **jointly** estimated. If every single dot represents a single hypothesis, and **if each dot has the same weight**, then the prior probability associated with each dot (hypothesis) is 1/10, or 0.1. This is a non-informative prior, and a probability mass function.

 ?? ## Step 3. Now what?

Answer: Collect data! **Suppose the cartel visits Vermont and learns that 10.2 million gallons were produced last year.**

 ?? ## Step 4. And then?

Answer: Step 4 is to determine how **likely** the observed data are, assuming each joint hypothesis is true.

Now, for each hypothesized combination of μ and σ, let's compute the likelihood of observing 10.2 million gallons.

At the beginning of this chapter, we showed that the normal **pdf** can be used to calculate the probability density of observing a value from a normal distribution whose parameters are given. This time, we have observed the value x, and now we are using the function to calculate the **likelihood** of observing the data under a combination of entries for μ and σ.

Just to give you an example, suppose one of 10 hypotheses was $\mu = 9.45$ and $\sigma = 1.98$ (the hypothesis highlighted by the arrow in Figure 12.4). Then we could ask: "What is the likelihood of observing 10.2 from a normal distribution whose mean is $\mu = 9.45$ and $\sigma = 1.98$." The answer is:

$$\mathcal{L}(x; \mu, \sigma) = \frac{1}{\sigma\sqrt{2\pi}} e^{-\frac{(x-\mu)^2}{2\sigma^2}} \tag{12.7}$$

$$\mathcal{L}(x = 10.2; \mu = 9.45, \sigma = 1.98) = \frac{1}{1.98\sqrt{2\pi}} e^{-\frac{(10.2-9.45)^2}{2 \cdot 1.98^2}} = 0.1875 \tag{12.8}$$

We can do similar calculations for other combinations as well. Table 12.1 shows 10 specific hypotheses; our highlighted example is hypothesis 6.

Table 12.1

Hypothesis	μ	σ	x	Prior (Pr)	Likelihood ($\mathcal{L}$)	$\mathcal{L} * \text{Pr}$	Denominator	Posterior
1	9.97	1.75	10.2	0.1	0.2260	0.0226	0.1842	0.1227
2	10.26	1.83	10.2	0.1	0.2179	0.0218	0.1842	0.1183
3	11.94	2.05	10.2	0.1	0.1357	0.0136	0.1842	0.0738
4	10.85	1.78	10.2	0.1	0.2097	0.0210	0.1842	0.1140
5	9.58	1.83	10.2	0.1	0.2058	0.0206	0.1842	0.1118
6	9.45	1.98	10.2	0.1	0.1875	0.0188	0.1842	0.1021
7	11.71	2.21	10.2	0.1	0.1429	0.0143	0.1842	0.0776
8	10.51	2.04	10.2	0.1	0.1933	0.0193	0.1842	0.1048
9	12.3	1.79	10.2	0.1	0.1120	0.0112	0.1842	0.0608
10	10.36	1.89	10.2	0.1	0.2103	0.0210	0.1842	0.1140

Let's walk through this table:

- The first three columns identify our 10 alternative hypotheses. Remember that a single hypothesis considers μ and σ jointly.
- Column 4 (x) represents our observed data (Vermont produced 10.2 million gallons of syrup last year).
- Column 5 is the prior probability for each hypothesis. In this example, each hypothesis has equal weight.
- Column 6 uses the normal pdf to compute the likelihood of observing the data under each hypothesis.

We'll stop there for now. It would be a good idea to take out a pencil and paper, calculator, or spreadsheet to verify at least a couple of results.

 ?? And Step 5?

Answer: The final step is to use Bayes' Theorem to compute the posterior probabilities for each combination of μ and σ. We won't present these calculations here.

?? Is this a realistic approach?

Answer: The main challenge is that we have an infinite number of hypotheses that represent each joint combination for the unknown parameters of the normal distribution. Accordingly, our prior and posterior distributions should reflect the **density** associated with each μ and σ combination.

 ?? Back to the drawing board?

Answer: Yes. In other words, a handful of dots is not sufficient. If we gave the cartel a zillion grains of sand, where each grain represents a joint hypothesis for μ and σ, their prior distribution for the parameters may look something like that shown in Figure 12.5.

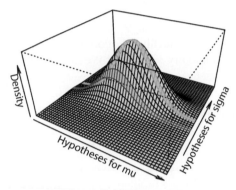

Figure 12.5

Here, the prior distribution has **volume**. For any given parameter combination, the corresponding height gives us the **density**, stressing the fact that we are dealing with a bivariate distribution in which both parameters are continuous. Our 10 original hypotheses are buried somewhere in this sandpile. This distribution really could have any shape—lumps indicate those hypotheses that have more weight than others, which implies that Figure 12.5 is an **informative prior** distribution.

It's plain to see that the shape of our prior distribution for μ **depends** on what level of σ you consider. For instance, Figure 12.5 illustrates that if we fix σ at the "red-level," we see plausible hypotheses for μ highlighted in red. And if we fix σ at the "green-level," we see plausible hypotheses for μ highlighted in green. These ribbons are called "slices" because they slice through the mound at a particular location. Given a particular slice for σ, you have a unique prior distribution for μ. Keep this point in the back of your mind—we will return to it shortly.

 ?? OK, I've got it. Now what?

Answer: Well, now that the cartel has a prior distribution, they would collect new data and use Bayes' Theorem to generate the posterior distribution. In other words, the pile of sand will change shape after new data are collected.

You might recall from previous chapters the generic version of Bayes' Theorem for **one** continuous variable:

$$P(\theta \mid \text{data}) = \frac{P(\text{data} \mid \theta) * P(\theta)}{\int P(\text{data} \mid \theta) * P(\theta) d\theta} \tag{12.9}$$

This is the generic version of Bayes' Theorem when the posterior distribution for a single parameter, given the observed data, is represented by a **pdf**. This posterior density is designated $P(\theta \mid \text{data})$, the prior density is designated $P(\theta)$, and the likelihood of observing the data under a given hypothesis for θ is designated as $P(\text{data} \mid \theta)$. The integral in the denominator accounts for the prior density times the likelihood for all possible hypotheses for θ and sums them.

Now, let's make this task even harder. For the maple syrup problem, there are **two** parameters to be estimated, μ and σ. With two parameters, Bayes' Theorem looks like this, where each hypothesis for μ and σ is considered (the posterior is shown in purple, the prior in blue, and the likelihood in red):

$$P(\mu,\sigma \mid \text{data}) = \frac{P(\text{data} \mid \mu,\sigma) * P(\mu,\sigma)}{\iint P(\text{data} \mid \mu,\sigma) * P(\mu,\sigma) d\mu d\sigma} \tag{12.10}$$

For any given prior hypothesis μ and σ, Bayes' Theorem will return the posterior density of the joint μ and σ by computing the likelihood of observing the data under hypothesis μ and σ, multiplied by the prior probability of μ and σ. We need to do this for an infinite number of hypotheses for each parameter.

 ?? ## How do we make headway?

Answer: As you might have guessed by the title of this chapter, there is a conjugate solution.

> The main approach is to simplify the problem by focusing on only one of the parameters and assuming the second parameter is known.

For a normal-normal conjugate, we use a normal distribution as our prior distribution for the mean, μ, and then collect data and update to the posterior distribution for the mean, μ, which will also be normally distributed. However, in this process, we must assume that σ is known; that is, we must choose a level for σ (such as the green value or red value for sigma as shown in Figure 12.5).

Let's assume we know that $\sigma = 2.0$ Then, we have alternative hypotheses for μ, which controls where the distribution is centered along the x-axis. For example, 5 alternative hypotheses for μ are shown in Figure 12.6.

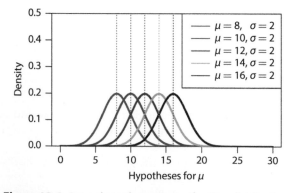

Figure 12.6 Annual maple syrup production distribution.

Now, are you paying attention? The normal-normal conjugate method uses a normal pdf to weight the alternative hypotheses for μ, like the one shown in Figure 12.7.

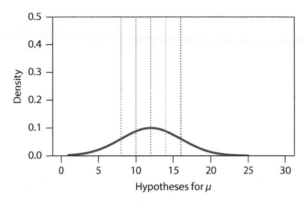

Figure 12.7 Annual maple syrup production, showing the prior distribution for μ.

Thus, the prior distribution for the unknown parameter, μ, is a **normal distribution**.

 ?? I'm confused!

Answer: That's completely normal. Figure 12.7 suggests that we have greater belief that the average annual syrup production is between 10 and 15 than, say, at 5 or 25. Remember that we have simplified the problem so we are only focusing on μ. We've already agreed that sigma, σ, is known and is 2 million gallons.

In fact, the blue distribution in Figure 12.7 is a normal distribution with $\mu_0 = 12$ and $\sigma_0 = 4$. Alternatively, this distribution can be specified in terms of the parameters μ_0 and τ_0, where $\mu_0 = 12$ and $\tau_0 = 0.0625$. These are formally called prior **hyperparameters**.

 ?? What is the parameter τ?

The spread of a normal distribution can be parameterized in a few ways:

- You can express the spread in terms of the standard deviation, σ.
- You can express the spread in terms of variance, σ^2, which is simply $\sigma * \sigma$. If $\sigma = 4$, then $\sigma^2 = 16.0$.
- You can express the spread in terms of precision using the Greek letter, tau (τ), which rhymes with "cow" and is $\frac{1}{\sigma^2}$. If $\sigma^2 = 16.0$, then $\tau = \frac{1}{16} = 0.0625$.

For reasons that will soon become apparent to you, we will define our prior distribution for the **unknown parameter** μ in terms of μ_0 and τ_0. The subscripts identify the two hyperparameters for the unknown μ. We will also define our **known parameter**, $\sigma = 2$, in terms of τ, which is then $1/(2^2) = 0.25$.

?? ## Why is this chapter called "The normal-normal conjugate"?

When you use a normal distribution as a prior distribution for the unknown parameter of a normal distribution, μ, assume that σ (or τ) is known, and then collect data (from a normal distribution). Then, you can use a conjugate shortcut to generate the posterior distribution for the unknown parameter μ, which will also be a normal distribution.

A diagram such as that shown in Figure 12.8 may help.

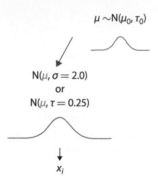

$\mu \sim N(\mu_0, \tau_0)$

$N(\mu, \sigma = 2.0)$
or
$N(\mu, \tau = 0.25)$

x_i

Figure 12.8

At the bottom of this diagram, we have our observed data, x_i, which we assume arises from a normal distribution with an unknown mean, μ, and known standard deviation, $\sigma = 2.0$. This is the same thing as a normal distribution with an unknown mean, μ, and known precision, $\tau = 0.25$. The parameter, μ, is the unknown parameter that we are trying to estimate. Here, we use a normal distribution as a prior distribution to set the "weights" on each and every hypothesis for the unknown parameter, μ. This normal distribution is defined by the hyperparameters μ_0 and τ_0.

 ?? ## So, what is the conjugate shortcut?

Answer: Ok then. Time for a concrete example. First, let's recall the steps for Bayesian analysis:

1. Identify your hypotheses—these would be the alternative hypotheses for the mean. (The conjugate shortcut requires that you assume either μ or σ is known; here we assume σ is known.)
2. Express your belief that each hypothesis is true in terms of a prior distribution (i.e., a pdf). For a normal distribution, μ can range from $-\infty$ to ∞. For our example, however, μ must be positive, so we will select a prior distribution such that the area under the curve for $\mu < 0$ is very, very small.
3. Gather the data.
4. Determine the likelihood of the observed data under each hypothesis.
5. Use Bayes' Theorem to compute the posterior distribution, which is a pdf.

Step 1. These would be the alternative hypotheses for the mean. Here, we will assume that τ is known and has a value of 0.25. Remember, this is **known.**

That leaves us with just one unknown parameter, μ, which has an infinite number of hypotheses, all of which are positive.

Step 2. Establish a **prior distribution** for the **unknown** parameter, μ. Remember that our prior distribution is a normal distribution. Although Vermont secretly collects data on syrup production, earlier in the chapter we used a dot exercise to "elicit" information from cartel members. Elicitation is a method for "boiling down" what experts think they know when other sources of data are not available (pun intended!).

Tip: "Use expert opinions in such a way that we stand to gain if they are correct and do not lose if they are wrong."—C.R. Rao

There are many different ways to elicit a prior distribution, and these are beyond the scope of this book. For the sake of this chapter, let's assume the cartel uses the normal distribution shown in blue in Figure 12.8; it has the following hyperparameters:

- $\mu_0 = 12$
- $\tau_0 = 0.0625$

Notice the subscript of 0 (naught) is used to indicate that these are prior hyperparameters.

Step 3. Collect data. **We have observed 10.2 million gallons of syrup produced.**
We now write our observed data as:

$$\sum_{i=1}^{n} x_i \tag{12.11}$$

Here, n is the number of samples (in this case, 1). Typically, we index each year with the letter i, where the index i goes from 1 to n. So, $x_i = x_1 = 10.2$.

If we collected seven years of data, the index i would be 1, 2,...,7. We would report our observed data as the sum of our seven observations, and n would be 7:

$$\sum_{n=1}^{7} x_i \tag{12.12}$$

Steps 4 and 5. Steps 4 and 5 are to determine the likelihood of the observed data under each hypothesis and then use Bayes' Theorem to compute the posterior probabilities for each hypothesis.

Here, though, we can update the prior normal distribution to the posterior normal distribution with the shortcut analytic solution and avoid the use of Bayes' Theorem directly. Here are the shortcut calculations—pay attention to subscripting!

The mean of the posterior distribution for the unknown parameter, μ, is calculated as:

$$\mu_{\text{posterior}} = \frac{\tau_0\mu_0 + \tau\sum_{i=1}^{n} x_i}{(\tau_0 + n * \tau)} = \frac{0.0625 * 12 + 0.25 * 10.2}{0.0625 + 1 * 0.25} = 10.56 \tag{12.13}$$

The precision of the posterior distribution (τ) for the unknown parameter, μ, is calculated as:

$$\tau_{\text{posterior}} = \tau_0 + n * \tau = 0.0625 + 1 * 0.25 = 0.31 \tag{12.14}$$

A graph of our prior and posterior is shown in Figure 12.9.

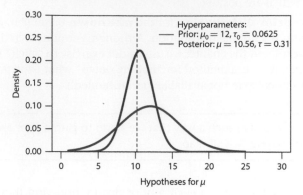

Figure 12.9 Annual maple syrup production, showing hypothesis for μ (in millions of gallons).

Our prior distribution for the unknown parameter, μ, is a normal distribution with $\mu_0 = 12$ and $\tau_0 = 0.0625$ (blue). After collecting one year of syrup data in Vermont (10.2 million gallons shown by the dashed line), our posterior distribution for the unknown parameter μ is a normal distribution with posterior hyperparameters $\mu = 10.56$ and $\tau = 0.31$ (purple). Notice how the posterior has tightened up a bit and shifted to the left.

 ?? **Why does the shortcut work?**

Answer: The answer is that the normal distribution is a conjugate distribution that can be updated with normal data (Raiffa and Schlaeffer, 1961). This is why we named this chapter "The normal-normal conjugate."

We mentioned the statistical definition of a conjugate distribution in Chapter 11, where we indicated that there are cases where you can use a particular **pdf** as a prior distribution, collect data of a specific flavor, and then derive the posterior **pdf** with a closed-form solution. In these special cases, the **pdf**'s of the prior and posterior distributions are the same probability density function, but their *parameters* may differ. The prior distribution is called a conjugate prior, and the effect of the data can then be interpreted in terms of changes in parameter values (Upton and Cook, 2014).

In the maple syrup problem, we used a **normal distribution** to set the priors for all hypotheses of μ, the mean number of gallons of syrup (millions). We used expert opinion to set the prior distribution. We then collected normal data, and then used a shortcut to generate the posterior distribution for μ.

Raiffa and Schlaeffer's (1961) conjugate solution is conveniently given in a Wikipedia page on Bayesian conjugates. A quick overview of that page's normal-normal conjugate can be depicted with the diagram shown in Figure 12.10.

Let's work our way through this figure. The **top panel** indicates that the model parameter of interest, μ, is a mean. The prior distribution is a normal distribution, defined by hyperparameters μ_0 and τ_0. The "naughts" signal that these are hyperparameters for the prior. The data collected are normal in flavor, with n being the total sample size. The likelihood to be used is the normal pdf, where τ is known. With this setup, the posterior distribution for the unknown parameter, μ, is a normal distribution with hyperparameters (the conjugate shortcuts) displayed.

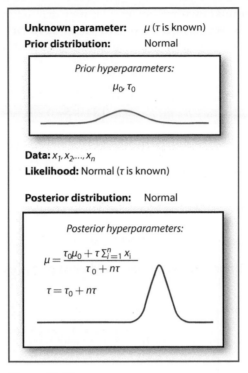

Unknown parameter: μ (τ is known)
Prior distribution: Normal

Prior hyperparameters:

μ_0, τ_0

Data: $x_1, x_2, \ldots, x_n$
Likelihood: Normal (τ is known)

Posterior distribution: Normal

Posterior hyperparameters:

$$\mu = \frac{\tau_0\mu_0 + \tau\sum_{i=1}^n x_i}{\tau_0 + n\tau}$$

$$\tau = \tau_0 + n\tau$$

Figure 12.10

?? Can you show me the proof?

Answer: All right. Have a look at Appendix 3. There, you can confirm that Bayes' Theorem is front and center, even though the results are astonishingly simple.

?? What happens to the posterior if we collect new data?

Answer: Let's try it! **Let's suppose you collect five more years of syrup production data (shown in Table 12.2).**

Table 12.2

Year	Gallons
1	7
2	10
3	10
4	8
5	4
	39

Note that we have five new observations from the Gaussian (normal) distribution, so $n = 5$. Also notice that the **sum** of the observed data is 39 million gallons produced. Our prior distribution for the unknown parameter, μ, is now defined by the hyperparameters $\mu_0 = 10.56$ and $\tau_0 = 0.31$; **we assume τ is still known to be 0.25**. If we repeat our conjugate calculations, the mean of the posterior distribution for the unknown parameter, μ, is calculated as:

$$\mu_{\text{posterior}} = \frac{\tau_0\mu_0 + \tau\sum_{i=1}^{n}x_i}{(\tau_0 + n*\tau)} = \frac{0.31*10.56 + 0.25*39}{0.31 + 5*0.25} = 8.35 \qquad (12.15)$$

And σ for the posterior distribution of the unknown parameter, μ, is calculated as:

$$\tau_{\text{posterior}} = \tau_0 + n*\tau = 0.31 + 5*0.25 = 1.56 \qquad (12.16)$$

$$\sigma^2_{\text{posterior}} = \frac{1}{\tau} = \frac{1}{1.56} = 0.641 \qquad (12.17)$$

$$\sigma_{\text{posterior}} = \sqrt{\sigma^2} = \sqrt{0.641} = 0.801 \qquad (12.18)$$

Let's add this new posterior **pdf** to our graph in dashed purple (see Figure 12.11).

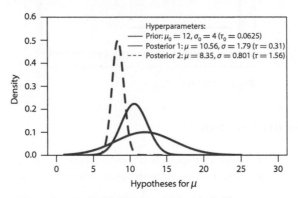

Figure 12.11 Annual maple syrup production, showing hypotheses for μ.

This is a formal documentation of **learning** about **average** maple syrup production in the Green Mountain State. A few things are worth pointing out here:

- We started off with an informative, subjective prior for μ shown in blue.
- We updated this based on a single year's data: 10.2 million gallons produced in one year in Vermont. This shifted the posterior to the left (solid purple), but with only a single year's data, the distribution is still quite broad.
- We then collected five more years of data, in which a total of 39 million gallons was produced. This further shifted the posterior to the left (dashed purple), and the distribution narrowed considerably.

That's what happens as you learn: the uncertainty associated with the true value of μ shrinks as we continue to collect data. We still have uncertainty about what the true mean is, but after some data collection we now have updated beliefs. **Remember that we have assumed that σ for maple syrup production is known with full certainty!**

?? Why do we use τ instead of σ directly?

Answer: We can work with a known variance (σ^2) OR known precision (τ, which is $1/\sigma^2$). Raiffa and Schlaeffer's (1961) conjugate solution is conveniently given in a Wikipedia page on Bayesian conjugates. Let's have another look at our figure, but expand it a bit (see Figure 12.12).

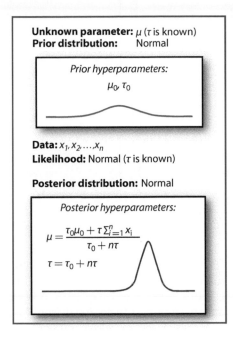
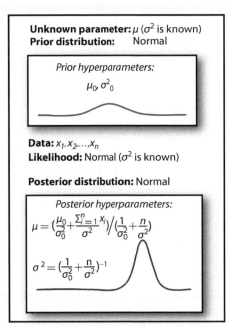

Figure 12.12

- In the left panel, we assume that τ is known, and we define our prior distribution for the unknown parameter, μ, with the hyperparameters μ_0 and τ_0. The conjugate calculations of the posterior hyperparameters are shown. We used this version here and we'll make use of it again in future chapters.
- In the right panel, we assume that σ^2 is known, and we define our prior distribution for the unknown parameter, μ, with the hyperparameters μ_0 and σ_0^2. The conjugate calculations of the posterior hyperparameters are shown.
- Both approaches will return the same result.

?? What if I know μ, but want to estimate σ or τ instead?

Answer: There is a conjugate shortcut for that too, but it wouldn't be called the **normal-normal conjugate** shortcut anymore! For example, Figure 12.13 shows the conjugate solutions for estimating σ^2 or τ given a known mean, μ.

An important difference is that we no longer use the normal **pdf** for our prior distribution for the unknown parameter. Instead, the prior distribution of σ^2 is either the inverse gamma distribution or the scaled inverse chi-square distribution, while the prior distribution for τ is the gamma distribution. We haven't introduced the first two distributions in

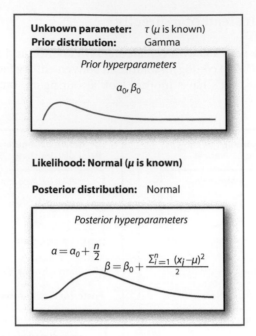

Figure 12.13

this book, but we learned about the gamma distribution in the Shark Attack Problem, and that is what is shown above. And we will make use of the gamma distribution to estimate τ with Gibbs sampling when we revisit the Maple Syrup Problem in Chapter 16. A key take-home point is that you have quite a bit of flexibility with respect to conjugate solutions for the parameters of the normal distribution.

 ?? **Can we summarize this chapter?**

Answer: Indeed.

- We started by mentioning that 60% of Quebec's maple syrup stockpile was stolen and that the cartel wishes to annex Vermont's maple syrup production.
- We introduced the normal probability density function as a way to quantify the average and variation in annual syrup production for the state of Vermont.
- We emphasized that the normal distribution has two parameters, the mean μ and the standard deviation σ, which need to be estimated jointly.
- We remembered that we have an infinite number of hypotheses for μ and σ; these can be described by continuous distributions. To complicate the matter, the estimate of one parameter depends on the value the second parameter.
- To make headway for this problem, we assumed that one of the parameters of the normal distribution was known. We said that σ was known to be 2.0.
- We talked about different measures of spread for the normal distribution: the standard deviation (σ), the variance (σ^2), and the precision ($\tau = \frac{1}{\sigma^2}$).
- We then set up a prior distribution to represent our hypotheses for the mean, μ, of the normal distribution. This prior distribution was also a normal distribution with specified hyperparameters μ_0 and τ_0.

- We then collected data. Vermont produced 10.2 million gallons of syrup last year.
- We used a shortcut, analytical solution that allowed us to update our prior normal distribution to a posterior normal distribution by using the data. The posterior distribution reflects our updated beliefs for each and every hypothesis for μ.
- We then collected five more years of data. This time, we used our newly updated posterior as the prior distribution, and then updated it again given the newly observed data. Thus, we are using Bayesian inference as a way to formalize "learning."

 ?? **What's next?**

Answer: This ends our section on Bayesian conjugates. You can see that these conjugate "shortcuts" are extremely handy for certain types of problems. In Section 5, we'll show you how to solve these same problems using a different approach called Markov Chain Monte Carlo, or MCMC for short.

SECTION 5

Markov Chain Monte Carlo

Overview

Welcome to Section 5! In this section, we will learn how to use MCMC (Markov Chain Monte Carlo) simulations to estimate the posterior distribution for a parameter that you are interested in estimating.

- Chapter 13 revisits the gamma-Poisson conjugate introduced in Chapter 11. There, our goal was to estimate the unknown parameter, λ. We set a gamma distribution as the prior distribution for λ, then collected Poisson-flavored data, and used a conjugate solution to update the prior distribution to a posterior distribution, which is also a gamma distribution. In other words, gamma prior + Poisson data $\rightarrow$ gamma posterior. In Chapter 13, we show you how to estimate the posterior distribution using MCMC with the Metropolis algorithm, and compare the results with the conjugate solution.
- Many things can go wrong with an MCMC analysis, and the analysist (you!) must put on your diagnostics hat to ensure that the posterior distribution that has been estimated really hits the mark. Chapter 14 introduces some common diagnostics.
- Chapter 15 revisits the beta-binomial conjugate introduced in Chapter 10. There, our goal was to estimate the unknown parameter, p. We set a beta distribution as the prior distribution for p, then collected binomial-flavored data, and used a conjugate solution to update the prior distribution to a posterior distribution, which is also a beta distribution. In other words, beta prior + binomial data $\rightarrow$ beta posterior. In Chapter 15, we demonstrate how to estimate the posterior distribution using MCMC and the Metropolis–Hastings algorithm, and compare the results with the conjugate solution.
- Chapter 16 revisits the normal-normal conjugate introduced in Chapter 12. There, our goal was to estimate one of the parameters of the normal distribution, μ, while assuming the second parameter of the normal distribution, τ was known. That is, normal prior + normal data $\rightarrow$ normal posterior. In Chapter 16, we assume **both** parameters are unknown, and thus we are trying to estimate a joint posterior distribution. Here, use MCMC with the Gibbs sampling.

CHAPTER 13

The Shark Attack Problem Revisited: MCMC with the Metropolis Algorithm

As you've seen, a major use of Bayesian inference involves estimating parameters.

Suppose we are trying to estimate a single parameter called theta (θ). You might recall that Bayes' Theorem in this case is specified as:

$$P(\theta \mid \text{data}) = \frac{P(\text{data} \mid \theta) * P(\theta)}{\int P(\text{data} \mid \theta) * P(\theta) d\theta} \tag{13.1}$$

This is the generic version of Bayes' Theorem, when the posterior distribution for a single parameter, given the observed data, is represented by a **pdf**. We've seen this many times before!

Do you remember the Shark Attack Problem in Chapter 11? In that chapter, we asked, "What is the average number of shark attacks in a given year?" In this problem, our goal was to estimate a parameter from the Poisson distribution called lambda (λ), which is an average rate of occurrences. Example rates include the number of car crashes at an intersection per year, the number of texts a person receives per day, or the number of photons reaching a telescope per second. In the Shark Attack Problem, λ represents the average rate of annual attacks, so it can range between 0 and, in theory, infinity. Of course, the number of shark attacks is not infinite, but the idea is that the upper limit can be any reasonable value for a given problem. Within the bounds you specify, there are an infinite number of alternative hypotheses. We can replace the parameter θ with the parameter λ and write Bayes' Theorem for this problem as:

$$P(\lambda \mid \text{data}) = \frac{P(\text{data} \mid \lambda) * P(\lambda)}{\int_0^\infty P(\text{data} \mid \lambda) * P(\lambda) d\lambda} \tag{13.2}$$

You might recall for the Shark Attack Problem that we set up an informative prior distribution (a gamma distribution) that reflected our prior beliefs in each alternative hypothesis for λ, collected some data (in the form of attacks per year), and then updated to the posterior distribution with a conjugate solution that allowed us to avoid the integration in the denominator of Bayes' Theorem. This posterior distribution was also a gamma distribution, and it represents our most up-to-date knowledge about the parameter, λ.

However, not all problems can be solved this way; Bayesian conjugates are special cases that can be solved analytically. But there is another approach, one that is so creative and

Bayesian Statistics for Beginners: A Step-by-Step Approach. Therese M. Donovan and Ruth M. Mickey,
Oxford University Press (2019). © Ruth M. Mickey 2019.
DOI: 10.1093/oso/9780198841296.001.0001

versatile that it can be used to solve almost any kind of parameter estimation problem. It involves building the posterior distribution from scratch using a process called a **Markov Chain Monte Carlo** simulation, or MCMC for short.

In this chapter, we introduce MCMC as a way to estimate the posterior distribution. We will quickly revisit the Shark Attack Problem by estimating the posterior distribution using the gamma-Poisson conjugate. **Then, we'll solve the same problem using the MCMC method so you can compare the two methods directly.**

MCMC makes use of another "version" (so to speak) of Bayes' Theorem. Recall that the integral of the denominator of Equation 13.2 is a constant. Because of that, Bayes' Theorem is sometimes written:

$$P(\lambda \,|\, \text{data}) \propto P(\text{data} \,|\, \lambda) * P(\lambda) \tag{13.3}$$

The symbol $\propto$ means "proportional to." This equation can be read "The posterior density of the parameter, λ, given the data, is proportional to the likelihood function of the data given the parameter λ, multiplied by the prior density of the parameter λ." We'll see this equation in action thousands of times in this chapter! By the end of this chapter, you should understand the following:

- Monte Carlo
- Markov chain
- Metropolis algorithm
- Tuning parameter
- MCMC inference
- Traceplot
- Moment matching

Let's start by reviewing the Shark Attack Problem.

 ?? What was the Shark Attack Problem again?

Answer: In that problem, the goal was to use a Bayesian inference approach to estimate the average number of shark attacks per year, λ. We used the same steps that we did in previous chapters:

1. Identify your hypotheses.
2. Express your belief that each hypothesis is true in terms of prior probabilities.
3. Gather the data.
4. Determine the **likelihood** of the observed data, assuming each hypothesis is true.
5. Use Bayes' Theorem to compute the posterior probability for each hypothesis.

Let's review these steps quickly:

 ?? Step 1. What are the hypotheses for λ?

Answer: We know that λ, the average rate in the Poisson distribution, can take on any value between 0 and some large number. Because the distribution of λ is **continuous**, there's nothing to stop us from considering the full range of hypotheses which are **infinite** ($\lambda = 0.01$, $\lambda = 0.011$, $\lambda = 1.234$, $\lambda = 6.076$, etc.).

?? Step 2. What were the prior probabilities for each hypothesis?

Answer: We used the **gamma** distribution to set a probability density for each and every hypothesis of λ. The gamma distribution is a continuous probability distribution; the x-axis is continuous, not discrete, and begins with 0 and goes to infinity. The shape of the gamma distribution is controlled by two parameters: the shape parameter, called alpha (α), and the rate parameter, called beta (β). Examples of a few alternative gamma probability distributions are shown in Figure 13.1, and we need to **choose one** to represent our prior beliefs for each and every λ. The higher the probability density, the more confidence we have in a particular hypothesis.

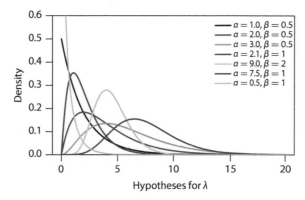

Figure 13.1 Gamma distributions.

Let's look at these graphs carefully. The x-axis for the gamma distribution ranges between 0 and 20, but it could stretch out between 0 and 100, or 0 and 1000, and so on. Our graph ends at 20 only because it's pretty clear that any value greater than 20 will have a density value very close to 0. The y-axis gives the probability density. The area under all of these distributions is 1.0. For any given graph, the higher the probability density, the more belief we have in a particular hypothesis.

For the Shark Attack Problem, we used an informative prior distribution of $\alpha_0 = 2.1$ and $\beta_0 = 1$, based on a previous (external) dataset of shark attacks (see Figure 13.2).

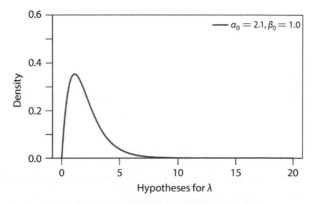

Figure 13.2 Prior distribution: $\alpha_0 = 2.1$, $\beta_0 = 1.0$.

This prior distribution suggests that 1 or 2 shark attacks per year is much more likely than, say, 6 or more shark attacks per year.

 ?? Step 3. Now what?

Answer: Collect data! **We have observed 5 shark attacks this year.**

 ?? Step 4. And then?

Answer: Step 4 is to determine the **likelihood** of the observed data, assuming each hypothesis for λ is true.

 ?? And step 5?

Answer: Step 5 is to generate the posterior distribution for λ by using the version of Bayes' Theorem in Equation 13.2. We mentioned that integration of the denominator is often tedious, and sometimes impossible. But, for problems where the prior distribution is a gamma distribution, and the data collected come in the form of Poisson data (number of events per unit time), an analytical approach can be used to generate the posterior distribution that has the same form as a prior distribution.

In short, prior gamma + Poisson data → posterior gamma.

This analytical shortcut (Raiffa and Schlaeffer, 1961) makes updating the prior to a posterior as easy as ABC. Remember that our prior distribution set $\alpha_0 = 2.1$ and $\beta_0 = 1$. Furthermore, we have 1 new observation: 5 shark attacks. Let's let the number of random variables that we observe be n. So here, $n = 1$, and its value is 5 (we observed 5 shark attacks in a year). **Here is the shortcut:**

The posterior α parameter is α_0 plus the sum of the Poisson observations. We have just one observation, which was 5 attacks:

$$\alpha_{\text{posterior}} = \alpha_0 + \sum_{i=1}^{n} x_i \tag{13.4}$$

$$\alpha_{\text{posterior}} = 2.1 + 5 = 7.1 \tag{13.5}$$

The posterior β parameter is β_0 plus n, the number of Poisson observations:

$$\beta_{\text{posterior}} = \beta_0 + n \tag{13.6}$$

$$\beta_{\text{posterior}} = 1 + 1 = 2 \tag{13.7}$$

Now we can look at the prior and posterior distributions for λ (see Figure 13.3).

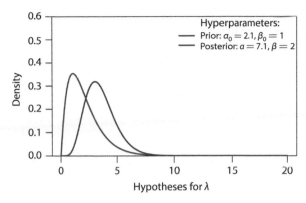

Figure 13.3 Gamma prior posterior.

We started off with a prior distribution shown in blue. Then we collected data in the form of a single Poisson random variable: five attacks. We then used the shortcut to generate the posterior distribution for all hypotheses of λ, shown in purple. As a result of our data, we now have **new** knowledge regarding the plausibility for each and every hypothesis of λ.

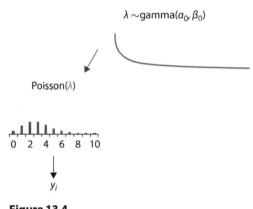

Figure 13.4

The graphical model diagram for this problem looked like the one shown in Figure 13.4.

?? **How would you solve the posterior distribution with a Markov Chain Monte Carlo (MCMC) approach?**

Answer: Ah, that is a great question. Here we go!

The main idea behind MCMC for Bayesian inference is that you "build" the posterior distribution by drawing samples from it. In other words, we are going to try to **build** the purple distribution shown in Figure 13.3 from scratch, using Bayes' Theorem in an entirely new way.

Remember, we had written Bayes' Theorem as:

$$P(\lambda \mid \text{data}) = \frac{P(\text{data} \mid \lambda) * P(\lambda)}{\int_0^\infty P(\text{data} \mid \lambda) * P(\lambda) d\lambda} \tag{13.8}$$

Now, knowing that the denominator is a constant (because it accounts for all possible hypotheses of λ), we can write Bayes' Theorem as:

$$P(\lambda \,|\, \text{data}) \propto P(\text{data} \,|\, \lambda) * P(\lambda) \tag{13.9}$$

In Bayesian parlance:

$$\text{posterior} \propto \text{likelihood} * \text{prior} \tag{13.10}$$

In words, the posterior density of a given hypothesis is **proportional** to the likelihood of observing the data under the hypothesis times the prior density of the given hypothesis. You'll see how this is used shortly. For now, let's color-code the pieces to help distinguish the terms, with the purple posterior being a blend of the red likelihood and blue prior:

$$\text{posterior} \propto \text{likelihood} * \text{prior} \tag{13.11}$$

Now, that idea of building a distribution from scratch may sound strange because you don't really know the distribution from which you are sampling. (Here, though, you have an advantage because you know we are trying to build the purple distribution in Figure 13.3). But come along with us and see. Let's first go through the steps and then discuss them after you have the big picture in mind.

In this chapter, we will be using an **algorithm** called the Metropolis algorithm to conduct our MCMC analysis. An algorithm is a self-contained step-by-step set of operations to be performed. Here is the sequence of operations. There are eight of them:

1. We start by proposing just one value from the posterior distribution. Any value will do as long as it is a possible value from the posterior. **Let's start with 3.100.** This is an arbitrary selection, and represents our **current** value of λ. We can say that 3.100 is a value of lambda (λ) that is drawn from the posterior distribution.

2. We then calculate the posterior density of observing the data under the current hypothesis that $\lambda = 3.100$. We just mentioned that the posterior density is proportional to the product of two terms, the likelihood and the prior:

$$\text{posterior} \propto \text{likelihood} * \text{prior} \tag{13.12}$$

- To get the result, first we ask, "What is the likelihood of observing 5 attacks given $\lambda = 3.100$?" We hope that the Poisson probability mass function springs to mind! Remember that this function has just one **parameter**, λ. The probability of observing five attacks given $\lambda = 3.1000$ is:

$$\mathcal{L}(x; \lambda) = \frac{\lambda^x e^{-\lambda}}{x!} \tag{13.13}$$

$$\mathcal{L}(5; 3.100) = \frac{3.100^5 e^{-3.100}}{5!} = 0.1075 \tag{13.14}$$

- Next, we ask "What is the probability density associated with 3.100 from the **prior distribution**?" We hope that the gamma probability density function springs to mind! We can use the gamma pdf to ask, "What is the probability density of drawing a random value of 3.100, given the parameters of the prior distribution?" Remember our prior distribution set $\alpha_0 = 2.1$ and $\beta_0 = 1$. Here, we use the traditional gamma pdf, and plug in λ for x:

$$g(x; \alpha, \beta) = \frac{\beta^\alpha x^{\alpha-1} e^{-\beta x}}{\Gamma(\alpha)} \tag{13.15}$$

$$g(3.100; 2.1, 1) = \frac{1^{2.1} 3.100^{2.1-1} e^{-1*3.100}}{\Gamma(2.1)} = 0.1494 \qquad (13.16)$$

- The posterior density associated with our first guess at λ (3.100) is proportional to the likelihood times the prior:

$$P(\lambda = 3.100 \mid 5) \propto 0.1075 * 0.1494 = 0.0161 \qquad (13.17)$$

- We will now refer to this as $P(\lambda_{current} \mid data)$. This is the posterior density value for a particular hypothesis of λ.

3. We then **propose** a second value for λ, drawn **at random** from a **symmetrical** distribution that is centered on the current value of λ. For this example, we will use a normal distribution as our symmetrical distribution, with $\mu = \lambda_{current}$, and $\sigma = 0.5$. Here, σ is called the **tuning parameter**, and we will talk about it later. **Suppose our randomly drawn proposed value is $\lambda = 4.200$.**

4. We now calculate the posterior for the proposed value of λ as follows:

$$posterior \propto likelihood * prior \qquad (13.18)$$

- First, we ask "what is the likelihood of observing five attacks given $\lambda = 4.200$?" We hope that the Poisson probability mass function springs to mind (again)! The probability of observing five attacks given $\lambda = 4.200$ is:

$$\mathcal{L}(x; \lambda) = \frac{\lambda^x e^{-\lambda}}{x!} \qquad (13.19)$$

$$\mathcal{L}(5; 4.200) = \frac{4.200^5 e^{-4.200}}{5!} = 0.1633 \qquad (13.20)$$

- Then we ask, "What is the probability density associated with 4.200 from the prior distribution?" And we hope once again that the gamma probability density function springs to mind! We can use the gamma pdf to ask "what is the probability density of drawing a random value of 4.200, given the parameters of the prior distribution?" Remember our prior distribution set $\alpha_0 = 2.1$ and $\beta_0 = 1$:

$$g(x; \alpha, \beta) = \frac{\beta^\alpha x^{\alpha-1} e^{-\beta x}}{\Gamma(\alpha)} \qquad (13.21)$$

$$g(4.200; 2.1, 1) = \frac{1^{2.1} 4.200^{2.1-1} e^{-1*4.200}}{\Gamma(2.1)} = 0.0695 \qquad (13.22)$$

- The posterior density associated with our proposed value of λ is proportional to the product of the likelihood and the prior:

$$P(\lambda = 4.200 \mid data) \propto 0.1633 * 0.0695 = 0.0113 \qquad (13.23)$$

- We will now refer to this as $P(\lambda_{proposed} \mid data)$.

5. Now you have two hypothesized λ values from the posterior distribution ($\lambda = 3.100$ and $\lambda = 4.200$), and you know their posterior densities. Never, ever, forget that Bayes' Theorem is at the heart of these values!

- The posterior densities for our two hypotheses are:

$$P(\lambda_{current} = 3.100 \mid data) \propto 0.0161 \qquad (13.24)$$

$$P(\lambda_{proposed} = 4.200 \mid data) \propto 0.0113 \qquad (13.25)$$

- The next step is to **throw one away**. Which one will you keep? Will you stay with your original value ($\lambda = 3.100$), or will you move to the proposed value ($\lambda = 4.200$)?
- To help you with that decision, the Metropolis algorithm specifies that you calculate the probability of accepting (or moving to) the proposed λ (which in our example is 4.200) as the smaller (minimum) of two values: 1 or the ratio of the two posterior densities associated with the proposed and current λ values:

$$p_{\text{move}} = min\left(\frac{P(\lambda_{\text{proposed}} \mid \text{data})}{P(\lambda_{\text{current}} \mid \text{data})}, 1\right) \tag{13.26}$$

$$p_{\text{move}} = min\left(\frac{0.0113}{0.0161}, 1\right) \tag{13.27}$$

$$p_{\text{move}} = min(0.7019, 1) = 0.7019 \tag{13.28}$$

- So, for this example, the probability of accepting the proposed value for λ is 0.7019.
6. Next, draw a random number from the **uniform distribution**, $U(0,1)$. This is a uniform distribution between 0 and 1. If the random number is less than the probability of moving, accept the proposed value of λ. If not, stay with the current value of λ. Notice that if the proposal has a greater posterior density than the current, you will **always** move to the proposal. But if your current posterior density is greater than your proposed, as is the case here, you may or may not move…it depends on what random number you draw. **Let's say our randomly drawn value is 0.8204.** This value is greater than 0.7019, so we now keep $\lambda = 3.100$. This now represents our **current** value of λ for MCMC trial 2.
7. Repeat, repeat, repeat hundreds or thousands of times!
8. Summarize the accepted values in the form of a distribution and its corresponding summary statistics.

By the way, the Metropolis algorithm is named for Nicholas Metropolis (see Figure 13.5), a physicist born in Chicago (Go Cubs!). However, his seminal paper had multiple coauthors, including Arianna W. Rosenbluth, Marshall N. Rosenbluth, Augusta H. Teller, and Edward Teller. Sometimes the algorithm is called the $M(RT)^2$ algorithm for the last names of the five authors. The history of this paper and the players involved is quite rich and intriguing.

Figure 13.5 Nicholas Metropolis (Original photograph courtesy of the Los Alamos National Laboratory).

?? Where can I learn about the mathematics of the Metropolis algorithm?

Answer: John Kruschke provides an excellent, non-technical discussion of the Metropolis algorithm in his book Doing Bayesian Data Analysis (2015). For those more mathematically inclined, Christophe Andrieu and co-authors (2003) provide "An introduction to MCMC for machine learning." Their first sentence reads, "A recent survey places the Metropolis algorithm among the ten algorithms that have had the greatest influence on the development and practice of science and engineering in the 20th century." Wow!

?? Can we see an example of these operations in action?

Answer: You bet. Operation 7 involves a loop where you repeat the operations over and over. This can be visualized with the map shown in Figure 13.6.

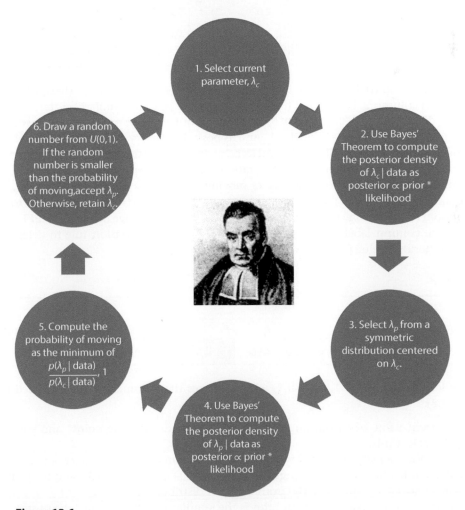

Figure 13.6

Let's carry out ten trials. **Remember that the standard deviation, σ, for our proposal distribution, is 0.5 (see Table 13.1).**

Table 13.1

	Data	Prior			Posterior				Decision			
Trial	Attacks	α_0	β_0	λ_c	P_c	λ_p	P_p	Ratio	P_{move}	Random	Accepted λ
1	5	2.1	1	3.100	0.0161	4.200	0.0113	0.7019	0.7019	0.8204	3.100
2	5	2.1	1	3.100	0.0161	2.360	0.0134	0.8323	0.8323	0.5716	2.360
3	5	2.1	1	2.360	0.0134	2.637	0.0151	1.1269	1.0000	0.2159	2.637
4	5	2.1	1	2.637	0.0151	2.306	0.0129	0.8543	0.8543	0.3645	2.306
5	5	2.1	1	2.306	0.0129	2.435	0.0139	1.0775	1.0000	0.0979	2.435
6	5	2.1	1	2.435	0.0139	2.674	0.0153	1.1007	1.0000	0.1749	2.674
7	5	2.1	1	2.674	0.0153	2.166	0.0117	0.7647	0.7647	0.5066	2.166
8	5	2.1	1	2.166	0.0117	2.629	0.0151	1.2906	1.0000	0.4551	2.629
9	5	2.1	1	2.629	0.0151	2.035	0.0104	0.6887	0.6887	0.0251	2.035
10	5	2.1	1	2.035	0.0104	2.616	0.0150	1.4423	1.0000	0.6145	2.616

In this table, our trials are given down column 1. The observed data, five shark attacks, is given down column 2. Notice that these are fixed for all 10 trials. Similarly, α_0 and β_0 from our prior distribution (columns 3 and 4) are fixed for all 10 trials. Under the four **Posterior** columns, the current λ value is labeled λ_c, and its posterior density (shaded purple) is labeled $\boldsymbol{P_c}$. The proposed λ value is labeled λ_p, and its posterior density (shaded purple) is labeled $\boldsymbol{P_p}$. Under the four **Decision** columns, the ratio of the two purple values is stored in the column labeled **Ratio**. The probability of moving to the proposed value (labeled $\boldsymbol{P_{move}}$) is the smaller of two values: the number 1.0000 or the ratio. The column labeled **Random** is a random number from a uniform distribution whose values range between 0 and 1. And, finally, the column labeled **Accepted** λ uses the Metropolis decision rule to determine which posterior λ value (λ_c or λ_p) we keep.

Table 13.2

	Data	Prior			Posterior				Decision			
Trial	Attacks	α_0	β_0	λ_c	P_c	λ_p	P_p	Ratio	P_{move}	Random	Accepted λ
1	5	2.1	1	3.100	0.0161	4.200	0.0113	0.7019	0.7019	0.8204	3.100

Now let's focus on row 1, or trial 1, which features the example we just discussed (see Table 13.2).

- Our current value of λ (column λ_c) is 3.100. This is our "starting point" and you, the analyst, select this.
- The posterior density associated with $\lambda = 3.100$ (column $\boldsymbol{P_c}$) is the product of two terms: the Poisson likelihood of observing the data (5 attacks), given $\lambda = 3.100$, and the gamma density associated with $\lambda = 3.100$ from the prior distribution. The answer is 0.0161. We calculated that by hand earlier, and this is the familiar Bayesian adage: "the posterior is proportional to the likelihood times the prior."

- We then consider a proposed value for lambda, $\lambda = 4.200$ (column λ_p). We obtain the proposal by drawing a random value from a normal distribution that is centered on λ_c and has a standard deviation of 0.5.
- The posterior density associated with $\lambda = 4.200$ (column P_p) is the product of two terms: the Poisson probability of observing the data (5 attacks), given $\lambda = 4.2$, and the gamma density associated with $\lambda = 4.2$ from the prior distribution. The answer is 0.0113 and is deeply rooted in Bayes' Theorem. We calculated that by hand earlier.
- The ratio of these two posterior densities is computed as the posterior density of the proposed λ divided by the posterior density of the current λ (column **Ratio**). We calculated that by hand earlier as 0.7019.
- The probability of moving (column $P_{\textbf{move}}$) is calculated as minimum of 1 and the ratio. Because the ratio 0.7019 is smaller than 1.0, the probability of moving (accepting the proposal) is 0.7019.
- We then draw a random number between 0 and 1. Our random number is 0.8204 (column **Random**).
- We then keep the λ value according to our random number. If the random number is less than the probability of moving, we accept the proposed value of λ_p; otherwise, we accept the current value of λ_c. Our random number is greater than 0.7019, so we keep $\lambda_c = 3.100$ (column **Accepted** λ). This value becomes λ_c for our next MCMC trial.

Now let's focus on row 2 (even if it is a bit tedious); see Table 13.3.

Table 13.3

	Data	Prior			Posterior			Decision			
Trial	Attacks	a_0	β_0	λ_c	P_c	λ_p	P_p	Ratio	P_{move}	Random	Accepted λ
1	5	2.1	1	3.100	0.0161	4.200	0.0113	0.7019	0.7019	0.8204	3.100
2	5	2.1	1	3.100	0.0161	2.360	0.0134	0.8323	0.8323	0.5716	2.360

- In trial 2, our current value of λ (column λ_c) is 3.100.
- The posterior density associated with $\lambda = 3.100$ (column P_c) is the product of two terms: the Poisson probability of observing the data (5 attacks) given $\lambda = 3.100$) and the gamma density associated with $\lambda = 3.100$ from the prior distribution. The answer is 0.0161. We calculated that by hand earlier.
- We then consider a proposed value, $\lambda = 2.360$ (column λ_p). We obtain the proposal by drawing a random value from a normal distribution that is centered on λ_c (3.100) and has a standard deviation of 0.5.
- The posterior density associated with $\lambda_p = 2.360$ (column P_p) is the product of two terms: the Poisson probability of observing the data (5 attacks), given $\lambda = 2.360$, and the gamma density associated with $\lambda = 2.360$ from the prior distribution. The answer is 0.0134.
- The ratio of these two posterior densities is computed as the posterior of the proposed λ_p divided by the posterior of the current λ_c (column **Ratio**), which is 0.8323.
- The probability of moving (column $P_{\textbf{move}}$) is calculated as minimum of 1 and the ratio. Since the ratio is 0.8323, the probability of moving (accepting the proposal) is 0.8323.
- We then draw a random number between 0 and 1 from a uniform distribution. Our random number is 0.5716 (column **Random**).

- We then keep the λ value according to our random number. If the random number is less than the probability of moving, we accept the proposed value for λ; otherwise, we accept the current value of λ. Our random number is less than 0.8323, so we accept $\lambda = 2.360$ (column **Accepted** λ). This value becomes λ_c for our next MCMC trial.

> Through this process, we "pit" two posterior densities against each other, and keep only the proposal (or proposed value) if its posterior density is higher than the current value. If it's not higher, we accept the proposal probabilistically. By repeating this process over and over again, we eventually end up converging on a reasonable estimate of the posterior distribution. Bam!

?? What are some key characteristics of the Metropolis algorithm?

Answer: The Metropolis algorithm has some unique requirements:

- The distribution that generates the proposed parameter must be symmetrical.
- Usually, the proposed parameter is drawn from a **normal distribution** that is centered on the current value. This means that the analyst must provide the standard deviation for the normal distribution to control what proposals can be drawn. In our example above, we used a standard deviation of 0.5.
- Setting the standard deviation can be a tricky business and requires **tuning** such that the probability of acceptance ranges between 20%–50% in order to converge on the posterior distribution. The parameter to be tuned is often called a **tuning parameter.** We'll touch more on this topic later.

?? Are there certain terms associated with this process?

Answer: Certainly. We're now ready to define a set of terms that you will see used in Bayesian analyses that use MCMC.

- **Starting point**. The starting point is the first value you designate in trial 1. We used an initial λ value of 3.100.
- **Monte Carlo**. Monte Carlo in this case refers to the Monte Carlo Casino, located in the Principality of Monaco. The Monte Carlo Casino is filled with games of chance. Encyclopedia Britannica tells us that Monte Carlo methods are "statistical methods of understanding complex physical or mathematical systems by using randomly generated numbers as input into those systems to generate a range of solutions...The Monte Carlo method is used in a wide range of subjects, including mathematics, physics, biology, engineering, and finance, and in problems in which determining an analytic solution would be too time-consuming." In this context, Monte Carlo refers to the notion that we use random sampling to generate a proposed parameter, and we also use random sampling to determine whether or not we accept the proposed value or retain the current parameter value.
- **Proposal Distribution**. The proposal distribution is the distribution that you use to draw the next, proposed value. In our example, we used a **normal distribution** whose mean is the current value in the chain and whose standard deviation (the **tuning parameter**) is set by you, the analyst.

- **Markov chain**. A Markov chain is named after the Russian mathematician, Andrey Markov. A Markov chain is "a sequence of random variables in which each variable depends only on its immediate predecessor, so the sequence moves at random, each move being determined by its previous position" (Upton and Cook, 2014). In our example, "chain" refers to the chain of accepted λ's. The construction of the chain is often described as "memorylessness" because the probability associated with the next link (in our case, λ_p) depends only on the current link (λ_c).
- **MCMC trials**. The number of MCMC trials is the number of trials you decide to run. In our example, we ran 10 trials.

 ?? ### How do we depict our results graphically?

Answer: Let's graph our **current λ values** as a line graph so that we can see how our values change with each iteration. In Figure 13.7, each point is a unique sample from the posterior distribution for λ.

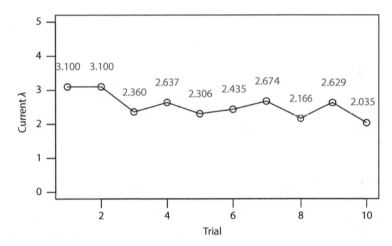

Figure 13.7

This graph shows the MCMC **chain** across trials and is called a **traceplot**. Here, trial number is along the x-axis, and we plot the accepted λ (the annual rate of shark attacks). We started off with $\lambda = 3.100$ but then retained $\lambda = 3.100$ (which is then our current value for trial 2). In trial 2, we proposed $\lambda = 2.360$ and kept the proposed value, which is the current value for trial 3. In trial 3, we proposed $\lambda = 2.637$ and then accepted that value (which is plotted). **Notice that for these 10 trials, each and every proposal was accepted with the exception of trial 1, where we retained $\lambda_c = 3.100$.**

When the proposal is not accepted, you will see the λ value remain constant from one trial to the next; that is, the line joining the accepted values will be horizontal.

The fact that we have a 90% acceptance rate suggests that our choice of standard deviation, 0.5, needs tuning, or adjusting. We will revisit this topic in Chapter 14.

?? **And how does this process lead to the posterior distribution?**

Answer: We create a frequency distribution of the current λ's. Let's try it for our 10 samples, grouping (binning) our λ's by 0.1. Notice that the range of λ's graphed is between 2 and 4 (see Figure 13.8).

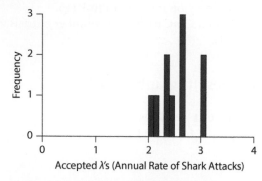

Figure 13.8 Histogram of MCMC results.

Here, we've let the current λ value run along the x-axis from 2.0 to 4.0, grouped into "bins" of 0.1. The y-axis is the frequency. For example, we observed one λ that ranged between 2.1 and 2.2. That would be trial 8, right? We observed two λ's that ranged between 2.3 and 2.4. Those would be trials 3 and 5 in Table 13.1. And so on.

This is an estimate of our posterior distribution for 10 trials, except that it is not normalized such that the sum of the bars equals 1.00.

 ?? **How many trials do you need to generate an adequate posterior distribution?**

Answer: Normally, hundreds or thousands. You've seen how easy it is to do an MCMC. The only limiting factor here is computing power and a software program where you can program the steps. If you are interested, check out Wolfram Demonstrations to see the algorithm in action on a different example.

Let's rerun our experiment, but this time let's run 10,000 trials to create our posterior distribution. But, instead of plotting frequency, let's plot probability density, which is the relative frequency (frequency of a bin divided by total trials) divided by the bin width (see Figure 13.9).

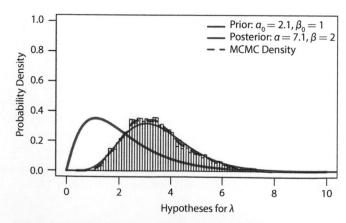

Figure 13.9 Prior and posterior density distributions.

Let's also plot the "smoothed" probability distribution in dashed purple. And, just for kicks, let's also add in the prior distribution (blue) as well as the posterior distribution we calculated from the conjugate shortcut method (in solid purple). Recall that this entire chapter demonstrates how MCMC with the Metropolis algorithm can be used to estimate the solid purple distribution!

> Mind-blowing, eh? Never forget that our MCMC distribution is estimated via a set of rules to be followed (in this case, the Metropolis algorithm). In each step, we have two competing hypotheses for a parameter of interest. For each hypothesis in each step, you compute the posterior density by multiplying the prior density by the likelihood. Then you retain one hypothesis via a rule in which random numbers are involved. If you repeat this process with new random numbers, you'll almost assuredly end up with a different purple distribution. However, as your number of trials becomes large, the differences should become negligible. You then use this distribution to base your inferences, or conclusions, about λ (the annual rate of shark attacks).

REV. T. BAYES
Improver of the Columnar Method developed by Barrett.

 ?? **How do we summarize the posterior distribution?**

Answer: With simple summary statistics! We have 10,000 datapoints now, which consist of our chain of accepted values of hypothesized λ's (the average rate of shark attacks per year). Now it is a matter of just computing some simple statistics:

- Mean: the average value of our values of λ
- Variance: the variance of the λ's
- Standard deviation: the standard deviation of our λ
- Minimum: the smallest value of λ
- Maximum: the largest value of λ

Here, we'll make our calculations based on the full set of 10,000 trials (see Table 13.4).

Table 13.4

Mean	Variance	Standard Deviation	Minimum	Maximum
3.476	1.448	1.203	0.893	7.549

 ??   ## Are these the statistics that we report for our analysis?

Answer: Possibly. But with an MCMC analysis, the Bayesian credible intervals are often reported. We'd like to show the credible range of λ values from our analysis...the minimum and maximum values would not be included in this interval. In Chapter 10, we mentioned alternative methods for defining a suitable Bayesian credible interval, including:

* Choosing the narrowest interval, which for a unimodal distribution will involve choosing those values of highest probability density including the mode. This is sometimes called the highest posterior density interval.
* Choosing the interval where the probability of being below the interval is as likely as being above it. This interval will include the median. This is sometimes called the equal-tailed interval.
* Assuming that the mean exists, choosing the interval for which the mean is the central point.

We touched on these in previous chapters. For this chapter, let's calculate the Bayesian credible intervals for λ using the second method, the equal-tailed interval method.

For this method, it's important that you understand the concept of a **quantile**. Quantiles are very flexible and easy to compute: just line up the datapoints from smallest to largest. In our case, the datapoints are the 10,000 λ values from the MCMC trials. Then, choose a quantile of interest, say 0.10. This means you'd like to find the datapoint where 10% of the data are smaller than that value, and 90% of the data are larger. If you'd like the 0.95 quantile, that means you'd like the find the datapoint where 95% of the data are smaller than that value, and 5% of the data are larger. Lining the data up in order is essential.

Table 13.5 shows some quantiles for our 10,000 MCMC results.

Table 13.5

Quantile	λ
0.010	1.41
0.025	1.57
0.050	1.76
0.250	2.59
0.500	3.33
0.750	4.17
0.950	5.75
0.975	6.24
0.990	6.87

Looking at these results, we can see that half of our 10,000 trials were less than 3.33, and half were greater than 3.33 (look at the 0.5 quantile, also called the **median**). Now look at the 0.99 quantile. The result (6.87) suggests that 99% of the λ estimates in our MCMC chain were less than 6.87, which means that 1% were greater than this value. Don't forget that λ here specifically means the average rate of shark attacks!

If we want to find the values of λ that represent 95% of the values (with 2.5% in each tail, or option 2 in the bulleted list above), we could just use the values associated with the 0.025 and 0.975 quantiles and report the following Bayesian credible interval:

$$1.57 < \lambda < 6.24 \tag{13.29}$$

In other words, given our prior distribution, and the fact that we observed five shark attacks this year, our MCMC results suggest that the average annual rate shark attacks is between 1.57 and 6.24.

 ?? But shouldn't we express the MCMC results as a gamma distribution?

Answer: You're right, of course! Our prior is a **gamma distribution**, and it makes sense that we'd like to express the posterior distribution as a gamma distribution as well. The method by which we convert our MCMC results to the gamma distribution is called **moment matching**. You might recall that the mean of a given gamma pdf is equal to:

$$\mu = \frac{\alpha}{\beta} \tag{13.30}$$

And the variance of a given gamma distribution is equal to:

$$\sigma^2 = \frac{\alpha}{\beta^2} \tag{13.31}$$

But our MCMC analysis provides us with the reverse: we have the mean and variance of the λ's from our MCMC trials. Can we go from these values to obtain the corresponding α and β parameters for the gamma distribution? We can use the first and second moment functions for the two unknowns (α and β), and solve in terms of μ and σ^2. This is done by solving one of the equations (it shouldn't matter which) for α and then substituting the result into the second equation for α. Then do the same for β. Try this yourself. You should end up with these solutions:

$$\alpha = \frac{\mu^2}{\sigma^2} \tag{13.32}$$

$$\beta = \frac{\mu}{\sigma^2} \tag{13.33}$$

Now we can estimate α and β from our MCMC results in Table 13.4:

$$\alpha = \frac{\mu^2}{\sigma^2} = \frac{3.476^2}{1.448} = 8.34 \tag{13.34}$$

$$\beta = \frac{\mu}{\sigma^2} = \frac{3.476}{1.448} = 2.40 \tag{13.35}$$

These are close to the conjugate solutions ($\alpha = 7.1$ and $\beta = 2$) but are not a perfect match. We've hinted that we can improve this estimate of the posterior distribution by adjusting the tuning parameter, which we'll cover in Chapter 14.

A graph of this distribution (from our MCMC analysis), along with our original prior in blue and the posterior from the conjugate solution in purple, can now be displayed (see Figure 13.10).

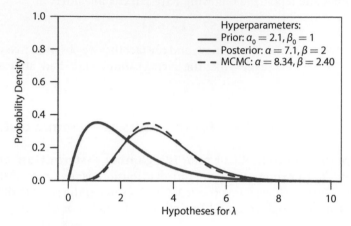

Figure 13.10

Cool! FYI, Tom Hobbs and Mevin Hooten provide an excellent discussion of moment matching in their book *Bayesian Models: A Statistical Primer for Ecologists* (Hobbs and Hooten, 2015).

 ?? Would we get different results if we used a different prior distribution?

Answer: As with all Bayesian analyses, the posterior distribution may be influenced by your choice of the prior distribution. In Chapter 11 (and here as well), we used an informative prior based on a 10-year dataset of annual shark attacks. This is justified: if you have previous knowledge or information that can be used to set the prior distribution, you should use it.

 ?? Are the other algorithms we can use in an MCMC analysis?

Answer: There are many others!

- In Chapter 15, we will revisit the White House Problem and introduce the more general **Metropolis–Hastings algorithm**, which relaxes the assumption that the proposed parameter is drawn from a symmetric distribution.
- In Chapter 16, we will revisit the Maple Syrup Problem, and use the **Gibbs Sampling algorithm**.
- There are others as well, and this is an area of active research...stay tuned! (Pun intended!)

?? **How do we know that our MCMC posterior really hits the spot?**

Answer: Ahh…That is a great question and one we will be visiting in Chapter 14. In this chapter, we knew all along that we were trying to build the "purple" posterior distribution—the one we found with the conjugate shortcut and is our "target distribution." But normally you would not have this advantage. Therefore, "diagnosing" the results of an MCMC analysis is a critical step in any Bayesian MCMC analysis. We already hinted that our tuning parameter, σ, needs adjustment. We won't cover it here because our aim in this chapter is to give you the big picture regarding MCMC. Aside from that, we're running out of steam!

 ?? **And what's the big picture?**

Answer: We can summarize this chapter in the following way:

- We are trying to estimate the average rate of shark attacks per year.
- The hypotheses range from 0 to something very large, and there are an infinite number of them.
- We used the gamma distribution to set our prior distribution; this gives our relative a priori probability (i.e., a density) for each alternative hypothesis for λ.
- We collected data: there were five shark attacks this year.
- We **computed** the posterior probability distribution by using the conjugate method.
- We then **estimated** the posterior probability distribution using a Markov Chain Monte Carlo simulation. This approach allows us to estimate a Bayesian posterior distribution and avoid the messy integration required in the denominator of Bayes' Theorem.
- We compiled the results of our MCMC analysis, computed the mean and variance, and then used the method of moments to define the posterior distribution as a gamma distribution.
- This posterior distribution provides the current state of knowledge about the annual rate of shark attacks. It is completely consistent with the scientific method in that we now have weights of plausibility for each and every alternative hypothesis for λ in light of new data. We could use this information to make future predictions or in decision-making.

 ?? **OK, what's next?**

Head over to Chapter 14, where we outline some of the things that can go wrong in the MCMC analysis.

CHAPTER 14

MCMC Diagnostic Approaches

In Chapter 13, we introduced the Markov Chain Monte Carlo approach as a means of estimating the posterior distribution for a parameter. With MCMC, the posterior distribution is created bit by bit (trial by trial).

In the end, you have a long list of output that tracks the value of the parameter of interest from trial to trial. With this long list, you can do the following:

- Create a **traceplot** of the parameter values across trials, which looks like a snake.
- Create a histogram of the current values in which the current values are "binned" and counted, producing a first glimpse at the shape of the posterior distribution.
- Compute summary statistics of the parameter values, such as the mean, variance, and quantiles. These can be used to create **Bayesian credible intervals**.
- Use **moment matching** to summarize the MCMC output in terms of the prior distribution's family, if applicable.

But, through this process, there has been one burning question:

 ?? **How do we know that our MCMC posterior really hits the spot?**

Answer: You often don't know! However, you can "diagnose" the results of your MCMC analysis to give you the best chance at inferring the true posterior distribution.

In this chapter, we will revisit the Shark Attack Problem, where our goal was to estimate the annual shark attack rate, λ. There, we set a prior distribution, collected new data, and then obtained the posterior distribution in two ways:

1. The first was a conjugate solution.
2. The second involved an MCMC analysis using the Metropolis algorithm.

Here, our goal is to illustrate some of the things that can go wrong in MCMC analysis and to introduce you to some diagnostic tools that help identify whether your results can be trusted or not.

By the time you have finished this chapter, you will understand the following diagnostic concepts:

- Tuning
- Acceptance rate
- Burn in
- Thinning
- Proposal issues

Bayesian Statistics for Beginners: A Step-by-Step Approach. Therese M. Donovan and Ruth M. Mickey,
Oxford University Press (2019). © Ruth M. Mickey 2019.
DOI: 10.1093/oso/9780198841296.001.0001

?? What was the Shark Attack Problem again?

Answer: In that problem, the goal was to use a Bayesian inference approach to estimate the average number of shark attacks per year. We called that rate λ. Since there is no such thing as a negative rate of shark attacks, λ ranges from 0 to some positive number. We used the same steps that we did in previous chapters:

1. Identify your hypotheses—these would be the alternative hypotheses for λ, and they range from 0 to ∞.
2. Express your belief that each hypothesis is true in terms of prior probabilities.
3. Gather the data—we observe a certain number of shark attacks in a given year.
4. Determine the **likelihood** of the observed data, assuming each hypothesis is true.
5. Use Bayes' Theorem to compute the posterior densities for each value of λ.

Let's review these steps quickly.

 ?? Step 1. What are the hypotheses for λ?

Answer: We know that λ, the average rate in the Poisson distribution, can take on any value between 0 and some large number. Since the distribution of λ is **continuous**, there's nothing to stop us from considering the full range of hypotheses which are **infinite** ($\lambda = 0.01$, $\lambda = 0.011$, $\lambda = 1.234$, $\lambda = 6.076$, etc.).

 ?? Step 2. What were the prior probabilities for each hypothesis?

Answer: We used the **gamma** distribution to set the probability density for each and every hypothesis for λ. For the Shark Attack Problem, we used an informative prior distribution of $\alpha = 2.1$ and $\beta = 1$. This prior distribution was determined based on a previous (external) dataset with 10 years of shark attack information (see Figure 14.1).

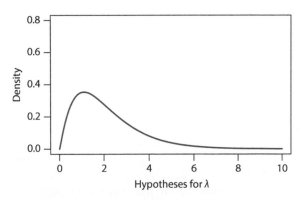

Figure 14.1 Gamma distribution: $a = 2.1$, $\beta = 1$.

This prior distribution suggests that one or two shark attacks per year is much more likely than, say, six or more shark attacks per year.

?? Step 3. Now what?

Answer: Collect data! **We have observed five shark attacks.**

?? Step 4. And then?

Answer: Step 4 is to determine the **likelihood** the observed data, assuming each hypothesis for λ is true.

?? And step 5?

Answer: Step 5 is to use Bayes' Theorem to generate the posterior distribution for λ, given the data:

$$P(\lambda \mid \text{data}) = \frac{P(\text{data} \mid \lambda) * P(\lambda)}{\int_0^\infty P(\text{data} \mid \lambda) * P(\lambda) d\lambda} \qquad (14.1)$$

We've repeatedly mentioned that integration of the denominator is often tedious, and sometimes intractable. However, this challenge can be solved in two ways.

First, for problems where the prior distribution is a gamma distribution, and the data collected come in the form of Poisson data (number of events per unit time), an analytical approach can be used to generate the posterior distribution that has the same form as a prior distribution, that is, a conjugate prior.

Alternatively, an MCMC approach can be used. Let's quickly review the MCMC steps we used in Chapter 13, which employed the Metropolis algorithm:

1. Start by proposing just one value of λ from the posterior distribution. Any reasonable value will do. We will call this the current hypothesis, λ_c.
2. Calculate the posterior density of observing the data under the **current** hypothesis, which is the likelihood of observing the data under the current hypothesis times the prior probability of the current hypothesis:

$$\text{posterior}_c \propto \text{likelihood}_c * \text{prior}_c \qquad (14.2)$$

$$P(\lambda_c|\text{data}) \propto P(\text{data}|\lambda_c) * P(\lambda_c) \qquad (14.3)$$

3. Propose a new value from the posterior distribution using a random process. We will call the randomly generated value the **proposal** hypothesis, λ_p. The distribution from which this proposal is drawn is called the **proposal distribution**. The Metropolis algorithm requires that the proposal distribution be a symmetric distribution (such as a normal distribution) that is centered on the current hypothesis. We used a normal distribution as our symmetric distribution and specified that the proposal distribution had a standard deviation of 0.5. The standard deviation here is called the **tuning parameter**. You, the analyst, set this parameter.

4. Calculate the posterior density of observing the data under the proposal hypothesis, which is proportional to the likelihood of observing the data under the proposal hypothesis times the prior probability of the proposal hypothesis:

$$\text{posterior}_p \propto \text{likelihood}_p * \text{prior}_p \tag{14.4}$$

$$P(\lambda_p \mid \text{data}) \propto P(\text{data}|\lambda_p) * P(\lambda_p) \tag{14.5}$$

5. You now have two competing hypotheses for λ from the posterior distribution. The next step is to **throw one away**. The Metropolis algorithm provides a rule for calculating the probability that you will move (transition) **from** the current hypothesis **to** the proposal. It is the smaller of two values: the ratio of the two posterior densities or 1:

$$p_{\text{move}} = min\left(\frac{P(\lambda_p \mid \text{data})}{P(\lambda_c \mid \text{data})}, 1\right) \tag{14.6}$$

6. Draw a random number between 0 and 1 from a uniform distribution. If the random number is less than the probability of moving, accept the proposed value, λ_p. If not, stay with the current value, λ_c.
7. Repeat, repeat, repeat.

In Chapter 13, we walked through both approaches in detail and then compared the resulting posteriors. **For the sake of discussion, the plot in Figure 14.2 is based on the first 1000 MCMC trials only**.

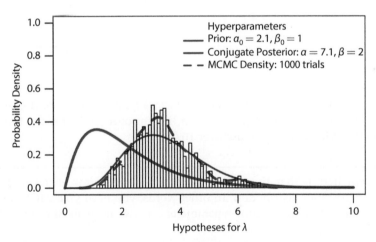

Figure 14.2 Prior and posterior probability density distributions with the tuning parameter $\sigma = 0.5$.

Here, the prior distribution is shown in blue, and the MCMC results with **1000 trials** are shown by the histogram. The two posterior distributions are also shown: a density plot fit to the MCMC results is shown in dashed purple, and the posterior distribution from the conjugate method is shown in solid purple. The goal here is to use MCMC to approximate the conjugate result. The conjugate result is the theoretically exact answer, but in many cases it's not feasible to obtain it. You can see that the results are not perfectly matched.

 ?? ## What can we do to hit our target posterior distribution?

Answer: There are several main considerations that you must keep in mind. First, let's focus on the number of MCMC trials. Let's see what our posterior distribution would look like if we ran only 100 trials (see Figure 14.3). Again, our aim is to match the solid purple posterior distribution.

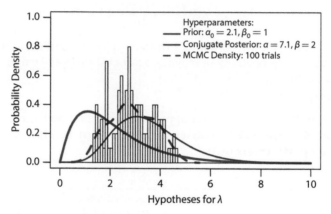

Figure 14.3 Prior and posterior probability density distributions with the tuning parameter $\sigma = 0.5$.

Clearly, increasing the number of trials produces a better estimate of the true posterior distribution. Even our 1000 trial example can be improved by increasing the number of trials. Frequently, MCMC analyses use over 10,000 trials.

A key concept here is that the MCMC approach should draw values that will fully represent the posterior distribution. In other words, you need your MCMC to explore the range of values that the unknown parameter can assume in an efficient and effective manner.

 ?? ## Is the number of trials the only thing to consider?

Answer: Ha ha!

Now let's turn to the concept of **tuning** the proposal distribution. In our example, we drew proposals from a normal distribution whose mean was the current value, λ_c, and whose standard deviation (our tuning parameter, σ) was 0.5. Let's look at our first 20 trials as a trace plot (see Figure 14.4).

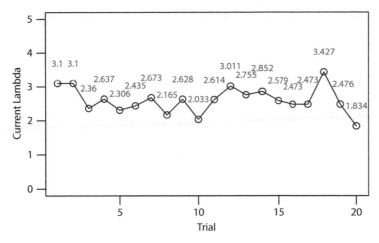

Figure 14.4

In a trace plot, points that are connected by a horizontal line indicate trials where the current value has been retained for the next MCMC trial. There is only 1 case here, so our **acceptance rate** is 95%. This is a problem.

Our friends at the SAS Institute explain it this way: "For a random walk Metropolis, high acceptance rate means that most new samples occur right around the current data point. Their frequent acceptance means that the Markov chain is moving rather slowly and not exploring the parameter space fully. On the other hand, a low acceptance rate means that the proposed samples are often rejected; hence the chain is not moving much. An efficient Metropolis sampler has an acceptance rate that is neither too high nor too low." In other words, if you accept the proposal all the time, your chain is moving towards the area of the posterior distribution that has significant weight, but it is moving like a turtle. If you never accept the proposal, your chain is stuck at some location, or the proposals are not very good proposals. Both of these situations will result in a posterior distribution that does not correspond with the true one you seek.

?? **What acceptance rate should we target?**

We can choose any standard deviation we wish, but you should aim for a standard deviation value such that the **acceptance rate** ranges roughly between 20%–50%. This is a heuristic, or rule of thumb, and the exact target depends on the problem you are trying to solve (see Gelman et al., 2004). For simple problems like the Shark Attack Problem, we could aim for a 25% acceptance rate. To reach this target, we need to **tune** our proposal distribution by twiddling the **tuning parameter**.

Since we have an acceptance rate that is too high, we should **increase** the standard deviation from its current setting of 0.5 to more fully explore the parameter space. But this might introduce yet another challenge.

?? **Uh…and what might that be?**

Answer: Drawing a proposal that is invalid. This may be shown with the following example (see Figure 14.5). Suppose our current value for λ_c is 1.00, near the left tail of the posterior distribution we are trying to describe (in purple). We center a normal distribution over the current value (shown in black) and draw one random variable from this proposal distribution.

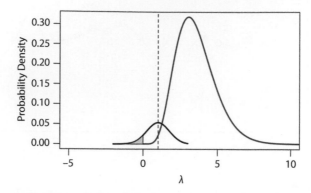

Figure 14.5 Proposal distribution centered on current λ.

If you draw a negative value from the blue shade area by chance, it will be invalid . . . the gamma distribution has a left boundary of 0. If this is the case, the computer program you are presumably using for the MCMC analysis will throw an error (i.e., the computer will give you an error message).

 ?? **What do we do in such cases?**

Answer: Well, there appear to be two approaches.

1. Stick with the current value and move to the next MCMC trial.
2. If your proposal cannot be negative, simply multiply any negative proposal by –1 and continue with the MCMC as usual.

In the graph in Figure 14.6, we reran our 1000 MCMC trials but this time set our tuning parameter, σ, to 1.

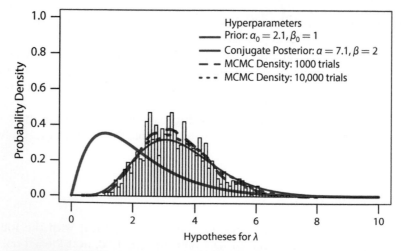

Figure 14.6 Prior and posterior probability distribution with the tuning parameter $\sigma = 1$.

If you compare these results with our first analysis of 1000 trials (Figure 14.2), you'll see that the tuning helped. Tuning is an iterative process; we would look at the acceptance rate under the condition that $\sigma = 1$ and then decide whether more fine-tuning is required. Figure 14.6 also shows the effect of increasing the number of MCMC trials as well—more is better!

 ?? **OK, once I have tuned my tuning parameter properly, am I finished with diagnostics?**

Answer:

One more biggie of concern deals with the starting values used in the MCMC analysis. Although the Markov chain should eventually converge to the desired distribution, the initial samples may follow a very different distribution, especially if the starting point is in a low density region of the posterior distribution. As a result, a **burn-in** period is typically necessary, where an initial number of samples (e.g., the first 1000 or so) are thrown away.

"**Burn in**" means you disregard a certain number of trials before building your posterior distribution. For instance, we may disregard our first ten trials and instead build the posterior using the remaining 990 trials. Both the starting value and the tuning parameter can dramatically affect the shape of your posterior distribution.

 ?? **Can we see an example of these challenges?**

Answer: But of course. You've already seen how the number of trials is an important MCMC consideration. Now, let's try a few experiments where we change the MCMC starting value and the proposal distribution's standard deviation (see Figure 14.7). To highlight the challenges involved, we'll run only 100 trials:

- Experiment 1 (the blue experiment): Set our **first** λ value at 2.1, and set the tuning parameter, σ, at 0.15 for the proposal distribution.
- Experiment 2 (the black experiment): Set our **first** λ value at 2.2, and set the tuning parameter, σ, at 0.8 for the proposal distribution.

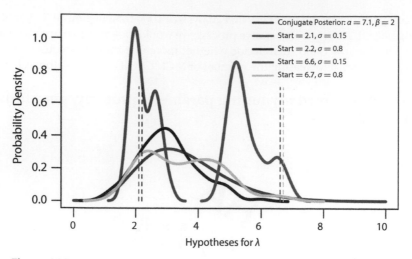

Figure 14.7

- Experiment 3 (the red experiment): Set our **first** λ value at 6.6, and set the tuning parameter, σ, at 0.15 for the proposal distribution.
- Experiment 4 (the gray experiment): Set our **first** λ value at 6.7, and set the tuning parameter, σ, at 0.8 for the proposal distribution.

The graphs are based on only 100 trials, but they demonstrate some of the issues that you, the analyst, must be aware of in using the Metropolis algorithm. The purple distribution is our conjugate posterior, that is, the posterior distribution we are trying to estimate with MCMC approaches. This is the gold standard. Comparing the four experiments, we see:

- The blue distribution (Experiment 1) is an approximation based on a starting λ_c of 2.1 and σ of 0.15. Notice that this posterior distribution is centered somewhat on our target posterior distribution (purple), but the distribution is much narrower than the target. The reason for this is that our tuning parameter, σ is too low … we started near the heart of the posterior distribution, our proposals never ventured very far from it. To address this issue, we should **increase** the standard deviation.
- The black distribution (Experiment 2) is an approximation based on a starting λ_c of 2.2 and σ of 0.8. Notice that this distribution is slightly to the right of the blue distribution and is a better match to our target (purple) distribution, even though the starting value was in the left tail of the distribution.
- The red distribution (Experiment 3) is an approximation based on a starting λ_c of 6.6 and σ of 0.15. Notice that it is shifted to the right of the other distributions. This is because the MCMC started in the far right portion of the target posterior distribution, and with a low standard deviation, it doesn't draw proposals towards the center of our target distribution.
- The gray distribution (Experiment 4) is an approximation based on a starting λ_c of 6.7 and σ of 0.8. Notice that it started near the tail of the posterior distribution, but with a higher tuning parameter, it approximates the conjugate posterior fairly well.

These issues can be further seen by examining the first 20 records of each experiment (see Figure 14.8).

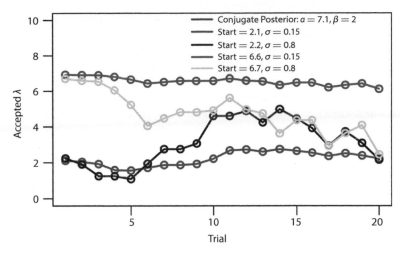

Figure 14.8

- Experiment 1 (blue) and Experiment 3 (red) had low values for the tuning parameter; thus, their traceplots are rather flat and do not venture far from the starting value.
- Experiment 2 (black) and Experiment 4 (gray) had high values for the tuning parameter; thus their traceplots "bounce."
- It's plain to see that these trials suggest that our tuning parameter needs attention. However, the acceptance rate is normally computed after the burn-in phase (which we ignored) and is based on a large number of trials (Figure 14.8 shows the first 20 trials.)

These experiments emphasize that tuning the standard deviation is important. In general, tune to get an acceptance rate of 20%–50%. And run as many trials as you are able but then **burn in** (discard) the first several samples to avoid creating a posterior distribution that is unduly influenced by your choice of the starting parameter. This, in a nutshell, is why you want to "Feel the Bern."

 ?? OK, I'm afraid to ask, but what else should concern me?

Answer: Due to the nature of how proposed values are selected, the samples from one trial to the next may be correlated. Can you see that this problem is magnified if the standard deviation is small? In MCMC analysis, **thinning** (also called **pruning**) has to do with omitting certain MCMC trials when building the posterior distribution, such as using every other trial, or every third trial, or every fifth trial. Thinning is the response to the correlation of nearby trials, which boils down to non-independence of samples being used to infer the shape of the posterior distribution. In order to map the posterior distribution, we must assume they are independent draws (realizations) from the posterior distribution of interest. There are diagnostics to ensure you have enough independent samples, and many of these topics are covered in more advanced books.

 ?? Is there anything else to worry about with MCMC analysis?

Answer: Another source of error involves how your computer stores numbers. Computers store numbers as bits in a base 2 system, and a number in your MCMC trial may require more bits than your software has allocated. For instance, the number 0.2 is stored in your computer as 00.0011001100110011..., the number 14 as 1110, and the number 133 as 10000101 (http://www.wolframalpha.com/examples/NumberBases.html). In calculating the posterior density, we multiply the prior density times the likelihood. If these are both small numbers, multiplying them results in a very small number. If the numbers are tiny enough, the number may not be stored properly by your computer. This is called numerical overflow. **In such cases, you should calculate the posterior densities using logs.**

 ??   Logs?

Answer: The idea is that instead of getting the posterior density by *multiplying* the likelihood by the prior density, you *add* the logs of these numbers.

A quick refresher on the rules of logs may help. You might recall that:

$$log_b(mn) = log_b(m) + log_b(n) \tag{14.7}$$

Here, if m is the likelihood, n is the prior density, and b is the base for the natural log (e, or Euler's number), then we have:

$$ln(mn) = ln(m) + ln(n) \tag{14.8}$$

This transformation allows us to **add** the natural log of the likelihood with the natural log of the prior, which is a much easier calculation to make (remember that multiplication is simply repeated addition). Here's a quick example. First, let's recall that:

$$P(\lambda_c \mid \text{data}) \propto P(\text{data} \mid \lambda_c) * P(\lambda_c) \tag{14.9}$$

Suppose the likelihood of observing the data under a current hypothesis for λ_c is 0.003, and the prior density for that hypothesis is 0.009. The posterior density is proportional to the product of these terms:

$$P(\lambda_c \mid \text{data}) \propto 0.003 * 0.009 = 2.7e - 05 \tag{14.10}$$

Suppose the proposed hypothesis for λ_p has a likelihood of 0.002 and the prior density is 0.004:

$$P(\lambda_p \mid \text{data}) \propto 0.002 * 0.004 = 8e - 06 \tag{14.11}$$

For this example, the probability of accepting the proposal is:

$$p_{\text{move}} = \frac{P(\lambda_p \mid \text{data})}{P(\lambda_c \mid \text{data})} = \frac{8e\text{-}06}{2.7e\text{-}05} = 0.2962963 \tag{14.12}$$

Keep this result in the back of your mind. **Now let's do the same calculations but work with natural logarithms instead.** Let's start with λ_c, noting the basic formula and log version:

$$P(\lambda_c \mid \text{data}) \propto P(\text{data} \mid \lambda_c) * P(\lambda_c) \tag{14.13}$$

$$ln(P(\lambda_c \mid \text{data})) \propto ln(P(\text{data} \mid \lambda_c)) + ln(P(\lambda_c)) \tag{14.14}$$

For the λ_c hypothesis, we have:

$$ln(P(\lambda_c \mid \text{data})) \propto ln(0.003) + ln(0.009) = -10.51967 \tag{14.15}$$

For the λ_p hypothesis, we have:

$$ln(P(\lambda_p \mid \text{data})) \propto ln(0.002) + ln(0.004) = -11.73607 \qquad (14.16)$$

Now we are in a position to calculate the probability of accepting the proposal. Remember the probability of moving to the proposal is the ratio of the two posterior densities. This now brings us to a second rule of logarithms:

$$log_b\left(\frac{u}{v}\right) = log_b(u) - log_b(v) \qquad (14.17)$$

In our case, this would be:

$$ln\left(\frac{u}{v}\right) = ln(P(\lambda_p \mid \text{data})) - ln(P(\lambda_c \mid \text{data})) \qquad (14.18)$$

$$ln(p_{\text{move}}) = -11.73607 - 10.51967 = -1.2164 \qquad (14.19)$$

But don't forget that this is now on the natural log scale. To get the probability of moving, we need to untransform the natural log:

$$p_{\text{move}} = exp(-1.2164) = 0.2962963 \qquad (14.20)$$

The end result for p_{move} is the same (compare with Equation 14.12), but your computer will appreciate the difference in calculation power.

?? Goodness! That's a lot of diagnosing!

Answer: Yes, it is. Good thing we aren't doctors! And there are other issues beyond the scope of this book. Gelman et al.'s (2004) work is the go-to book for Bayesian analysis. The concepts of burn in and thinning apply to every MCMC analysis. The concept of tuning depends on whether you need to have a tuning parameter in your MCMC analysis.

?? Can we summarize this chapter?

Answer: Why not? The goal of a Bayesian MCMC analysis is to estimate the posterior distribution while skipping the integration required in the denominator of Bayes' Theorem. The MCMC approach does this by breaking the problem into small, bite-sized pieces, allowing us to build the posterior distribution bit by bit (sand grain by sand grain). The main challenge, however, is that several things might go wrong in the process, and you must run several diagnostic tests to ensure that your MCMC analysis provides an adequate estimate of the posterior distribution. These diagnostics are required of all MCMC analyses. This is a big topic, and certainly an area in which you will want to dig deeper.

?? What's next?

Answer: This chapter ends our quick introduction to MCMC diagnostics. Our first analysis featured an MCMC approach with the Metropolis algorithm. But there are alternative algorithms that can be used. In Chapter 14, we revisit the White House Problem and introduce you to the Metropolis–Hastings algorithm.

CHAPTER 15

The White House Problem Revisited: MCMC with the Metropolis–Hastings Algorithm

In Chapters 13 and 14, we introduced the basic principles of Markov Chain Monte Carlo (MCMC) and the diagnosis of MCMC results. These chapters were chock-full of new terms, new ideas, and new calculations, and focused on the Metropolis algorithm.

In this chapter, we'll work through another MCMC example to help solidify your understanding. This time, we'll revisit the White House Problem that we studied in Chapter 10. The main twist for this chapter's material is that we will use the **Metropolis–Hastings algorithm**, which is a generalized version of the Metropolis algorithm.

By the end of this chapter, you will have a firm understanding of the following:

* Monte Carlo
* Markov chain
* Metropolis–Hastings algorithm
* Metropolis–Hastings proposal distribution
* Metropolis–Hastings correction factor.

This is a long chapter, so make sure to get up and stretch every now and then!

Now then, remember the White House Problem? We asked, "What is the probability that any famous person (like Shaq) can drop by the White House without an invitation?" In that problem, we were trying to estimate a parameter from the **binomial distribution** called p, which is the probability of success and ranges between 0 and 1. We used a Bayesian inference approach to estimate p, as follows:

$$P(p \mid \text{data}) = \frac{P(\text{data} \mid p) * P(p)}{\int_0^1 P(\text{data} \mid p) * P(p) dp} \tag{15.1}$$

For the White House Problem, we set up a prior distribution (a beta distribution), collected some data (in the form of successful or failed attempts to visit the White House without an invitation), and then updated to the posterior beta distribution with an analytical (conjugate) shortcut that allowed us to avoid the integration in the denominator of Bayes' Theorem.

However, not all problems can be solved this way . . . Bayesian conjugates are special cases that can be solved analytically. But, there is another approach, one that is so creative and versatile that it can be used to solve almost any kind of parameter estimation problem. It involves building the posterior distribution from scratch using a process called a Markov Chain Monte Carlo simulation, or MCMC for short.

Bayesian Statistics for Beginners: A Step-by-Step Approach. Therese M. Donovan and Ruth M. Mickey, Oxford University Press (2019). © Ruth M. Mickey 2019.
DOI: 10.1093/oso/9780198841296.001.0001

In this chapter, we will use an MCMC as a way to estimate the posterior distribution of p. We will quickly revisit the White House Problem, starting off by setting the prior distribution associated with alternative hypotheses for p and then finding the posterior distribution using the beta-binomial conjugate. **Then, we'll solve the same problem using MCMC with the Metropolis–Hastings algorithm so you can compare the two methods directly.**

Let's start by reviewing the White House Problem.

 ?? What were the analytic steps again?

Answer: We used the same steps that we did in previous chapters:

1. Identify your hypotheses—these would be the alternative hypotheses for p, ranging from 0 to 1.
2. Express your belief that each hypothesis is true in terms of prior densities.
3. Gather the data—Shaq makes his attempt(s), and will either fail or succeed.
4. Determine the **likelihood** of the observed data, assuming each hypothesis is true.
5. Use Bayes' Theorem to compute the posterior densities for each hypothesis of p.

Let's review these steps quickly:

 ?? Step 1. What are the hypotheses for p?

Answer: We know that p, the probability of success in the binomial distribution, can take on any value between 0 and 1. Since p is a **continuous variable**, there's nothing to stop us from considering the full range of hypotheses between 0 and 1, which are **infinite** ($p=0.01$, $p=0.011$, $p=0.0111$, etc.).

 ?? Step 2. What were the prior probabilities for each hypothesis?

Answer: We used the **beta** distribution to set relative prior probabilities for each and every hypothesis for p. Examples of a few alternative beta probability distributions are shown in Figure 15.1, and we need to **choose one** to represent our prior beliefs for each and every p.

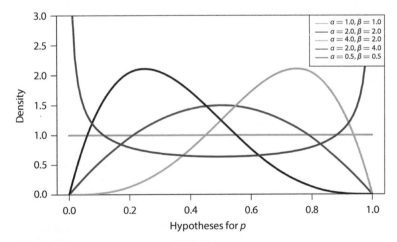

Figure 15.1

Let's look at these graphs carefully. The x-axis for the beta distribution ranges between 0 and 1. The y-axis gives the probability density (relative probability). For any given graph, the higher the probability density, the more belief we have in a particular hypothesis.

Because Shaq and his friend have totally different ideas of what p should be (with Shaq believing his chances were high and his friend believing his chances were low), we imagined that they settled on a beta distribution with $\alpha = 0.5$ and $\beta = 0.5$ as a prior distribution, which gives the U-shaped result shown in blue. This is a subjective prior. For reasons discussed in Chapter 10, it is also a vague prior. **So, the prior distribution for this example is a beta distribution with $\alpha_0 = 0.5$ and $\beta_0 = 0.5$.** Remember that the *naught* subscripts are used to indicate that these are hyperparameters, or parameters of a prior distribution.

 ## Step 3. Now what?

Answer: Collect data! **Let's assume that Shaq makes one attempt at visiting the White House without an invitation, and fails to get in**. In the binomial function terms, the number of trials $n = 1$, and the number of successes $y = 0$.

 ## Step 4. And then?

Answer: Step 4 is to determine the **likelihood** of the observed data, assuming each hypothesis for p is true.

 ## And step 5?

Answer: Step 5 is to generate the posterior distribution for p by using Bayes' Theorem. We mentioned that integration of the denominator is often tedious, and sometimes impossible.

But for problems where the prior distribution is a beta distribution, and the data collected come in the form of binomial data (number of successes out of a given number of trials), an analytical approach can be used to generate the posterior distribution.

The analytical shortcut makes updating to the posterior a snap. Here it is:

- posterior $\alpha = \alpha_0 + y$
- posterior $\beta = \beta_0 + n - y$

For the White House Problem, our prior distribution was a beta distribution with $\alpha_0 = 0.5$ and $\beta_0 = 0.5$. Shaq made one attempt, so $n = 1$. He failed to get into the White House, so $y = 0$. We can now use this shortcut to calculate the parameters of the posterior distribution:

- posterior $\alpha = \alpha_0 + y = 0.5 + 0 = 0.5$
- posterior $\beta = \beta_0 + n - y = 0.5 + 1 - 0 = 1.5$

Now we can look at the prior and posterior distributions for p (see Figure 15.2).

In short, we started off with a prior distribution shown in blue. Then we collected binomial data: one failure out of one trial. We then used an analytical approach to generate the posterior distribution for all hypotheses of p (shown in purple). As a result of Shaq's failed attempt, we now have **new** beliefs in each and every hypothesis of p. Notice how the posterior really shifted toward the lower ends of the p spectrum after just one new observation.

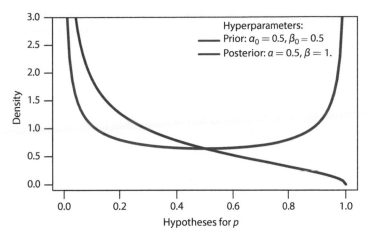

Figure 15.2

 ?? **How would you solve the posterior distribution with an MCMC approach?**

Answer: Here we go!

The main idea behind MCMC for Bayesian inference is that you "build" the posterior distribution by drawing samples from it. In other words, we are going to try to **build** the purple distribution in Figure 15.2 from scratch. Bayes' Theorem is still at the heart of the analysis.

Earlier, we had written Bayes' Theorem as:

$$P(p|\text{data}) = \frac{P(\text{data}\,|\,p) * P(p)}{\displaystyle\int_{0}^{1} P(\text{data}\,|\,p) * P(p)dp} \tag{15.2}$$

Now, knowing that the posterior probability is a proportion, and that the denominator is fixed across all hypotheses, we can write Bayes' Theorem as:

$$\text{posterior} \propto \text{likelihood} * \text{prior} \tag{15.3}$$

In words, the posterior density of a given hypothesis is **proportional** to the likelihood of observing the data under the hypothesis times the prior density of the given hypothesis. The symbol $\propto$ means "proportional to." Our color-coding is intended to help remind you that the posterior (purple) is a blend of the likelihood (red) and prior (blue).

Now, the idea of building a distribution from scratch may sound strange because you don't really know the distribution from which you are sampling. But in Chapter 14 we demonstrated it is possible using the **Metropolis algorithm**. We used the following set of operations in that chapter:

1. Start by proposing just one value from the posterior distribution. We will call this the current hypothesis, p_c.
2. Calculate the posterior probability density of observing the data under the **current** hypothesis, which is the likelihood of observing the data under the current hypothesis times the prior density of the current hypothesis:

$$P(p_c\,|\,\text{data}) \propto P(\text{data}\,|\,p_c) * P(p_c) \tag{15.4}$$

3. Propose a new value from the posterior distribution using a random process. We will call the randomly generated value the **proposal** hypothesis, and the distribution from

which it is drawn the **proposal distribution**. The Metropolis algorithm requires that the proposal distribution be a symmetric distribution (such as a normal distribution) that is centered on the current hypothesis.

4. Calculate the posterior probability density of observing the data under the proposal hypothesis, which is the likelihood of observing the data under the proposal hypothesis times the prior density of the proposal hypothesis:

$$P(p_p \mid \text{data}) \propto P(\text{data} \mid p_p) * P(p_p) \tag{15.5}$$

5. You now have two competing hypotheses for p from the posterior. The next step is to **throw one away**. The Metropolis algorithm provides a rule for calculating the probability that you will move (transition) **from** the current hypothesis p_c **to** the proposal p_p. It is the smaller of two values: the ratio of the two posterior densities, or 1:

$$p_{\text{move}} = min\left(\frac{P(p_p \mid \text{data})}{P(p_c \mid \text{data})}, 1\right) \tag{15.6}$$

Thus, if the posterior density of the proposed p is greater than posterior density of the current p, you will always move to the proposed p because the probability of moving will be 1. If not, whether you move or not depends on a random draw. In case you were wondering, the probability of moving or accepting the proposal is also called a **transition kernel**.

---

6. Draw a random number between 0 and 1 from a uniform distribution. If the random number is less than the probability of moving, accept the proposed value of p. If not, stay with the current value of p.

7. Repeat and repeat hundreds or thousands of times!

We mentioned previously that the Metropolis algorithm is a special case of a more general algorithm called the **Metropolis–Hastings** algorithm. We introduced **Nicholas Metropolis** in Chapter 14. Here, say hello to W. Keith Hastings (1930–2016), a Canadian (see Figure 15.3).

Figure 15.3 Keith Hastings (Original photograph courtesy of Gary Bishop).

?? **OK then, what is the Metropolis–Hastings algorithm?**

It is virtually identical to the Metropolis algorithm in operations, with one important difference: the probability of accepting the proposal is calculated differently.

In generalized terms, if $P(\theta_c|\text{data})$ is the posterior density associated with a current parameter (θ_c), and $P(\theta_p|\text{data})$ is the posterior density associated a proposed parameter (θ_p), then the Metropolis probability of accepting the proposal is:

$$p_{\text{move}} = min\left(\frac{P(\theta_p \mid \text{data})}{P(\theta_c \mid \text{data})}, 1\right) \tag{15.7}$$

The Metropolis–Hastings algorithm is virtually identical, but with an added **correction factor** that is used to compute the probability of accepting the proposal:

$$p_{\text{move}} = min\left(\frac{P(\theta_p \mid \text{data}) * g(\theta_c \mid \theta_p)}{P(\theta_c \mid \text{data}) * g(\theta_p \mid \theta_c)}, 1\right) \tag{15.8}$$

The correction factor is:

$$\frac{g(\theta_c \mid \theta_p)}{g(\theta_p \mid \theta_c)} \tag{15.9}$$

 ?? **Why do we need a correction factor?**

Answer: The correction factor is needed if the **proposal distribution** is **not** symmetric. If it is symmetric, the correction factor equals 1, so we are back to the Metropolis algorithm. This is why the Metropolis algorithm is said to be a specific case of the Metropolis–Hastings algorithm.

 ?? **How does the correction factor work?**

Answer: The correction factor focuses on the symmetry, or lack thereof, of the **proposal distribution**, that is, the distribution from which we draw a new, proposed hypothesis for a parameter of interest. In the Metropolis algorithm, we draw a random variable from a normal distribution whose mean is centered on the current parameter—the analyst sets the standard deviation for this distribution. For example, if θ_c is 0.5, and we use a standard deviation of 0.05, then our proposal would be drawn from a normal distribution with $\mu = 0.5$ and $\sigma = 0.05$. You might recall that σ in this example is referred to as the **tuning parameter**.

The Metropolis–Hastings correction factor is the probability density of drawing the **current** value (θ_c) from a distribution that is centered on the **proposed** value (θ_p) divided by the probability density of drawing the **proposed** value (θ_p) from a distribution centered on the **current** value (θ_c). The denominator is what actually happened, while the numerator is the reverse. It is written this way, with color-coding that will hopefully help us keep track of the numerator and denominator. The **n**umerator will be coded in **n**avy, and the **d**enominator will be coded in **d**irt brown (or blue sky above and dirt below):

$$\frac{g(\theta_c \mid \theta_p)}{g(\theta_p \mid \theta_c)} \tag{15.10}$$

First, let's see how this works when our proposal distribution is the **symmetric**, normal distribution. Suppose θ_c is 0.5, and θ_p is 0.6. Let's also assume that our proposal distribution has a tuning parameter, σ, of 0.05:

- For the **denominator** of the correction factor, which is written $g(\theta_p \mid \theta_c)$ and is what actually happened, we compute the probability density of 0.6 from a normal distribution with $\mu = 0.5$ and with $\sigma = 0.05$. The normal probability density function can be used to calculate this answer, which is 1.08. This is shown by the dirt-brown dashed line in Figure 15.4.

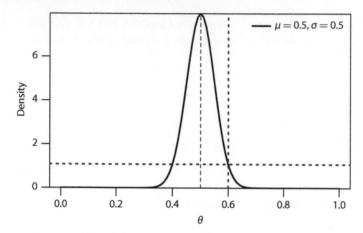

Figure 15.4 Proposal distribution centered on θ_c.

For the **numerator** of the correction factor, which is written $g(\theta_c \mid \theta_p)$, we compute the probability density of 0.5 from a normal distribution whose mean is 0.6 and whose standard deviation is 0.05. The normal probability density function can be used to calculate this answer, which is 1.08. This is shown by the navy blue dashed line in Figure 15.5.

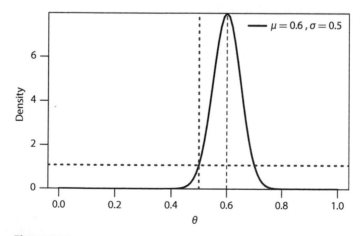

Figure 15.5 Proposal distribution centered on θ_p.

With a symmetric distribution, you can see that no matter what proposal we make, the probability density of drawing the proposal centered on the current will be identical to the probability density of drawing the current centered on the proposal.

The correction factor for this example is:

$$\frac{g(\theta_c \mid \theta_p)}{g(\theta_p \mid \theta_c)} = \frac{1.08}{1.08} = 1.0 \tag{15.11}$$

Thus, the probability of moving from the current to the proposal for this problem is:

$$p_{\text{move}} = min\left(\frac{P(\theta_p)}{P(\theta_c)} * 1.0, 1\right) \tag{15.12}$$

which is just the Metropolis algorithm. Don't forget that the term $P(\theta_p)$ is the posterior density of the proposal, while $P(\theta_c)$ is the posterior density of the current, and Bayes' Theorem is at the heart of the calculations.

 ?? What if we want to use a proposal distribution that is not symmetrical?

Hastings' giant insight was that you can use **any** proposal distribution you want as long as you correct for the fact that the proposal distribution is non-symmetric. The same principles apply.

Let's try a new example, returning to our parameter of interest for this chapter, p, the probability that a celebrity can visit the White House without an invitation. Suppose that our current value of $p_c = 0.5$ and that, instead of drawing a proposal from a normal distribution, we use a beta distribution instead. **Pay attention now: we are now using the beta distribution as our proposal distribution... this is completely separate from the use of a beta distribution to represent our prior hypotheses that a famous celebrity can visit the White House without an invitation.**

Remember that the beta distribution has two parameters named α and β. Just as we had to specify the tuning parameter, σ, to draw from a normal distribution, here we have to specify either α or β to draw from a beta distribution. **Let's assume that we specify that $\beta = 3$ for our proposal distribution. This is our tuning parameter for the proposal distribution.**

We now need to calculate α so that we end up with a beta distribution that is centered on $p_c = 0.5$. One way to do this is to calculate the mean of the beta distribution, which is:

$$\mu = \frac{\alpha}{\alpha + \beta} \tag{15.13}$$

If we set the mean to 0.5, with a little rearranging we can solve for α with the following:

$$\alpha = \frac{\beta * \mu}{-\mu + 1} = \frac{3 * 0.5}{-0.5 + 1} = 3. \tag{15.14}$$

Now we have specified a proposal beta distribution with parameters $\alpha = 3.000$ and $\beta = 3$; this distribution is centered over p_c. We can draw a random proposal from it. **Suppose we draw a proposed p_p of 0.2**. So,

- $p_c = 0.5$
- $p_p = 0.2$

Now we can calculate the correction factor, starting with the denominator (you can do this in any order, but the denominator gives what actually happened, so it may be more logical to start there):

- For the **denominator** of the Metropolis–Hastings correction factor, we compute the probability density of 0.2 from a beta distribution whose α is 3 and whose β is 3 (see Figure 15.6). The beta probability density function can be used to calculate this answer, which is 0.768. Here, our proposal distribution is colored black and is centered on the current value (it happens to look symmetrical but this is often not the case). The dirt colored dashed lines represents our random draw and its corresponding density.

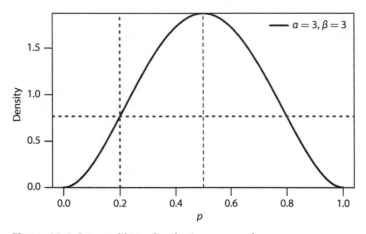

Figure 15.6 Proposal beta distribution centered on p_c.

- For the **numerator** of the correction factor, we now consider what *might* have happened instead of what actually happened. We center a beta distribution so that the mean of the beta distribution is the **proposed** value ($p_p = 0.2$) and ask, "What is the probability density of drawing the current value ($p_c = 0.5$)?" Again, we'll need to solve for α to center this new beta distribution, knowing that our tuning parameter $\beta = 3$:

$$\mu = \frac{\alpha}{\alpha + \beta} \tag{15.15}$$

$$\alpha = \frac{\beta * \mu}{-\mu + 1} = \frac{3 * 0.2}{-0.2 + 1} = 0.75 \tag{15.16}$$

- Now we need to compute the probability density of drawing p_c (0.5), given a beta distribution that is centered on the p_p (0.2), which is a beta distribution with $\alpha = 0.75$ and $\beta = 3$. The beta probability density function can be used to calculate this answer, which is 0.537. This is shown by the navy dashed lines in Figure 15.7.

Intuitively, you can see that the probability density of drawing the proposal from a distribution that is centered on the current value is nowhere near the probability density of drawing the current value from a distribution that is centered on the proposal.

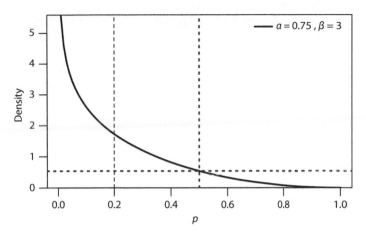

Figure 15.7 Proposal beta distribution centered on reverse.

The Metropolis–Hastings correction factor for this single example would be:

$$\frac{g(p_c|p_p)}{g(p_p|p_c)} = \frac{0.537}{0.768} = 0.7 \tag{15.17}$$

Don't lose sight that all of this was to compute a correction factor because our proposal distribution is not symmetrical!

Additionally, don't lose sight that this correction factor is used to calculate the probability that we accept the proposal:

$$p_{move} = min\left(\frac{P(p_p)g(p_c|p_p)}{P(p_c)g(p_p|p_c)}, 1\right) \tag{15.18}$$

?? Do you have to center the distribution on the mean?

Answer: You can center it on the mean, mode, or median if you'd like. You just need to be able to recalculate the parameters of the beta distribution for each step to center it. Alternatively, you can avoid the recentering and draw from a fixed distribution where you specify the parameters of the proposal distribution, and these remain constant throughout the MCMC analysis. This is known as the Independence Metropolis–Hastings Sampler.

?? Why can't you just use the normal distribution as the proposal distribution?

Answer: Sometimes, the parameter that we are trying to estimate is bounded. For instance, in this problem, we are trying to estimate p, the probability that Shaq can visit the White House without an invitation, which is bounded between 0 and 1. If we use a normal distribution for our proposal distribution, we can end up with proposal draws that are outside of this boundary. Incidentally, we could have run into this issue in Chapter 13, where we were trying to estimate λ via a Metropolis MCMC but somehow avoided it. Additionally, sometimes we really understand the distribution from which proposals should be drawn, and it might not be a symmetric distribution. With the Metropolis–Hastings algorithm, we can use any proposal distribution we want. The only limiting factor is that the correction factor must be computed for each MCMC trial, and this will increase your computing time.

 ?? ## OK, I think I've got it. Can we walk through the White House MCMC analysis using the Metropolis–Hastings algorithm?

Answer: OK, back to Shaq! Remember that our **prior distribution** for this problem is a beta distribution with hyperparameters $\alpha_0 = 0.5$ and $\beta_0 = 0.5$. Also recall that the **data** for this problem consists of one failure out of one attempt (Shaq made an attempt but failed, so $n = 1$ trial and $y = 0$ successes). With a beta prior, and binomial data (likelihood), we can use the conjugate shortcut to obtain the beta posterior (see Figure 15.8).

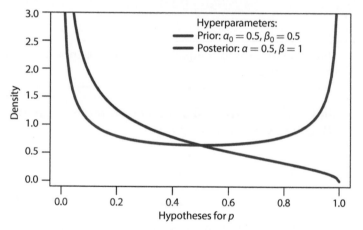

Hyperparameters:
— Prior: $\alpha_0 = 0.5, \beta_0 = 0.5$
— Posterior: $\alpha = 0.5, \beta = 1$

Figure 15.8

The conjugate posterior distribution is our gold standard. Now let's estimate the purple posterior distribution in Figure 15.8 with an MCMC analysis that uses the Metropolis–Hastings algorithm. But before we get started, let's remind ourselves that Bayes' Theorem is a central part of the MCMC analysis:

$$\text{posterior} \propto \text{likelihood} * \text{prior} \tag{15.19}$$

Let's recall the MCMC operations:

1. Start by providing a current value of p. We will call this p_c.
2. Calculate the posterior density of observing the data under the hypothesis p_c using Bayes' Theorem:

$$P(p_c \,|\, \text{data}) \propto P(\text{data} \,|\, p_c) * P(p_c) \tag{15.20}$$

3. Draw a random value from a proposal distribution that is centered on the current p. We will call this the proposed p_p.

4. Calculate the posterior density of observing the data under p_p using Bayes' Theorem:

$$P(p_p \mid \text{data}) \propto P(\text{data} \mid p_p) * P(p_p) \qquad (15.21)$$

5. You now have two competing hypotheses for p from the posterior. The next step is to **throw one away**. We will use the Metropolis–Hastings rule to calculate the probability that you will move from the current hypothesis p_c to the proposal p_p. It is the smaller of two values, 1 or the ratio of the two posterior probabilities multiplied by the correction factor:

$$p_{\text{move}} = min\left(\frac{P(p_p \mid \text{data})g(p_c \mid p_p)}{P(p_c \mid \text{data})g(p_p \mid p_c)}, 1\right) \qquad (15.22)$$

6. Draw a random number between 0 and 1 from a uniform distribution. If the random number is less than the probability of moving, accept p_p. If not, stay with p_c.
7. Repeat hundreds or thousands of times!

?? Should we start with a table?

Answer: Good idea. Let's set up a table that will store the results of our MCMC trials (see Table 15.1). Remember that we are using the Metropolis–Hastings algorithm, so we have a few extra steps in our MCMC algorithm due to the correction factor (CF).

Table 15.1

	Data		Prior		Current				Proposed			Decision				
Trial	n	y	a_0	β_0	P_c	$P(p_p \mid \text{data})$	a_c	p_p	$P(p_p \mid \text{data})$	a_p	Ratio	CF	k	Random	Accept
1	1	0	0.5	0.5	—	—	—	—	—	—	—	—	—	—	—	—
2	1	0	0.5	0.5	—	—	—	—	—	—	—	—	—	—	—	—
3	1	0	0.5	0.5	—	—	—	—	—	—	—	—	—	—	—	—
4	1	0	0.5	0.5	—	—	—	—	—	—	—	—	—	—	—	—
5	1	0	0.5	0.5	—	—	—	—	—	—	—	—	—	—	—	—
6	1	0	0.5	0.5	—	—	—	—	—	—	—	—	—	—	—	—
7	1	0	0.5	0.5	—	—	—	—	—	—	—	—	—	—	—	—
8	1	0	0.5	0.5	—	—	—	—	—	—	—	—	—	—	—	—
9	1	0	0.5	0.5	—	—	—	—	—	—	—	—	—	—	—	—
10	1	0	0.5	0.5	—	—	—	—	—	—	—	—	—	—	—	—

In this table, our MCMC trials are shown in column 1. The observed data, $n = 1$ attempt by Shaq and $y = 0$ successes, are given in columns 2 and 3, respectively. Notice that these are fixed for all 10 trials. Similarly, a_0 and β_0 from our **prior distribution** (columns 4 and 5, respectively) are fixed for all 10 trials.

?? What is operation 1?

Answer: We need to start by proposing just one value from the posterior distribution. Any value will do as long as there is a non-0 density associated with that value. Let's go with 0.5, as in our example. This represents our **current** value, p_c, which is the probability that a famous person like Shaq can visit the White House without an invitation. We can say that $p_c = 0.5$ is a random value that is drawn from the posterior distribution.

 ?? **What is operation 2?**

Answer: We then calculate the **posterior density** of observing the data under the hypothesis that $p_c = 0.5$. The posterior density is proportional to the product of **two terms**, the likelihood times the prior density:

$$\text{posterior} \propto \text{likelihood} * \text{prior} \tag{15.23}$$

$$P(p_c \mid \text{data}) \propto P(\text{data} \mid p_c) * P(p_c) \tag{15.24}$$

- For the first term, we ask, "What is the likelihood of observing 0 successes out of 1 trial, given $p_c = 0.5$?" Let's hope the binomial probability mass function springs to mind! Remember that this function has two fixed **parameters**: n and p. The parameter n = the total number of trials (in our case, $n = 1$ attempt to get into the White House, or 1 trial). The parameter p = the probability of success (in our case, $p = p_c = 0.5$). The likelihood of observing 1 failure in 1 trial is:

$$\mathcal{L}(y; n, p) = \binom{n}{y} p^y (1-p)^{(n-y)} \tag{15.25}$$

$$\mathcal{L}(0; 1, 0.5) = \binom{1}{0} 0.5^0 (1-0.5)^{(1-0)} = 0.5 \tag{15.26}$$

- For the second term, we ask, "What is the probability density associated with $p_c = 0.5$ from the **prior distribution**?" Again, we hope the beta probability density function springs to mind! We can use the beta pdf to ask, "What is the probability density of drawing a random variable of 0.5, given the parameters of the prior distribution?" Remember that our prior distribution set α_0 and β_0 to 0.5. The beta distribution is written below, where B is the beta function that ensures the curve integrates to 1.0. Here, we plug in p_c for x in the beta pdf:

$$f(x; \alpha, \beta) = \frac{1}{B(\alpha, \beta)} x^{\alpha-1} (1-x)^{\beta-1} \tag{15.27}$$

$$f(0.5; \alpha = 0.5, \beta = 0.5) = \frac{1}{B(0.5, 0.5)} 0.5^{0.5-1} (1-0.5)^{0.5-1} = 0.637 \tag{15.28}$$

- The posterior density associated with our first guess at p_c is the product of these two results:

$$P(p_c \mid \text{data}) \propto P(\text{data} \mid p_c) * P(p_c) \tag{15.29}$$

$$P(p_c \mid \text{data}) \propto 0.5 * 0.637 = 0.318 \tag{15.30}$$

Let's fill these values into our MCMC table for trial 1 (see Table 15.2).

Table 15.2

	Data		Prior		Current				Proposed				Decision				
Trial	n	y	α_0	β_0	p_c	$P(p_c \mid \text{data})$	α_c	p_p	$P(p_p \mid \text{data})$	α_p	Ratio	CF	k	Random	Accept
1	1	0	0.5	0.5	0.500	0.318	—	—	—		—	—	—	—	—	—

?? What is operation 3?

Answer: We then **propose** a second value for p, **drawn at random** from a proposal distribution. Let's hope this won't be too confusing, but we'll use a second beta distribution for our **proposal distribution**, which is totally distinct from our prior distribution, which also happens to be a beta distribution. **The proposal distribution will set the tuning parameter $\beta = 3$, as in our previous narrative**.

- We need to find α for the proposal distribution so that the proposal beta distribution has a mean of 0.5, our current p_c:

$$\mu = \frac{\alpha}{\alpha + \beta} \tag{15.31}$$

$$\alpha = \frac{\beta * \mu}{-\mu + 1} = \frac{3 * 0.5}{-0.5 + 1} = 3.000 \tag{15.32}$$

- Now that we have a beta distribution that is centered over our current value, we can draw a random value from it. That is, we will draw a random value of p from a proposal beta distribution where $\alpha = 3.000$ and $\beta = 3$. **Suppose we draw a proposed p of 0.2, as in our narrative.**

Let's now update our MCMC table (see Table 15.3).

Table 15.3

	Data		Prior		Current			Proposed			Decision				
Trial	n	y	α_0	β_0	p_c	$P(p_c \mid \text{data})$	α_c	p_p	$P(p_p \mid \text{data})$	α_p	Ratio	CF	k	Random	Accept
1	1	0	0.5	0.5	0.500	0.318	3.000	0.200	—		—	—	—	—	—	—

?? What is operation 4?

Answer: We then calculate the **posterior density** under the hypotheses that $p = 0.2$. This result is a product of two terms, the likelihood times the prior density:

$$P(p_p \mid \text{data}) \propto P(\text{data} \mid p_p) * P(p_p) \tag{15.33}$$

- For the first term, we ask, "What is the likelihood of observing 0 successes out of 1 trial given $p = 0.2$?" Again, the binomial probability mass function is used:

$$\mathcal{L}(y; n, p) = \binom{n}{y} p^y (1-p)^{(n-y)} \tag{15.34}$$

$$\mathcal{L}(0; 1, 0.2) = \binom{1}{0} 0.2^0 (1 - 0.2)^{(1-0)} = 0.8 \tag{15.35}$$

- For the second term, we ask, "What is the probability density associated with $p_p = 0.2$ from the **prior distribution**?" Again, the beta probability density function is used:

$$f(x; \alpha, \beta) = \frac{1}{B(\alpha, \beta)} x^{\alpha-1} (1 - x)^{\beta-1} \tag{15.36}$$

$$f(0.2; \alpha = 0.5, \beta = 0.5) = \frac{1}{B(0.5, 0.5)} 0.2^{0.5-1} (1 - 0.2)^{0.5-1} = 0.796 \tag{15.37}$$

- The posterior density associated with our second hypothesis for p (p_p) is the product of these two results:

$$P(p_p \mid \text{data}) \propto P(\text{data} \mid p_p) * P(p_p) \tag{15.38}$$

$$P(p_p \mid \text{data}) \propto 0.8 * 0.796 = 0.637 \tag{15.39}$$

Let's now update our MCMC table (see Table 15.4).

Table 15.4

	Data		Prior			Current		Proposed			Decision				
Trial	n	y	α_0	β_0	p_c	$P(p_c \mid \text{data})$	α_c	p_p	$P(p_p \mid \text{data})$	α_p	Ratio	CF	k	Random	Accept
1	1	0	0.5	0.5	0.500	0.318	3.000	0.200	0.637	—	—	—	—	—	—

 ?? What is operation 5?

Answer: Now you have two proposed p values from the posterior distribution ($p_c = 0.5$ and $p_p = 0.2$). You also have the posterior density associated with each (0.318 and 0.637). Note that the posteriors are not normalized (they are just the likelihood times the prior density.).

In this step we will **throw one away**. Which one will you keep for your next MCMC trial? Will you stay with your original value ($p_c = 0.5$), or will you move to the proposed value ($p_p = 0.2$)? As you've guessed, we'll use the Metropolis–Hastings algorithm to compute the probability of moving to the proposed value (i.e., the probability of accepting the proposal).

- Calculate the probability of accepting a move to the proposed p_p, which in our example is 0.2, as the smaller of two values: 1 or the ratio of the two posterior probabilities multiplied by the correction factor. Here, to avoid the use of yet another letter p, we will let k be the probability of moving (accepting) to the proposal:

$$k = min\left(\frac{P(p_p \mid \text{data}) * g(p_c \mid p_p)}{P(p_c \mid \text{data}) * g(p_p \mid p_c)}, 1 \right) \tag{15.40}$$

$$k = min\left(\frac{0.637 * g(p_c \mid p_p)}{0.318 * g(p_p \mid p_c)}, 1 \right) \tag{15.41}$$

- The ratio of the two posterior densities is $0.637/0.318 = 2.003$. The only thing left to do is to calculate the correction factor. Remember, this is the probability density of $p_c = 0.5$ if the proposal distribution was centered on $p_p = 0.2$ divided by the probability density of $p_p = 0.2$ if the proposal distribution is centered on $p_c = 0.5$. To calculate the correction factor, we need the α and β parameters for each case, and in both cases our tuning parameter $\beta = 3$.
- For the distribution centered on p_c:

$$\mu = \frac{\alpha}{\alpha + \beta} \tag{15.42}$$

$$\alpha = \frac{\beta * \mu}{-\mu + 1} = \frac{3 * 0.5}{-0.5 + 1} = 3 \tag{15.43}$$

- For the distribution centered on p_p:

$$\mu = \frac{\alpha}{\alpha + \beta} \tag{15.44}$$

$$\alpha = \frac{\beta * \mu}{-\mu + 1} = \frac{3 * 0.2}{-0.2 + 1} = 0.75 \tag{15.45}$$

- Now we can use the beta probability density function to get the density in each case (as we did earlier):

$$\text{numerator} = p(0.5 \mid \alpha = 0.75, \beta = 3) = 0.537 \tag{15.46}$$

$$\text{denominator} = p(0.2 \mid \alpha = 3, \beta = 3) = 0.768 \tag{15.47}$$

- Now that we have the correction factor numerator and denominator, we can compute the correction factor as $0.537/0.768 = 0.699$. Our probability of accepting the proposal is calculated as:

$$k = min\left(\frac{0.637 * 0.537}{0.318 * 0.768}, 1\right) \tag{15.48}$$

$$k = min\left(\frac{0.342069}{0.24576}, 1\right) \tag{15.49}$$

$$k = min(1.39, 1) \tag{15.50}$$

$$k = 1 \tag{15.51}$$

Let's now update our MCMC table before we lose track of things (see Table 15.5).

Table 15.5

	Data		Prior			Current			Proposed				Decision		
Trial	n	y	α_0	β_0	p_c	$P(p_c \mid \text{data})$	α_c	p_p	$P(p_p \mid \text{data})$	α_p	Ratio	CF	k	Random	Accept
1	1	0	0.5	0.5	0.500	0.318	3.000	0.200	0.637	0.750	2.003	0.699	1.000	—	—

?? What is operation 6?

Answer: Draw a random number between 0 and 1. If the random number is less than the probability of moving, accept p_c. If not, stay with p_c. Let's say our randomly drawn value is 0.372 (see Table 15.6). This value is less than the probability of moving, so we accept our proposed value of p_p, which is 0.2. This will then serve as p_c for trial 2. Note that since $k = 1$, you already know that you will accept p_p, so the draw is optional!

Table 15.6

	Data		Prior			Current			Proposed					Decision		
Trial	n	y	α_0	β_0	p_c	$P(p_c \mid \text{data})$	α_c	p_p	$P(p_p \mid \text{data})$	α_p	Ratio	CF	k	Random	Accept
1	1	0	0.5	0.5	0.500	0.318	3.000	0.200	0.637	0.750	2.003	0.699	1.000	0.372	0.200

?? What is operation 7?

Answer: Repeat hundreds or thousands of times!

?? What is operation 8?

Answer: Summarize the p_c's in the form of a distribution and its corresponding summary statistics.

?? Can we see an example of operations 6 and 7, please?

Answer: You bet.

Let's carry out ten trials, keeping in mind that we are drawing proposals from a beta distribution that has a tuning parameter $\beta = 3.0$ (see Table 15.7).

Table 15.7

	Data		Prior			Current			Proposed					Decision		
Trial	n	y	α_0	β_0	p_c	$P(p_p \mid \text{data})$	α_c	p_p	$P(p_p \mid \text{data})$	α_p	Ratio	CF	k	Random	Accept
1	1	0	0.5	0.5	0.500	0.318	3.000	0.200	0.637	0.750	2.003	0.699	1.000	0.372	0.200
2	1	0	0.5	0.5	0.200	0.637	0.750	0.006	4.097	0.018	6.432	0.009	0.058	0.572	0.200
3	1	0	0.5	0.5	0.200	0.637	0.750	0.070	1.161	0.226	1.823	0.226	0.412	0.216	0.070
4	1	0	0.5	0.5	0.070	1.161	0.226	0.089	1.018	0.293	0.877	1.479	1.000	0.364	0.089
5	1	0	0.5	0.5	0.089	1.018	0.293	0.023	2.074	0.071	2.037	0.104	0.212	0.098	0.023
6	1	0	0.5	0.5	0.023	2.074	0.071	0.001	10.061	0.003	4.851	0.003	0.015	0.175	0.023
7	1	0	0.5	0.5	0.023	2.074	0.071	0.124	0.846	0.425	0.408	14.602	1.000	0.507	0.124
8	1	0	0.5	0.5	0.124	0.846	0.425	0.001	10.061	0.003	2.003	0.699	1.000	0.372	0.124
9	1	0	0.5	0.5	0.124	0.846	0.425	0.032	1.751	0.099	6.432	0.009	0.058	0.572	0.032
10	1	0	0.5	0.5	0.032	1.751	0.099	0.044	1.484	0.138	1.823	0.226	0.412	0.216	0.044

In this table, our MCMC trials are shown in column 1. The observed data, $n = 1$ attempt by Shaq and $y = 0$ successes, are given in columns 2 and 3. Notice that these are fixed for all 10 trials. Similarly, α_0 and β_0 from our **prior distribution** (columns 4 and 5) are fixed for all 10 trials.

Now let's focus on row 1, or trial 1, which is the trial we tediously walked through earlier. It is easy to get lost in the weeds, so let's work our way through it carefully. Take a look at the column group labeled **Current**:

- Our current value of p (column P_c) is 0.5.
- The posterior probability density associated with $p_c = 0.5$ is 0.318, stored in column 7. This column is shaded purple to remind us that this is the posterior density from Bayes' Theorem.
- We now need to center the beta distribution on p_c. Remember that tuning parameter β for the proposal distribution was fixed at 3. We then calculate what α must be so that the beta distribution is centered on our 0.5. This result, 3.00, is stored in column 8 (α_c).

Let's move on to the column group labeled **Proposed**:

- We then drew a proposed value $p_p = 0.2$ (column P_p) from the beta distribution centered on p_c. This distribution has parameters $\alpha = 3.00$ and $\beta = 3$.
- The posterior probability density associated with $p_p = 0.2$ is 0.637, stored in column 10.
- To prepare for the correction factor calculations, we need the α value for a beta distribution centered p_p, which was 0.750. This is stored in column α_p.

Now take a look at the column group labeled **Decision**:

- The **Ratio** of the two posteriors is computed as posterior density of p_p divided by the posterior density of p_c, which is 2.003.
- The correction factor, **CF**, is the probability density of the current value of p from a beta distribution centered on the proposal divided by the probability density of the proposed value of p from a beta distribution centered on the current. The result is 0.699.
- The probability of moving to p_c (column k) is calculated as minimum of 1 and the ratio times the correction factor. In this example, the result is 1.
- We then draw a random number between 0 and 1. Our random number is 0.372 (column **Random**).
- We then keep the p value according to our random number. If the random number is less than the probability of moving, we accept p_c; otherwise, we accept p_c. Our random number is less than 1, so we accept $p_p = 0.2$ (column **Accept**). This now becomes p_c in trial 2, and you can trace it in the table. Whew!

> Through this process, we "pit" two posterior densities against each other and keep only the greater of the two, according to a rule. The rule we have used in this example is the Metropolis–Hastings rule. By repeating this process over and over again, we eventually end up converging on a reasonable estimate of the posterior distribution.

 ?? Can anything go wrong with this approach?

Answer: Well, yes, there are a few places where things can go awry in any MCMC analysis. The same issues we dealt with in Chapter 14 apply here as well.

?? Can we review the terms associated with this process?

Answer: Yes. A lot of jump-starting your Bayesian education involves learning a new vocabulary. Let's review the terms first introduced in Chapter 10:

- **Starting point**. The starting point is the first value you designate in trial 1. We used a starting p value of 0.5.
- **Monte Carlo**. Monte Carlo in this case refers to the Monte Carlo Casino, located in the Principality of Monaco. In statistics, the Monte Carlo methods "are statistical methods of understanding complex physical or mathematical systems by using randomly generated numbers as input into those systems to generate a range of solutions... The Monte Carlo method is used in a wide range of subjects, including mathematics, physics, biology, engineering, and finance, and in problems in which determining an analytic solution would be too time-consuming." In this context, Monte Carlo refers to the notion that we use random sampling to generate a proposed parameter, and we also use random sampling to determine whether or not we accept the proposed value or retain the current parameter value.
- **Proposal distribution**. The proposal distribution is the distribution that you use to draw the next, proposed value. In our example, we used a **beta distribution** that is centered on p_c.
- **Markov chain**. A Markov chain is named after the Russian mathematician, Andrey Markov. A Markov chain is a sequence of random variables in which each variable depends only on its immediate predecessor, so the sequence moves at random, each move being determined by its previous position (Upton and Cook, 2014). The Markovian process pertains to the value of the accepted p. The accepted p in trial t will become p_c in trial $t + 1$ (see Table 15.7).
- **MCMC trials**. The number of MCMC trials is the number of trials you decide to run. In our example, we ran 10 trials.

Now, let's graph our accepted values as a line graph so that we can see how our values change with each iteration (see Figure 15.9).

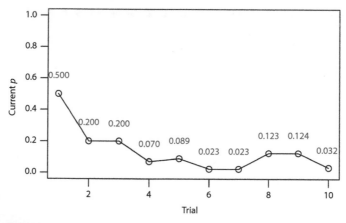

Figure 15.9

This graph shows the MCMC **chain** across 10 trials. This graph is called a **traceplot**. Here, trial number is along the x-axis, and we plot p_c (the probability that a famous person can make it into the White House without an invitation). See if you can match these results with those in our table of 10 trials. Horizontal stretches in this plot are trials where the proposal was **not** accepted. You can see that we have a 100% acceptance rate for these 10 trials, which may suggest that our tuning parameter, $\beta = 3.00$, may need adjusting. This topic was discussed in Chapter 13 as well.

?? Can we see the posterior distribution after multiple trials?

Answer: Let's rerun our experiment, but this time let's run **10,000** trials to create our posterior distribution. With 10,000 trials, we can "bin" our results in tight units (say, 0.01) and then tally the number of times p_c falls into each bin. A frequency histogram of these results could be shown. But, instead of plotting frequency, let's plot probability density, which is the relative frequency (frequency of a bin divided by total trials) divided by the bin width (see Figure 15.10).

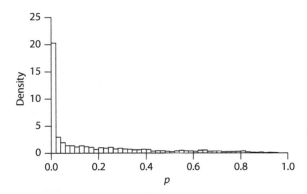

Figure 15.10 Posterior probability distribution.

> Never forget that this distribution is estimated via a set of rules to be followed (in this case, the Metropolis–Hastings algorithm) and that random draws are involved. If you repeat this process with new random numbers, you'll almost certainly end up with a different distribution. However, if your number of trials is large, the differences should be negligible.

?? How do we summarize the posterior distribution?

Answer: With simple summary statistics! We have 10,000 datapoints now, which consist of our chain of accepted hypotheses for p. Now it is a matter of just computing some simple statistics:

- Mean: the average value of p
- Variance: the variance of p
- Standard deviation: the standard deviation of p
- Minimum: the smallest value of p
- Maximum: the largest value of p

Here, we'll make our calculations based on the full set of 10,000 trials (see Table 15.8).

Table 15.8

Mean	Variance	Standard Deviation	Minimum	Maximum
0.18	0.057	0.238	0.001	0.986

We can also report the 95% Bayesian credible interval for p by providing the 0.025 and 0.975 quantiles (see Table 15.9).

Table 15.9

Quantile	p
0.010	0.001
0.025	0.001
0.050	0.001
0.250	0.001
0.500	0.058
0.750	0.290
0.950	0.711
0.975	0.807
0.990	0.880

Here, with the prior distribution we selected and data consisting of 1 failed attempt at entering the White House without an invitation, our MCMC results suggest that:

$$0.001 < p < 0.807 \tag{15.52}$$

That's quite a range and suggests quite a bit of uncertainty for future celebrities. That's what you get with a sample size of 1!

?? But shouldn't we express the MCMC results as a beta distribution?

Answer: How right you are! Our prior is a **beta distribution**, and it makes sense that we'd like to express the posterior distribution as a beta distribution as well. After all, we're trying to replicate the conjugate solution with an MCMC approach.

The method by which we convert our MCMC results to the beta distribution is called **moment matching**. We touched on this in previous chapters. Let's quickly recall some general information about the beta distribution. The mean of the beta distribution can be calculated as:

$$\mu = \frac{\alpha}{\alpha + \beta} \tag{15.53}$$

The variance of the beta distribution can be calculated as:

$$\sigma^2 = \frac{\alpha\beta}{(\alpha + \beta)^2(\alpha + \beta + 1)} \tag{15.54}$$

This is great, but our MCMC analysis provides us with the reverse: we have the mean and variance of the p's from our MCMC trials, which are $\mu = 0.18$ and $\sigma^2 = 0.057$, respectively. Can we use these values to obtain the corresponding α and β parameters for the beta distribution? We can by solving these two equations with two unknowns, α and β. The solution is:

$$\alpha = -\frac{\mu(\sigma^2 + \mu^2 - \mu)}{\sigma^2} \tag{15.55}$$

$$\alpha = -\frac{0.18(0.057 + 0.18^2 - 0.18)}{0.057} = \frac{0.0163}{0.057} = 0.2861 \tag{15.56}$$

$$\beta = \frac{(\sigma^2 + \mu^2 - \mu)(\mu - 1)}{\sigma^2} \tag{15.57}$$

$$\beta = \frac{(0.057 + 0.18^2 - 0.18)(0.18 - 1)}{0.057} = \frac{0.074292}{0.057} = 1.303 \tag{15.58}$$

Note that Equation 15.55 (15.57) is equivalent to Equation 10.10 (10.9); for details see supplemental appendix at https://global.oup.com/booksites/content/9780198841296. Figure 15.11 shows a graph of this distribution (from our MCMC analysis), along with our original prior in blue, the posterior from the conjugate solution in solid purple (the one we are trying to match), and the moment-matching result from the MCMC analysis in dashed purple.

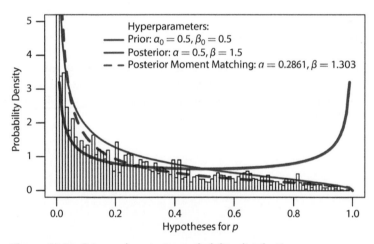

Figure 15.11 Prior and posterior probability distributions.

Not bad! As we mentioned, we could probably do a better job with a bit of tuning, in addition to thinning and a healthy burn-in. But, let's keep the big picture in mind here! We can summarize this chapter as follows:

- We are trying to estimate the probability that a famous person (like Shaq) can visit the White House without an invitation.
- The hypotheses range from 0 to 1, and there are an infinite number of them.
- We used the beta distribution to set our prior distribution (blue), which gives our relative *a priori* probability for each alternative hypothesis between 0 and 1.
- We collected data: Shaq tries to visit the White House without an invitation and fails.
- We **analytically computed** the posterior probability distribution by using the conjugate method (solid purple).
- We then **estimated** the posterior probability distribution using a Markov Chain Monte Carlo simulation that incorporates the Metropolis–Hastings algorithm (bars). This

approach allows us to estimate a Bayesian posterior distribution and avoid the messy integration required in the denominator of Bayes' Theorem.

• We used **moment matching** to convert our MCMC results into a beta distribution (dashed purple).

 ?? **Where can I find more information on this algorithm?**

Answer: The original citation is:

W. K. Hastings. "Monte Carlo sampling methods using Markov chains and their applications." *Biometrika* 57.1 (1970): 97–109. doi:10.1093/biomet/57.1.97. JSTOR 2334940. Zbl 0219.65008.

The Oxford Dictionary of Statistics (Upton and Cook, 2014) provides this background: "Hastings was a student at U. Toronto where his Ph.D. (1962) was supervised by Geoffrey Watson. After two years at U. Canterbury in New Zealand, and two at Bell Labs, he returned in 1966 to U. Toronto where he wrote the paper (in Biometrika) that placed an algorithm due to Metropolis in a statistical context."

 ?? **Are there other algorithms we can use?**

Answer: There are many others! We've touched on two so far but you may learn about other approaches in other texts. In our next chapter, we will use the **Gibbs sampling algorithm**. As we'll see, Gibbs sampling is another special case of the Metropolis–Hastings algorithm in which the proposal is always accepted.

CHAPTER 16

The Maple Syrup Problem Revisited: MCMC with Gibbs Sampling

As you've seen, a major use of Bayesian inference involves estimating parameters. You might recall from previous chapters the generic version of Bayes' Theorem for **one** continuous parameter:

$$P(\theta \mid \text{data}) = \frac{P(\text{data} \mid \theta) * P(\theta)}{\int P(\text{data} \mid \theta) * P(\theta) d\theta} \tag{16.1}$$

This is the generic version of Bayes' Theorem, when the posterior distribution for a single parameter, given the observed data, is represented by a **pdf**. This is designated $P(\theta \mid \text{data})$, **where P is probability density**. This is the left side of the equation. On the right side of the equation, the numerator multiplies the prior probability **density** of θ, which is written $P(\theta)$, by the likelihood of observing the data under a given hypothesis for θ, which is written $P(\text{data} \mid \theta)$. In the denominator, we see the same terms, but this time we also see a few more symbols. The symbol $\int$ means "integrate," which roughly means "sum up all the pieces" for each tiny change in θ, which is written $d\theta$. In other words, the denominator accounts for the prior density times the likelihood for all possible hypotheses for theta and sums them.

Do you remember the Maple Syrup Problem in Chapter 12? In that chapter, we assumed that the Federation of Quebec Maple Syrup Producers was interested in annexing Vermont's syrup production. We sought to estimate the mean, μ, and precision, τ, for annual maple syrup production in the great state of Vermont. These are the two parameters of a normal distribution. Both of the parameters are continuous in nature. With two continuous parameters, Bayes' Theorem looks like this, where each hypothesis for μ and τ is considered:

$$P(\mu, \tau \mid \text{data}) = \frac{P(\text{data} \mid \mu, \tau) * P(\mu, \tau)}{\int \int P(\text{data} \mid \mu, \tau) * P(\mu, \tau) d\mu d\tau} \tag{16.2}$$

For any given prior hypothesis for μ and τ, Bayes' Theorem will return the posterior joint density of μ and τ by computing the likelihood of observing the data under the joint hypothesis μ and τ, multiplied by the prior density of μ and τ. We need to do this for an **infinite** number of joint hypotheses.

What might the prior distribution of the parameters look like? If we gave the Federation a zillion grains of sand, where each grain represents a joint hypothesis for μ and τ, their prior distribution for the parameters may look something like the graph shown in Figure 16.1.

Bayesian Statistics for Beginners: A Step-by-Step Approach. Therese M. Donovan and Ruth M. Mickey, Oxford University Press (2019). © Ruth M. Mickey 2019.
DOI: 10.1093/oso/9780198841296.001.0001

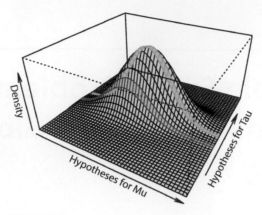

Figure 16.1

This distribution can take on many shapes, but the key thing to keep in mind is that it is a **surface**. You might recall that, with one parameter, we were concerned with the **area** under the curve. Here, with two parameters, we are concerned with the **volume** under the surface. For any given parameter combination, the corresponding height gives us the **probability density**, stressing the fact that we are dealing with a bivariate distribution in which both parameters are continuous. With the collection of new data, Bayes' Theorem could be used to update the shape of our sandpile.

> The function that sets the prior probabilities must integrate to 1. That is, the volume under the surface must integrate to 1.0. This can be a tough nut to crack!

In Chapter 12, we introduced a conjugate shortcut that will allow us to generate a posterior distribution that eliminates the need for integration. **The approach was to simplify the problem by focusing on only one of the parameters and assuming the second parameter is known. For a normal-normal conjugate solution, we assumed that τ was known and so focused on the unknown parameter, μ. We used a normal distribution as our prior distribution for the unknown mean, μ, and then collected data and updated the posterior distribution for the unknown mean, μ, which was also normally distributed. However, in this process, we assumed that τ was known; that is, we chose a level for τ (such as the green value or red value for τ in Figure 16.1).**

If you recall, the graphical diagram for this problem looked like the one shown in Figure 16.2.

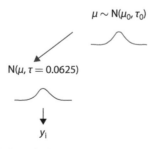

Figure 16.2

Here, the observed data, y_i, are at the bottom of the diagram. We assume the data arise from a normal distribution with parameters μ and $\tau = 0.0625$. The parameter μ is unknown and is the focus of estimation. The prior distribution for the unknown parameter μ is a normal distribution with hyperparameters μ_0 and τ_0.

Did you feel a sense of dissatisfaction with the conjugate approach? Seriously, how often do you **KNOW** one of the parameters (such as τ) of the normal distribution?

Seriously.

With two unknown parameters, the diagram is more realistically depicted as shown in Figure 16.3.

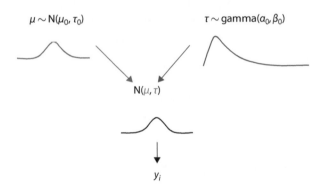

Figure 16.3

Here, the observed data, y_i, are at the bottom of the diagram. We assume the data arise from a normal distribution with parameters μ and τ, both of which are unknown and must be estimated. The prior distribution for the unknown parameter, μ, is a normal distribution with hyperparameters μ_0 and τ_0. The prior distribution for the unknown parameter, τ, is a gamma distribution with hyperparameters α_0 and β_0.

 ?? ## Can we make headway on this problem?

Answer: Fear not! We can! There is another approach, one that is so creative and versatile that it can be used to solve almost any kind of parameter estimation problem. It involves building the posterior distribution from scratch using a Markov Chain Monte Carlo simulation, or MCMC for short. The MCMC approach we will introduce in this chapter uses **Gibbs sampling**, which makes use of the conjugate shortcuts.

By the end of this chapter, you will have a good, general understanding of the following:

- Monte Carlo
- Markov chain
- Gibbs sampling

 ?? What is Gibbs sampling?

Answer: From the SAS Institute: "The Gibbs sampler . . . is a special case of the Metropolis sampler in which the proposal distributions exactly match the posterior conditional distributions and proposals are accepted 100% of the time."

So, Gibbs sampling is another MCMC algorithm, just like the Metropolis algorithm we used in the Shark Attack Problem (Chapter 13) and the Metropolis–Hastings algorithm we used in the White House Problem (Chapter 15). Remember that an algorithm is a step-by-step procedure for solving a problem or accomplishing some end, especially by a computer.

 ?? What is so special about Gibbs sampling?

Answer: A key feature is that Gibbs sampling allows one to estimate **multiple parameters**. In this method, in each trial, you update the prior distribution to the posterior distribution for a given parameter, **conditional** on the values of the remaining parameters. You then draw a proposal from this trial's posterior distribution for the target parameter and accept it. Then, you focus on the next parameter of interest, repeating the process for each unknown parameter in turn for each trial.

The algorithm generates a sample from the posterior distribution one sand grain at a time, conditional on the current values of the other variables. After many trials, the results provide an approximation of the joint distribution as well as the marginal distributions for each parameter separately.

 ?? What would this look like visually?

Answer: A figure displayed in Wikipedia beautifully illustrates the idea (see Figure 16.4).

In Figure 16.4, there are two variables of interest (X and Y). The individual dots on the grid represent a single hypothesis for a joint combination of X and Y. We can create a joint density distribution similar to Figure 16.1 from these dots by counting the number of dots in each grid cell, where the height of the distribution at each cell is proportional to the total dots in each grid cell. This is a joint distribution.

The marginal distributions for each variable are also shown in Figure 16.4 (blue for X, and red for Y). The heights of these distributions can be estimated by counting the dots in each row while ignoring the columns (for, say, X) or by counting the dots in each column while ignoring the rows (for Y). We'll be building a distribution similar to Figure 16.4, but it won't be a bivariate normal distribution. Generally speaking, though, if we let μ stand in for X, and τ stand in for Y, our goal for this chapter will be to estimate the posterior distributions for μ and τ, given the data. We can estimate both marginal distributions and the joint distribution.

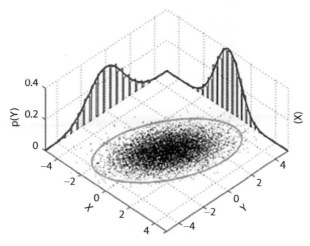

Figure 16.4 Many samples from a bivariate normal distribution. The marginal distributions are shown in red and blue. The marginal distribution of X is also approximated by creating a histogram of the X coordinates without consideration of the Y coordinates.

 ?? **So we'll be collecting dots or sand grains across many trials?**

Answer: In a manner of speaking, yes. Each trial will generate a dot. The informal pseudocode for the algorithm looks something like this:

1. Define the unknown parameters of interest. For the Maple Syrup Problem, they are μ and τ.
2. Specify the prior distribution for all parameters (e.g., Figure 16.3).
3. Choose initial values for the unknown parameters. For example, let's initialize τ with some reasonable value (e.g., 40.123).
4. Begin trial 1. Our goal now is to obtain a sample from the posterior distribution for μ, **conditional on** the previous trial's value for τ. Because we assume that τ is known (fixed) at some level (e.g., 40.123), we can use a conjugate solution to update the prior distribution for μ to the posterior distribution for μ for trial 1. We will draw a proposal from this distribution, designated p_μ, and accept it. This is our value for μ for trial 1.
5. Still in trial 1, we now focus our attention on the unknown parameter, τ. Our goal now is to obtain a sample from the posterior distribution for τ, **conditional on** this trial's value for p_μ. Because we assume that μ is known (fixed), we can use a conjugate solution to update the prior distribution for τ to the posterior distribution for τ for trial 1. We will draw a proposal from this distribution, designated p_τ and accept it. This is our value for τ for trial 1.
6. Repeat for numerous trials, each time accepting the proposal.

 ?? **OK! Can we start now?**

Let's start by reviewing the steps in Bayesian inference:

1. Identify your hypotheses.
2. Express your belief that each hypothesis is true in terms of prior probabilities.

3. Gather the data.
4. Determine the likelihood of the observed data under each hypothesis.
5. Use Bayes' Theorem to compute the posterior distribution for the unknown parameters.

 ?? What is step 1?

Answer: In this step, we identify the alternative hypotheses for the mean, μ, and alternative hypotheses for the precision, τ.

 ?? And step 2?

Answer: We establish prior distributions for **each** of the unknown parameters. Here, as in Chapter 12, we will assume that the prior distributions are set based on the expert opinion of federation members.

- The prior distribution for the unknown parameter, μ, will be a normal distribution. **Suppose the federation has specified the hyperparameters $\mu_0 = 12$, and $\tau_0 = 0.0625$** (i.e., $\sigma_0 = 4$).
- The prior distribution for the unknown parameter, τ, will be a gamma distribution. We learned about this distribution when estimating the rate of shark attacks. A quick reminder may be helpful. From Wikipedia (accessed August 19, 2017): "in probability theory and statistics, the gamma distribution is a two-parameter family of continuous probability distributions. There are three parameterizations in common use:
  - With a **shape** parameter k and a **scale** parameter θ.
  - With a **shape** parameter $\alpha = k$ and an inverse scale parameter $\beta = \frac{1}{\theta}$, called a **rate** parameter.
  - With a shape parameter k and a **mean** parameter $\mu = \frac{k}{\beta}$."
- We would like to add the following advice: **Beware!!** Different textbooks and statistical programs refer to β in different ways. For instance, some refer to the rate parameter as β, while others refer to the scale parameter as β. Rate and scale are reciprocals of each other, so rate = 1/scale or scale = 1/rate. You will need to be fully aware of how your computing program of choice refers to these parameters.
- Wikipedia goes on to explain the parameterization, with α (shape) and β (rate) being more common in Bayesian statistics. We need a gamma distribution that will reflect our beliefs and uncertainty for the unknown parameter, τ. **We will assume that the Federation has identified a gamma distribution with a shape parameter $\alpha_0 = 25$ and the rate parameter $\beta_0 = 0.5$. Remember that these are hyperparameters: in this case, they are parameters that describe the prior distribution for a parameter of interest. We will add subscripts to the hyperparameters to help them stand out.**

Now let's graph the prior distributions for **each** of the unknown parameters of the normal distribution (see Figure 16.5).

Here, we have color-coded the prior distribution for μ in blue, and τ in red, and we will use this color convention to help keep track of the two parameters throughout this chapter. Take a moment to firmly fix in your mind the two unknown parameters and their corresponding prior distributions that are identified by hyperparameters.

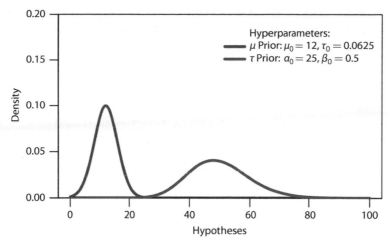

Figure 16.5 Annual maple syrup production, showing hypotheses for μ and τ.

 ?? Is it time for step 3?

Answer: Yes. Step 3 is to gather data. **We have observed 1 year of syrup production in Vermont, wherein 10.2 million gallons of the delicious sticky stuff was produced.**

 ?? And steps 4 and 5?

Answer: Steps 4 and 5 involve computing the likelihood of observing the data under each joint hypothesis for μ and τ using Bayes' Theorem. But, as you know, this is fairly intractable.

Instead, we'll use MCMC with Gibbs sampling to estimate the marginal and joint distributions of the unknown parameters, μ and τ.

 ?? How do we begin?

Answer: As in Chapters 13, 14, and 15, there will be a series of operations that will guide us.

1. First, let's set up our MCMC table that will store our calculations (Table 16.1). Here, we'll do a short MCMC that features five trials.

Table 16.1

Trial	Data			Proposal Distribution for μ			Proposal Distribution for τ		
				Hyperparameters			Hyperparameters		
$[t]$	Total Gallons	n	$\mu_{[t]}$	$\tau_{[t]}$		$p_{\mu[t]}$	$a_{[t]}$	$\beta_{[t]}$	$p_{\tau[t]}$
0	10.2	1	12	0.0625			25	0.5	
1	10.2	1							
2	10.2	1							
3	10.2	1							
4	10.2	1							
5	10.2	1							

- Notice that the trials are listed in column 1, and that the observed data (10.2 total gallons of maple syrup with $n = 1$ year of data) will go down columns 2 and 3. The data are fixed for all trials.
- The hyperparameters associated with the unknown parameter, μ, are given in columns 4 and 5. These are parameters that describe either the prior distribution for μ (fixed for all trials) or the posterior distribution for μ in any given trial. The columns labeled $\mu_{[t]}$ and $\tau_{[t]}$ will store trial-specific values, where the subscript, $[t]$, designates the trial. **For this exercise, the hyperparameters for the prior distribution (i.e., μ_0, τ_0) will be stored in trial 0. We will use these values repeatedly.** For the remaining trials, $\mu_{[t]}$ and $\tau_{[t]}$ are the updated trial-specific hyperparameters for the posterior distribution of μ. A randomly drawn proposal from the trial's posterior distribution will be stored in the blue-shaded column 6 labeled $p_{\mu[t]}$, where p stands for proposal.
- The hyperparameters associated with the unknown parameter, τ, are given in columns 7 and 8. The columns labeled $\alpha_{[t]}$ and $\beta_{[t]}$ will store trial-specific values, where $[t]$ designates the trial. **Again, for this exercise, the hyperparameters for the prior distribution (i.e., α_0, β_0) will be stored in trial 0. We will use these values repeatedly to calculate the updated parameters of the posterior distribution for each and every trial.** For the remaining trials, $\alpha_{[t]}$ and $\beta_{[t]}$ are the updated trial-specific parameters for the posterior distribution. A randomly drawn proposal from the trial's posterior distribution will be stored in the red-shaded column labeled $p_{\tau[t]}$, where p stands for proposal.
- Take a moment now to locate the hyperparameters of the prior distribution for each of our two unknown parameters and match them with Figure 16.5.

2. We then set a starting value for one of the unknown parameters. Either one will do. **Let's initialize τ with a value of 40.123.** This "primes" our MCMC chain and is entered for trial 0 (see Table 16.2).

Table 16.2

Trial	Data			Proposal Distribution for μ			Proposal Distribution for τ		
				Hyperparameters			Hyperparameters		
$[t]$	Total Gallons	n	$\mu_{[t]}$	$\tau_{[t]}$	$p_{\mu[t]}$		$\alpha_{[t]}$	$\beta_{[t]}$	$p_{\tau[t]}$
0	10.2	1	12	0.0625			25	0.5	40.123
1	10.2	1							
2	10.2	1							
3	10.2	1							
4	10.2	1							
5	10.2	1							

3. Now we are ready for trial 1, focusing on the parameters of the distribution of the unknown parameter, μ. More specifically, we are interested in obtaining $p_\mu \mid p_\tau$ for this trial.
   - We now assume that τ is **known** to be 40.123. This lets us **calculate** the hyperparameters for the posterior distribution of the unknown parameter, μ, using one of the conjugate shortcuts we've come to know and love (below in red). Pay attention to our subscripting please; parameters that are subscripted with a 0 are hyperparameters of the prior distribution (see Figure 16.6).

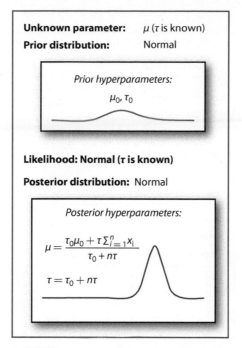

Figure 16.6

- The posterior distribution of the unknown parameter, μ, now has trial 1 hyperparameters:

$$\mu_{\text{posterior}[1]} = \frac{(\tau_0\mu_0 + \tau\sum_{i=1}^{n}x_i)}{(\tau_0 + n*\tau)} = \frac{0.0625*12 + 40.123*10.2}{0.0625 + 1*40.123} = \frac{410.0346}{40.1855} = 10.2028$$

(16.3)

and

$$\tau_{\text{posterior}[1]} = \tau_0 + n*\tau = 0.0625 + 1*40.123 = 40.1855$$

(16.4)

- Let's now draw a random proposal from our updated distribution for μ, that is, the normal distribution with parameters $\mu_1 = 10.2028$ and $\tau_1 = 40.1855$. Let's call this proposal $p_{\mu[1]}$. **Let's say this proposal's value is 10.5678. Again, we accept this proposal.**
- Now let's fill these values into our table as the trial 1 posterior parameters for the unknown parameter, μ, and a single proposal from this new distribution (see Table 16.3).

4. Now we are ready to focus on the unknown parameter, τ for trial 1. Specifically, we are interested in obtaining p_τ for trial 1.

- We now assume that μ is **known** to be 10.5678. This lets us **update** the parameters for the posterior distribution of the unknown parameter, τ, using the conjugate shortcut (below in red; also see Figure 16.7). Remember that α_0 and β_0 are prior hyperparameters for the gamma distribution. We added subscripting to make this more clear.

Table 16.3

Trial	Data		Proposal Distribution for μ			Proposal Distribution for τ		
			Hyperparameters			Hyperparameters		
[t]	Total Gallons	n	$\mu_{[t]}$	$\tau_{[t]}$	$p_{\mu[t]}$	$\alpha_{[t]}$	$\beta_{[t]}$	$p_{\tau[t]}$
0	10.2	1	12	0.0625		25	0.5	40.123
1	10.2	1	10.2028	40.1855	10.5678			
2	10.2	1						
3	10.2	1						
4	10.2	1						
5	10.2	1						

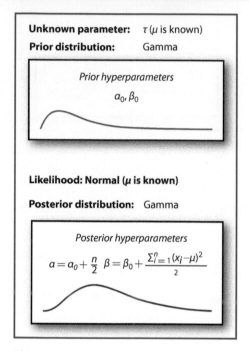

Unknown parameter: τ (μ is known)

Prior distribution: Gamma

Prior hyperparameters

α_0, β_0

Likelihood: Normal (μ is known)

Posterior distribution: Gamma

Posterior hyperparameters

$$\alpha = \alpha_0 + \frac{n}{2} \quad \beta = \beta_0 + \frac{\sum_{i=1}^{n}(x_i-\mu)^2}{2}$$

Figure 16.7

- The unknown parameter, τ, now has trial 1 posterior hyperparameters:

$$\alpha_{\text{posterior}[1]} = \alpha_0 + \frac{n}{2} = 25 + \frac{1}{2} = 25.5 \tag{16.5}$$

and

$$\beta_{\text{posterior}[1]} = \beta_0 + \frac{\sum_{i=1}^{n}(x_i-\mu)^2}{2} = 0.5 + \frac{(10.2-10.5678)^2}{2} = 0.5 + \frac{0.1352768}{2} = 0.5676 \tag{16.6}$$

- Let's now draw a random proposal from this trial's posterior distribution, and call it $p_{\tau[1]}$. **Let's say this proposal's value is 45.678.** We accept it.
- And now let's fill these values into our table as the trial 1 hyperparameters for the distribution of the unknown parameter, τ, and a single proposal from this new distribution (see Table 16.4).

Table 16.4

Trial	Data		Proposal Distribution for μ			Proposal Distribution for τ		
			Hyperparameters			Hyperparameters		
[t]	Total Gallons	n	$\mu_{[t]}$	$\tau_{[t]}$	$p_{\mu[t]}$	$\alpha_{[t]}$	$\beta_{[t]}$	$p_{\tau[t]}$
0	10.2	1	12	0.0625		25	0.5	40.123
1	10.2	1	10.2028	40.1855	10.5678	25.5	0.5676	45.678
2	10.2	1						
3	10.2	1						
4	10.2	1						
5	10.2	1						

5. And now we are ready for trial 2. As painful as it may be, we'll walk through the calculations once more, just to make sure you nail it. Trial 2 will begin by assuming that the τ is known to be 45.678. We can now use this value to update the hyperparameters for posterior distribution of μ in trial 2.

   • The posterior distribution of the unknown parameter, μ, now has trial 2 hyperparameters:

$$\mu_{\text{posterior}[2]} = \frac{(\tau_0\mu_0 + \tau\sum_{i=1}^{n}x_i)}{(\tau_0 + n*\tau)} = \frac{0.0625 * 12 + 45.678 * 10.2}{0.0625 + 1 * 45.678} = \frac{466.6656}{45.7405} = 10.2025$$

(16.7)

and

$$\tau_{\text{posterior}[2]} = \tau_0 + n*\tau = 0.0625 + 1 * 45.678 = 45.7405 \qquad (16.8)$$

   • Let's now draw a random proposal from our updated distribution for μ, that is, the normal distribution with hyperparameters $\mu_{[2]} = 10.2025$ and $\tau_{[2]} = 45.7405$. Let's call this proposal $p_{\mu[2]}$. **Let's say this proposal value is 10.0266.**
   • And now let's fill these values into our table as updated hyperparameters for the posterior distribution of the unknown parameter, μ, and a single proposal from this new distribution (see Table 16.5).

Table 16.5

Trial	Data		Proposal Distribution for μ			Proposal Distribution for τ		
			Hyperparameters			Hyperparameters		
[t]	Total Gallons	n	$\mu_{[t]}$	$\tau_{[t]}$	$p_{\mu[t]}$	$\alpha_{[t]}$	$\beta_{[t]}$	$p_{\tau[t]}$
0	10.2	1	12	0.0625		25	0.5	40.123
1	10.2	1	10.2028	40.1855	10.5678	25.5	0.5676	45.678
2	10.2	1	10.2025	45.7405	10.0266			
3	10.2	1						
4	10.2	1						
5	10.2	1						

6. Now let's fill in some entries under trial 2 for the unknown parameter, τ.

- We now assume that μ is **known** to be 10.0266. This lets us **update** our hyperparameters for the posterior distribution of the unknown parameter, τ, using the conjugate shortcut.
- The unknown parameter, τ, now has trial 2 hyperparameters:

$$\alpha_{posterior[2]} = \alpha_0 + \frac{n}{2} = 25 + \frac{1}{2} = 25.5 \tag{16.9}$$

and

$$\beta_{posterior[2]} = \beta_0 + \frac{\sum_{i=1}^{n}(x_i - \mu)^2}{2} = 0.5 + \frac{(10.2 - 10.0266)^2}{2} = 0.5 + \frac{0.03006756}{2} = 0.5150 \tag{16.10}$$

- Let's now draw a random proposal from this distribution and call it $p_{\tau[2]}$. **Let's say this proposal value is 52.39**. We accept it.
- And now let's fill these values into our table as updated hyperparameters for the posterior distribution of the unknown parameter, τ, and a single proposal from this new distribution (see Table 16.6).

Table 16.6

Trial	Data			Proposal Distribution for μ			Proposal Distribution for τ		
				Hyperparameters			Hyperparameters		
[t]	Total Gallons	n	$\mu_{[t]}$	$\tau_{[t]}$	$p_{\mu[t]}$	$\alpha_{[t]}$	$\beta_{[t]}$	$p_{\tau[t]}$
0	10.2	1	12	0.0625		25.0	0.5000	40.12
1	10.2	1	10.2028	40.1855	10.5678	25.5	0.5676	45.68
2	10.2	1	10.2025	45.7405	10.0266	25.5	0.5150	52.39
3	10.2	1						
4	10.2	1						
5	10.2	1						

This brings us to trial 3...

Isn't this brilliant? The Gibbs sampler is similar to the Metropolis and Metropolis–Hastings algorithms we previously studied in that we collect a sample from the posterior distribution in each trial. With these algorithms, we drew a sample and then either accepted or rejected it. However, with Gibbs sampling, there is no reject–accept criterion. Instead, we draw a proposal from the distribution **conditional** on the values for the remaining parameters and accept it. Gibbs sampling requires that you (1) can compute a posterior distribution of a parameter **conditional** on the values of other parameters and (2) can draw a sample from it. Here, we made use of the conjugate solutions. If one of the parameters is assumed to be known for a given trial (e.g., the red or green ribbon in Figure 16.1), we can use a conjugate solution to derive the posterior distribution for the parameter of interest.

Before we go further, a few questions may have leapt to mind.

 ?? Who the heck is Gibbs?

Answer: From Wikipedia: the Gibbs sampler is named after Josiah Willard Gibbs, "a theoretical physicist and chemist who was one of the greatest scientists in the United States in the 19th century."

Figure 16.8 Josiah Gibbs.

 ?? **Did Josiah Gibbs come up with the Gibbs sampler?**

Answer: Actually, no. Wikipedia again: "The algorithm was described by brothers Stuart and Donald Geman in 1984, some eight decades after the death of Gibbs."

Figure 16.9 Stuart Geman (left) and Donald Geman (right) (Original photograph courtesy of Stuart Geman).

That's Stuart on the left, and Donald on the right, working on their paper in Paris. The brothers came up with the method to help restore old or blurred images.

 ?? Why did they name their algorithm the Gibbs sampler?

Answer: The answer is explained by Sharon Bertsch McGrayne (2011) in The Theory That Would Not Die: How Bayes' Rule Cracked the Enigma Code, Hunted Down Russian Submarines, and Emerged Triumphant from Two Centuries of Controversy:

"Sitting at a table in Paris, Donald Geman thought about naming their system. A popular Mother's Day gift at the time was a Whitman's Sampler assortment of chocolate bonbons (see Figure 16.10); a diagram inside the box top identified the filling hidden inside each candy. To Geman, the diagram was a matrix of unknown but enticing variables. 'Let's call it Gibbs sampler,' he said, after Josiah Willard Gibbs, a nineteenth century American physicist who applied statistical methods to physical systems."

Figure 16.10 A Whitman's Sampler.

 ?? Do you have a reference for the Geman brother's paper?

Answer: The brothers published their method in 1984.

> S. Geman and D. Geman. 1984. "Stochastic relaxation, Gibbs distributions, and the Bayesian restoration of images." *IEEE Transactions on Pattern Analysis and Machine Intelligence* 6(1984): 721–41.

You can find the abstract here.

 ?? Can we finish our trials please?

Answer: Of course! See if you can trace your way through Table 16.7.

Table 16.7

[t]	Total Gallons	n	$\mu_{[t]}$	$\tau_{[t]}$	$p_{\mu[t]}$	$a_{[t]}$	$\beta_{[t]}$	$p_{\tau[t]}$
Trial	Data		Proposal Distribution for μ Hyperparameters			Proposal Distribution for τ Hyperparameters		
0	10.2	1	12.0	0.0625	NA	25.0	0.5000	40.12
1	10.2	1	10.2	40.1855	10.57	25.5	0.5676	45.68
2	10.2	1	10.2	45.7405	10.03	25.5	0.5150	52.39
3	10.2	1	10.2	52.4521	10.15	25.5	0.5010	44.58
4	10.2	1	10.2	44.6445	10.29	25.5	0.5043	48.33
5	10.2	1	10.2	48.3911	10.10	25.5	0.5055	46.92

 ?? How do we display the results of a Gibbs MCMC?

Answer: We show the traceplots of the MCMC proposals for each unknown parameter, μ and τ, as in Figure 16.11.

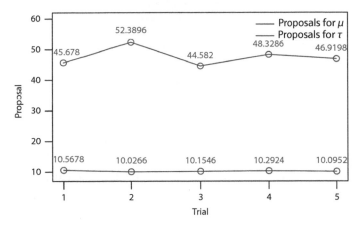

Figure 16.11

Of course, this is just 5 trials. Let's run the analysis again and show the traceplots for 1000 trials (see Figure 16.12).

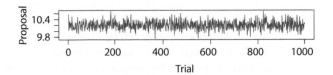

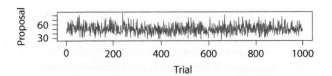

Figure 16.12 Traceplots for μ (top) and τ (bottom).

Here, you can see the familiar "bouncing" as new proposals are drawn. We talked about how to evaluate these chains in Chapter 14.

 ?? How do we use the MCMC results to draw conclusions about the posterior distributions for μ and τ?

Answer: We summarize the chains. A histogram that shows the frequencies of proposals is a good start. Figure 16.3 shows the frequency histogram of μ proposals in our 1000 trials. Remember these are marginalized over values of the other parameter.

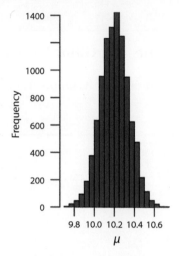

Figure 16.13 Posterior distribution of μ.

It is also common to present summary statistics of the data, such as various quantiles, as shown in Table 16.8.

Table 16.8

Minimum	25th Percentile	50th Percentile	75th Percentile	Maximum
9.77	10.11	10.2	10.3	10.64

Based on your analysis, you can now report that you believe the mean annual maple syrup production ranges between 9.77 and 10.64 thousand gallons. You could also present the Bayesian credible intervals, as we've done in previous chapters.

But the mean (10.2) and standard deviation (0.14) of the trials are very useful because these can now serve as hyperparameters for μ for your next analysis. Cool!

 ?? What conclusions do I draw about τ?

Answer: Same idea. Here, we show the histogram for the τ proposals. But, we also convert τ back to σ and show the histogram for the σ's as well. Remember that:

$$\tau = \frac{1}{\sigma^2} \tag{16.11}$$

$$\sigma = \frac{1}{\sqrt{\tau}} \tag{16.12}$$

This is probably what you are most familiar with (see Figure 16.14).

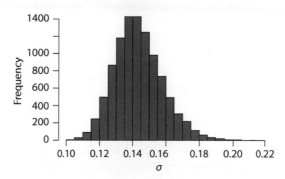

Figure 16.14 Posterior distribution of σ.

Again we'll present summary statistics of the data, including the mean, the standard deviation, and quantiles, as shown in Table 16.9.

Table 16.9

Minimum	25th Percentile	50th Percentile	75th Percentile	Maximum
0.11	0.13	0.14	0.15	0.2

If you were to use this information for your **next** analysis, you would need to use **moment matching** to convert the mean and standard deviation of the MCMC results to α and β from a gamma distribution. We discussed this in Chapter 13, so we won't repeat the calculations here.

 ?? **What's the difference between our maple syrup estimation conjugate approach and the Gibbs sampling approach?**

Answer: The main difference is that Chapter 12's normal-normal conjugate required that we **KNEW** σ...a bit unrealistic. With Gibbs sampling, we took advantage of the Bayesian conjugate solutions in a different (and wickedly creative) way. This approach relaxes the assumption that we have to know σ (or τ) and let us estimate **both** unknown parameters. We had a look at the marginal distributions for each.

 ?? **Can we look at the joint posterior distribution too?**

Answer: Yes. Figure 16.15 shows a scatter plot using proposals from every 10th trial out of 10,000 trials.

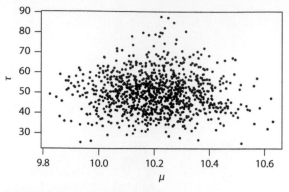

Figure 16.15 Annual maple syrup production, showing joint hypotheses for μ and τ.

Compare Figure 16.15 with Figure 16.4. Each dot is a joint hypothesis.

Imagine now that we place a grid of hexagons over the dots and count the dots within each hexagon so that darker hexagons indicate more values in that bin. Our resulting graph is an approximation of the joint posterior distribution (see Figure 16.16).

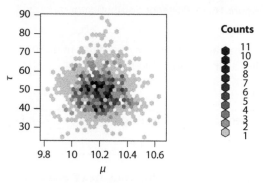

Figure 16.16 Joint hypotheses for μ and τ.

And, even slicker, if we "bin" our results as shown in Figure 16.4, we can plot both the joint and marginal distributions for the parameters μ and τ, as shown in Figure 16.17.

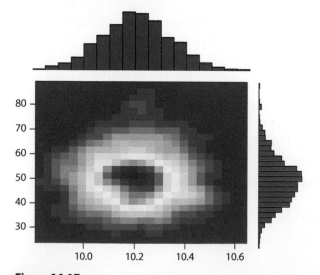

Figure 16.17

?? **You don't do these calculations by hand, do you?**

Answer: You can! Carrying out a few trials by hand may well be the best way to internalize this material. But after you understand the method, you certainly don't want to do the calculations by hand... computers live for tasks such as these. There are several software programs available, but here we mention two that specifically highlight the use of Gibbs sampling by name:

- The BUGS (Bayesian inference Using Gibbs Sampling) project: "BUGS is concerned with flexible software for the Bayesian analysis of complex statistical models using Markov chain Monte Carlo (MCMC) methods. The project began in 1989 in the MRC Biostatistics Unit, Cambridge, and led initially to the 'Classic' BUGS program, and then onto the WinBUGS software developed jointly with the Imperial College School of Medicine at St Mary's, London. Development is now focused on the OpenBUGS project."
- JAGS (Just Another Gibbs Sampler) (Plummer, 2003): "JAGS is Just Another Gibbs Sampler. It is a program for analysis of Bayesian hierarchical models using Markov Chain Monte Carlo (MCMC) simulation not wholly unlike BUGS. JAGS was written with three aims in mind:
  - To have a cross-platform engine for the BUGS language
  - To be extensible, allowing users to write their own functions, distributions and samplers.
  - To be a platform for experimentation with ideas in Bayesian modeling."

There are several excellent guides written specifically to teach you how to use these programs.

 ?? **Why is the Gibbs sampler known as a special case of the Metropolis–Hastings algorithm?**

Answer: Great question! It has to do with the proposal, how it is drawn, and whether it is accepted or not. You might recall that, in the Metropolis algorithm, the proposal is drawn from a **symmetric** distribution. The Metropolis–Hastings algorithm, in contrast, relaxes this assumption so that the proposal can be drawn from an asymmetric distribution. (You might recall that a correction factor is used to help determine whether the proposal is accepted or not.) Similarly, with the Gibbs sampling algorithm, the proposal can be drawn from any distribution (symmetric or non-symmetric). We saw an example of this when we drew proposals from a symmetric normal distribution as well as from a non-symmetric gamma distribution. The difference, however, is that we **always accept** the proposal with Gibbs sampling because it is a sample from the trial's posterior distribution, conditional on the values of the remaining parameters for that trial.

The SAS Institute's tutorial introduction to MCMC describes the procedure this way: "Gibbs sampling requires you to decompose the joint posterior distribution into full conditional distributions for each parameter in the model and then sample from them. The sampler can be efficient when the parameters are not highly dependent on each other and the full conditional distributions are easy to sample from. Some researchers favor this algorithm because it does not require an instrumental proposal distribution as Metropolis methods do. However, while deriving the conditional distributions can be relatively easy, it is not always possible to find an efficient way to sample from these conditional distributions."

?? Can we solve the Maple Syrup Problem using the Metropolis–Hastings algorithm instead?

Answer: You can! There is nothing to stop you from drawing proposals for μ and τ as we did in previous chapters. The key is, still, that for any given trial, you draw a proposal conditional on the values for the remaining parameters.

?? Why not just use the Metropolis–Hastings algorithm instead?

Answer: The simplest answer for the Maple Syrup Problem is computing speed...using the conjugate shortcuts is much faster than carrying out all the steps for the Metropolis–Hastings algorithm.

?? How do we know if our MCMC results are any good?

Answer: As with previous MCMC chapters, you MUST carefully consider the resulting posterior distributions and run some diagnostic tests. These include the concepts of "burn-in" and "pruning." We covered MCMC diagnostics in Chapter 14.

?? Can we summarize the main points of this chapter?

Answer: Sure. It's easy to lose sight of the big picture. Keep in mind that the goal of a Bayesian MCMC analysis is to estimate the posterior distribution while skipping the integration required in the denominator of Bayes' Theorem. For a problem with two or more parameters, the approach is to sample from the posterior distribution for a given parameter, *conditional* on the values of the remaining parameters. Thus, the MCMC approach allows us to build the posterior distribution by breaking the problem into small, bite-sized pieces, allowing us to build the posterior distribution bit by bit (sand grain by sand grain). In this chapter, we introduce the Gibbs sampler, which is a special case of the Metropolis–Hastings algorithm. Gibbs sampling requires that you (1) can compute a posterior distribution of a parameter **conditional** on the values of other parameters and (2) can draw a sample from it. Here, we made use of the conjugate solutions, which enabled us decompose the joint posterior distribution into full conditional distributions for each parameter in the model and then sample from them.

?? What's next?

Answer: This ends our section on MCMC. The next section of the book highlights some applications of Bayes' Theorem. Chapter 17, in particular, makes use of MCMC with Gibbs sampling.

SECTION 6

Applications

Overview

You've made it! Analysts across the world are using Bayes' Theorem to tackle a great variety of problems. In this final section, we illustrate a variety of methods in which Bayes' Theorem is used to solve pressing problems of the day.

- One of the most common uses of Bayes' Theorem is in the statistical analysis of a dataset (i.e., statistical modeling). In Chapter 17, we consider another application of Gibbs sampling, that of parameter estimation for simple linear regression. Our example considers the relationship between how many days a contestant lasts in a Survivor game as a function of how many years of formal education they have. It is a bit more complicated than the previous chapter because it involves estimation of the joint posterior distribution of three parameters. As in earlier chapters, we describe the estimation process in detail and then illustrate its application on a step-by-step basis. Finally, we discuss how to estimate the posterior predictive distribution.
- Chapter 18 provides a very brief introduction to Bayesian model selection. We still use the Survivor game example but now want to compare two models: one that uses years of formal education, and a second one that uses grit as a predictor of number of days a contestant lasts on Survivor. We use Gibbs sampling for parameter estimation and introduce Deviance Information Critera (DIC) as a guide for model selection. We describe in detail how this measure is computed.
- Bayesian belief networks comprise the focus of Chapter 19, and provide a very powerful tool to aid in decision-making when there is uncertainty. The ideas are first explained in terms of a small, standard example that explores two alternative hypotheses for why the grass is wet: the sprinkler is on versus it is raining. We describe how to depict causal models graphically with the use of influence diagrams and directed acyclic graphs, and show where Bayes' Theorem comes into play for computing conditional probabilities and for updating probabilities once new information is obtained or assumed. We illustrate the software program, Netica. Finally, we provide a second example based on The Lorax by Dr. Seuss.
- Chapter 20 introduces decision trees, a very useful technique that can be used to answer a variety of questions and assist in making decisions. We've assumed you've read our previous chapter on Bayesian networks. If so, you already know that Bayes' nets can be

an aid in decision-making. We show that a decision tree is a graphical representation of the alternatives in a decision-making problem and are very closely related to Bayes' networks, except that they take the shape of a tree instead. The tree itself consists of decision nodes, chance nodes, and end nodes, which provide an outcome. In the tree, we point out the probabilities associated with chance nodes are conditional probabilities, which Bayes' Theorem can be used to estimate or update. We describe the calculation of expected values (or expected utility) of competing alternative decisions. We illustrate the use of decision trees with another application from The Lorax by Dr. Seuss.

If this book makes it to a second edition, this section will likely grow, as there are so many fun applications of Bayes' Theorem!

CHAPTER 17

The Survivor Problem: Simple Linear Regression with MCMC

Welcome to this section on applications of Bayes' Theorem. One of the most common uses of Bayes' Theorem is in the statistical analysis of a dataset (i.e., statistical modeling). This chapter, and the one that follows, introduce you to statistical modeling à la Bayes. This is not meant to be a course on statistics, but hopefully it will give you the general sense of how things are done. It's a long chapter, so make sure to get up and stretch every now and then. When all is said and done, you will have a basic understanding of the following concepts:

- Linear equation
- Sums of squares
- Linear regression with MCMC and Gibbs sampling
- Posterior predictive distribution

Our chapter title suggests we will be learning about simple linear regression, which is a statistical method that allows us to formalize the relationship between two variables as a function, or model.

We thought it might be useful to begin with a refresher, so...

 ?? What is a function?

Answer: In math, a **function** relates an input to an output, and the classic way of writing a function is:

$$f(x) = \ldots \tag{17.1}$$

For instance, consider the function shown in Figure 17.1.

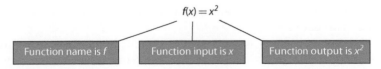

Figure 17.1

The function name is f. The **inputs** (also called arguments) go within the parentheses. Here, we have a single input, or argument, called x. The function itself is a set of instructions that tell us what to do with the input. The **output** is what the instructions return. Here, we input a value for x; the instructions tell us that we should square x to give us the output.

Bayesian Statistics for Beginners: A Step-by-Step Approach. Therese M. Donovan and Ruth M. Mickey, Oxford University Press (2019). © Ruth M. Mickey 2019.
DOI: 10.1093/oso/9780198841296.001.0001

In other words, $f(x)$ **maps** the input x to the output. We can visualize this as shown in Figure 17.2.

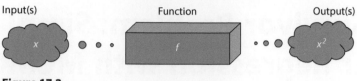

Input(s)   Function   Output(s)

x f x^2

Figure 17.2

The Oxford Dictionary Plus Science and Technology (Upton and Cook, 2016) tells us that a function is "...a relationship between the elements of one set and one element of another or the same set, such that one or many inputs are related to only one output. For example, a function might relate a real number to the cube of that number. This relation can be written as $f(x) = x^3$; f is the name of the function, x is the input, and x^3 is the output, and $f(x)$ is spoken as 'f of x'."

 ?? Can we see another example?

Answer: Of course. How about this one:

$$g(x) = mx + b. \tag{17.2}$$

The function's name is g, and it has a single argument (input), called x. You input a value for x, and the function will multiply it by m and then add b to it. That is the function output: $mx + b$. Here, m and b are called **parameters**; they are part of the function's definition.

 ?? This function looks vaguely familiar . . . is this a linear function?

Answer: It is! The linear function above might be more familiar to you in this form:

$$y = mx + b. \tag{17.3}$$

This is still a function, but it is nameless and lacks definition. Regardless, this function has an input, an output, and instructions. Here, we assume that x is the input. You input a value for x; the instructions tell you to multiply it by the parameter m and then add the result to the parameter b. The final result, or output, is called y. Both x and y are called **variables** because if you change the value of x, the value of y may also change. Encyclopedia Britannica defines variable as "a symbol (usually a letter) standing in for an unknown numerical value in an equation."

A function like this may have great utility. For example, if you speak in Fahrenheit but your Canadian neighbor wants to know the average temperature in February, you can convert temperature in degrees Fahrenheit (the input) to temperature in degrees Celsius (the output) with the following linear equation (also see Figure 17.3):

$$y = mx + b \tag{17.4}$$

$$\text{Celsius} = 5/9 * \text{Fahrenheit} + (-32 * 5/9). \tag{17.5}$$

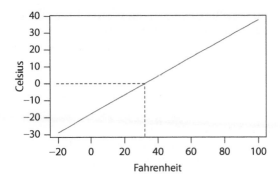

Figure 17.3 Conversion of Fahrenheit to Celsius scale.

Here, the variables are degrees in Celsius and degrees in Fahrenheit. Notice that all of the points fall exactly on the line itself. We say that this is a **deterministic** function, or a deterministic model. This model lets you plug in a value of Fahrenheit and output the *exact* degrees in Celsius. For example, if you input 32 degrees Fahrenheit (shown by the red dashed line), the output would be 0 degrees Celsius, which, of course, you already knew.

 ?? **So what does this have to do with regression analysis?**

Answer: Regression analysis is a statistical process for estimating the relationships among variables.

The goal of a regression analysis is to identify a function, also called a **statistical model**, that describes how two or more variables are related. The main difference between a statistical model and a deterministic model (e.g., the exact relationship between Celsius and Fahrenheit) is that the datapoints won't necessary fall on the line directly, and the functional relationship must be **estimated**.

 ?? **Can we see an example?**

Answer: By all means. Consider the hypothetical dataset shown in Table 17.1.

Table 17.1

Var1	Var2	Error	Var3
−1.0	1.0	−0.15	0.85
−0.8	1.2	0.79	1.99
−0.6	1.4	0.27	1.67
−0.4	1.6	0.45	2.05
−0.2	1.8	0.23	2.03
0.0	2.0	−0.36	1.64
0.2	2.2	0.01	2.21
0.4	2.4	−0.22	2.18
0.6	2.6	−0.48	2.12
0.8	2.8	0.07	2.87
1.0	3.0	−0.12	2.88

Here, our dataset has four variables, and you can see that they can be positive or negative. In column 1, we generated Var1 by setting up a series of numbers from −1 to +1 in increments of 0.2. In column 2, Var2 was calculated with the deterministic model:

$$Var2 = 1 * Var1 + 2. \tag{17.6}$$

This is the equation of a line. Here, the slope is 1, and the intercept is 2:

$$y = mx + b \tag{17.7}$$

Now, let's rename the intercept b as b_0, and let's rename the slope m as b_1. Statisticians conventionally use these names for the intercept and slope. With minor rearrangement, our linear function can be written:

$$y = b_0 + b_1 * x \tag{17.8}$$

$$y = 2 + 1 * x. \tag{17.9}$$

Continuing with our table, we then generated some random numbers associated with each datapoint in column 3, which we are calling "Error." Finally, in column 4, Var3 is Var2 plus Error. Notice that Var3 contains both a deterministic component (which we can call the **signal**) and an error component (which we can call the **noise**).

Now then. Suppose we don't know how Var3 was created, and we would like to know if there is a **relationship** between **Var1** and **Var3**. To begin, we can plot the two variables against each other (see Figure 17.4).

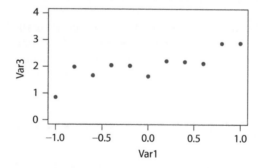

Figure 17.4 Scatterplot of Var1 and Var3.

You can see that our two variables are **positively** related: as one variable goes up, the other tends that direction too. In other words, there is a **pattern** in the data, and we would like to formalize this pattern as a statistical model.

 ?? **Can you summarize the relationship between Var1 and Var3 with a straight line?**

Answer: Let's try it. For this example, we will only use a straight line to summarize the relationship (see Figure 17.5).

See the green circles? These are our datapoints. See the blue line? That is the linear relationship between Var1 and Var3 (i.e., the signal). Each datapoint has a vertical line that measures its distance to the line itself. This distance is called "error," and it matches up

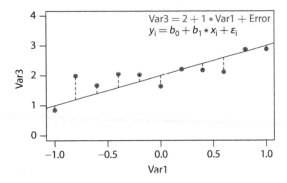

Figure 17.5

with the "Error" column in Table 17.1. It is not an error as in a "mistake," but rather it is information (noise) that cannot be explained by the line. An "error" for a given datapoint i is written as ϵ_i. Notice that some of the errors are positive, and some are negative. Also note that some datapoints have large error (i.e., they are far from the line) while others have small error (they are close to the line, or even on it).

See the equation in blue at the top of Figure 17.5? That is our expression for the y_i's. Notice that the equation itself specifies the **signal** (the linear equation $b_0 + b_1 x_i$) and **error** (noise), indicating the linear equation by itself does not perfectly fit the data. Here, we say that Var3 and Var1 are stochastically related.

 ?? **What's the second equation written in black?**

Answer: That is a general way of expressing the stochastic relationship between Var3 and Var1. With new color-coding, our equation has the general form:

$$y_i = b_0 + b_1 * x_i + \epsilon_i. \tag{17.10}$$

Encylopedia Britannica calls ϵ "a probabilistic error term that accounts for the variability in y that cannot be explained by the linear relationship with x. If the error term were not present, the model would be deterministic; in that case, knowledge of the value of x would be sufficient to determine the value of y."

Speaking of ϵ_i (the error term associated with datapoint i), it is not measured by the analyst (although we provided it in Equation 17.1). Here, it can be determined as the difference between y_i and its model signal, but normally this term is estimated. We can rearrange the terms in our model to highlight the error term:

$$\epsilon_i = y_i - (b_0 + b_1 * x_i). \tag{17.11}$$

 ?? **So our goal is to find the signal within the data?**

Answer: Yes! As such, it is a major part of science. After all, understanding **pattern** and **process** (the underlying causes of patterns) is what science is all about. This goes way back to our section on Bayesian inference, where we provided a short introduction to the scientific method. Let's quickly revisit these concepts now.

?? What exactly is science?

Answer: There are many definitions, formal and informal. Generally speaking, science refers to a system of acquiring knowledge.

Answer: (From NASA): Science is curiosity in thoughtful action about the world and how it behaves.

Answer: (From Wikipedia): Science (from Latin *scientia*, meaning "knowledge") is a systematic enterprise that builds and organizes knowledge in the form of testable explanations and predictions about the universe.

 ?? How do we go about actually conducting science?

Answer: We normally use what is called the scientific method. The Oxford English Dictionary (Stevenson, 2010) says that the scientific method is "a method or procedure that has characterized natural science since the 17th century, consisting in systematic observation, measurement, and experiment, and the formulation, testing, and modification of hypotheses."

A key concept in scientific endeavors is formulating testable, alternative explanations about how the universe works. The scientific method actually consists of two types of inquiry: **induction** and **deduction**, which, when used in concert, produce knowledge.

The scientific process is nicely captured in the diagram in Figure 17.6 (adapted from Rao, 1997). Let's walk through this diagram, noting that it is a (diamond-shaped) circle at heart and has neither beginning nor end. It can be thought of as a race track or a wheel.

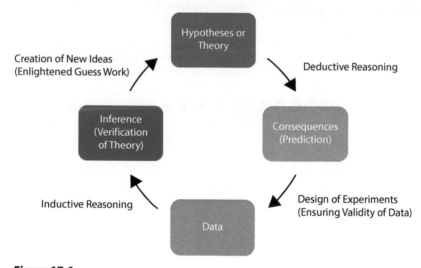

Figure 17.6

We need to start somewhere in this diagram, which contains four boxes and four arrows. Let's start with the upper box:

1. **Hypothesis or Theory Box**. A hypothesis is a proposed explanation for a phenomenon. A scientific theory is a coherent group of propositions formulated to explain a group of facts or phenomena in the natural world and repeatedly confirmed through experiment or observation. We will also note that process-based models, such as models of global climate circulation, are hypotheses at heart.

- A theory: Darwin's Theory of Evolution
- A hypothesis: The earth is warming due to increased CO_2 levels in the atmosphere

2. **Deductive Reasoning Arrow**. The Oxford Reference tells us that deductive reasoning is reasoning from the general to the particular. Here, you start with a hypothesis or theory and test it from an examination of facts.

3. **Consequences or Predictions Box**. Dr. Sylvia Wassertheil-Smoller, a research professor at Albert Einstein College of Medicine, explains "In deductive inference, we hold a theory and based on it we make a prediction of its consequences. That is, we predict what the observations should be if the theory were correct" (source: http://www.livescience.com/21569-deduction-vs-induction.html; accessed August 17, 2017). A prediction is the result of deduction.

4. **Design of Experiments Arrow**. In this step, you plan an experiment that will allow you to collect data to test your hypothesis or hypotheses. A well-designed experiment will ensure that the data you collect are valid.

5. **Data Box**. After the experiment is designed, we then collect some data. We can also use existing datasets if they are appropriate.

6. **Inductive Reasoning Arrow**. The Oxford Reference tells us that inductive reasoning involves inferring general principles from specific examples.

7. **Inference Box**. Inductive reasoning makes broad generalizations from specific observations. Dr. Sylvia Wassertheil-Smoller explains, "In inductive inference, we go from the specific to the general. We make many observations, discern a pattern, make a generalization, and infer an explanation or a theory" (source: http://www.livescience.com/21569-deduction-vs-induction.html; accessed August 17, 2017). At this point, we may verify the pattern, or falsify our hypothesis or theory.

8. **Creativity Arrow**. The final process involves creativity, in which we bring to bear creative ideas that may explain a pattern or a phenomenon, which brings us full circle.

C. R. Rao (1997) notes that the inference-through-consequences portion of the diagram comes under the subject of research and the creative role played by scientists, while the design of experiments through inductive reasoning comes under the realm of statistics.

Note that you can go through a portion of the scientific race track or wheel (e.g., collect data in the lower box, analyze the data to infer a generalized pattern, and then stop). However, going around the race track once or multiple times builds **knowledge** about a system. As Dr. Wassertheil-Smoller states, "In science there is a constant interplay between inductive inference (based on observations) and deductive inference (based on theory), until we get closer and closer to the 'truth,' which we can only approach but not ascertain with complete certainty."

> "Our search is only for working hypotheses which are supported by observational facts and which, in course of time, may be replaced by better working hypotheses with more supporting evidence from a wider set of data and provide wider applicability."— C. R. Rao

 ?? What flavor of science will we explore in this chapter?

Answer: Well, we have data in hand, and we can use a statistical analysis to take the **specific** observations in our dataset and infer a pattern. Thus, many types of statistical analysis are firmly rooted in the induction side of the diagram in Figure 17.7.

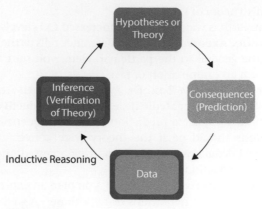

Figure 17.7

Statistical inference is the process of identifying patterns from data. The Oxford Dictionary of Statistics defines statistical inference as "the process of drawing conclusions about the nature of some system on the basis of data subject to random variation."

In this chapter, we will use a statistical approach called **simple linear regression** to generate a general pattern from our specific observations. Our approach will be to use **Bayesian analysis with MCMC**.

 ?? **Great! What will we be analyzing?**

Answer: For this chapter (and the next), we will be analyzing a small, hypothetical dataset that is (sort of) based on research by Angela Duckworth (University of Pennsylvania).

Duckworth and her talented colleagues are keenly interested in answering the following question: **"Why do some individuals accomplish more than others of equal intelligence?"**

They explain: "In addition to cognitive ability, a list of attributes of high-achieving individuals would likely include creativity, vigor, emotional intelligence, charisma, self-confidence, emotional stability, physical attractiveness, and other positive qualities. *A priori*, some traits seem more crucial than others for particular vocations. Extraversion may be fundamental to a career in sales, for instance, but irrelevant to a career in creative writing. However, some traits might be essential to success no matter the domain. We suggest that one personal quality is shared by the most prominent leaders in every field: grit."

 ?? **Grit?**

Answer: The researchers define grit as "perseverance and passion for long-term goals." They continue: "Our hypothesis that grit is essential to high achievement evolved during interviews with professionals in investment banking, painting, journalism, academia, medicine, and law. Asked what quality distinguishes star performers in their respective fields, these individuals cited grit or a close synonym as often as talent. In fact, many were awed by the achievements of peers who did not at first seem as gifted as others but whose sustained commitment to their ambitions was exceptional. Likewise, many noted with surprise that prodigiously gifted peers did not end up in the upper echelons of their field."

They put forth the "grit" hypothesis that "perseverance and passion for long-term goals" is a characteristic that leads some people to accomplish more than others. The grit hypothesis is at the top of our scientific wheel.

Duckworth and colleagues predicted that a person's grit, if it can be measured, is one factor that can explain variability in accomplishment among a group of individuals. To test this hypothesis, they collected data on thousands of individuals, which moves us to the box at the bottom of Figure 17.7.

 ?? Can you really measure a person's grittiness?

Answer: You can! In fact, you can find your own grit score by taking the **grit test**, found at http://angeladuckworth.com/grit-scale/. And you can read Duckworth and colleagues' seminal work here:

A. L. Duckworth, C. Peterson, M. D. Matthews, et al. (2007). Grit: Perseverance and passion for long-term goals. *Journal of Personality and Social Psychology* 92.6, 1087–101.

 ?? Can we have a peek at their dataset?

Answer: Well, no. For this chapter, we have **completely fabricated a tiny dataset** in the spirit of the Duckworth et al. analysis so that we can illustrate a simple regression analysis with Bayesian techniques. As with other chapters, we'll be solving things by hand, hence the need to keep the dataset tiny. Let's have a look at Table 17.2.

Table 17.2

ID	Success	IQ	Years in School	Grit
1	33.48	112	12	2.2
2	42.53	94	14	3.2
3	48.53	118	18	3.4
4	30.21	87	10	1.8
5	38.76	96	13	2.8
6	38.59	106	22	0.2
7	52.93	71	17	4.4
8	32.65	91	15	1.0
9	52.42	95	16	4.6
10	22.22	94	9	0.4
11	41.40	100	19	1.6
12	16.28	98	8	0.0
13	40.83	94	20	1.2
14	24.43	113	11	0.6
15	56.38	85	21	4.2

In this table, the column labeled **ID** identifies an individual. The column labeled **Success** is some generic measure of success. We hate to identify what this measure might be, but it could be something like "Number of days a person makes it in a Survivor competition," where the decimal portion of the score would indicate the exact time in which a person exited the competition.

In case you aren't familiar with Survivor, Wikipedia tells us: "Survivor is a reality competition television franchise produced in many countries throughout the world. The show features a group of contestants who are marooned in an isolated location, where they must provide food, water, fire, and shelter for themselves. The contestants compete in challenges for rewards and immunity from elimination. The contestants are progressively eliminated from the game as they are voted out by their fellow contestants until only one remains and is declared the winner" (article accessed August 20, 2017).

Our goal for this chapter is to analyze the dataset and look for **patterns** in the data. Specifically, we will be looking to see if success can be predicted based on the following variables:

1. **IQ**, which is a meaure of a contestant's Incessant Quibbling.
2. **Years in School**, which is the number of years that a participant attended a formal school.
3. **Grit**, measured on a scale between 0 and 5.

It's reasonable to think that all three factors might influence how long somebody lasts in a game of "Survivor." As an aside, notice that, in Table 17.2, there is no column called "Error."

 ??   **How do I get started with my analysis?**

Answer: Typically, the first step in data analysis is to examine each column of the dataset to get an overview. Here, we will generate a **box plot** for each of our four variables (see Figure 17.8). Box plots are oh-so-very helpful in looking at a column of numeric data. The Oxford Dictionary of Statistics defines a box plot (also known as a box-and-whisker diagram) as: "a graphical representation of numerical data, introduced by [John] Tukey and based on the five-number summary. The diagram has a scale in one direction only. A rectangular box is drawn, extending from the lower quartile to the upper quartile, with the median shown dividing the box. 'Whiskers' are then drawn extending from the end of the box to the greatest and least values."

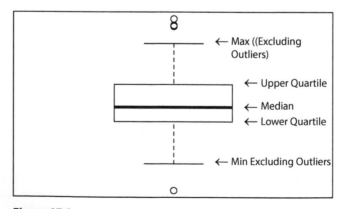

Figure 17.8

Before we interpret this plot, imagine that we took all 15 individuals and ordered them from lowest score to highest score for a particular variable, like IQ. This ordering is critical. Now look at the box plot. The thick line in the middle of the plot is the median value. The median is the datapoint in the middle of the IQ lineup: 50% of the data have values greater than the median, and 50% have values less than the median. Then we see the upper and lower quartiles in our box plot. The word "quartiles" sounds like "quarters." You'd be right in assuming that the upper quartile is the datapoint (in our ordered vector) where 25% of the data have a greater value, and 75% have a lower value. Same goes for the lower quartile, but in reverse. The tails of our box plot give the maximum and minimum values, excluding the outliers. And the outliers themselves are identified as dots. The outliers are identified by first calculating the Inter-Quartile Range, or IQR, which is the difference between the upper and lower quartiles. Observations less than the lower quartile minus 1.5 times IQR are outliers, as are observations greater than the upper quartile plus 1.5 times IQR.

Now let's have a look at the box plots for **each** of the four variables in our dataset (see Figure 17.9).

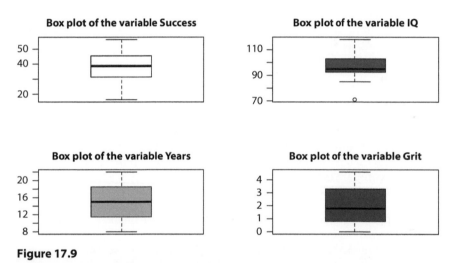

Figure 17.9

Looks like we have one outlier with respect to IQ, but you can (and should) see if the four box plots make sense based on the raw data. Remember, this is a pattern of the **raw data** considered for **each variable** separately.

 ??   OK, so how do I find patterns *between* variables?

Answer: The first step is to generate a scatter plot of the data like we did at the beginning of the chapter. A scatterplot requires that one variable is selected to be graphed on the y-axis and that a second variable is selected to be graphed on the x-axis. Usually, the variable on the y-axis is called the **dependent variable** or **response variable**—it is the variable that you think might respond to changes in the other variables. We are interested in the number of days a person can last in a game of Survivor, so **Success** will be our response variable. The x-axis is occupied by an **independent variable** or a **predictor variable**. This is a variable that can predict the response. For instance, can we predict the success of a person in a game of Survivor based on their IQ? Can we predict their success based on the number of years they attended school? Is grit a predictor?

Let's look at the scatterplots of the **specific observations** in our dataset (see Figure 17.10).

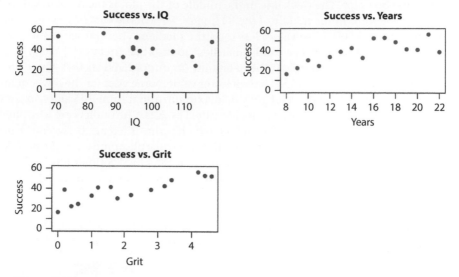

Figure 17.10

Now we are ready to begin our analysis.

For this chapter, we will estimate the linear relationship between success and years in school. In Chapter 18, we will estimate the linear relationship between success and grit score and then compare the two models.

Before we dive in, it might be a good time to stretch or take a break.

Ahhh. That's more like it!

?? **Where were we?**

In Equation 17.10, we mentioned that an individual's success can be expressed as:

$$y_i = b_0 + b_1 x_i + c_i \qquad i = 1, \ldots, n \tag{17.12}$$

where y_i is **Success** in Survivor (number of days) and x_i is **Years** of formal education.

 ?? **Is this a statistical model?**

While we have an equation in 17.12, it is incomplete as a model.

 ?? **Why is it incomplete? What exactly is a statistical model?**

Answer: Wikipedia tells us that "a statistical model is a class of mathematical model, which embodies a set of assumptions concerning the generation of some sample data, and similar data from a larger population. A statistical model represents, often in considerably idealized form, the data-generating process. The assumptions embodied by a statistical model describe a set of probability distributions, some of which are assumed to adequately approximate the distribution from which a particular data set is sampled. The probability distributions inherent in statistical models are what distinguishes statistical models from other, non-statistical, mathematical models" (Article accessed August 20, 2017).

 ?? **What assumptions do we make about how the Survivor data were generated?**

Answer: Fantastic question! You may recall the notation used in previous examples in this book. For example, the random variable X arises from a normal distribution with $\mu = 2.0$ and $\tau = 0.001$ can be written:

$$X \sim N(\mu, \tau) \tag{17.13}$$

$$X \sim N(\mu = 2.0, \tau = 0.001). \tag{17.14}$$

A specific value of X could be designated as x.

We need to similarly describe the process for our Survivor data. In this dataset, we have n random variables: $Y_1, Y_2, \ldots, Y_n$, where Y_1 is a variable that represents the success for the first person in the dataset. We have observed that $y_1 = 33.48$ (Table 17.2). This person had 12 years of formal education, which we can write as $x_1 = 12$. We have observed that $y_2 = 42.53$ (Table 17.2). This person had 14 years of formal education, which we can write as $x_2 = 14$, and so on. Thus, we **pair** the observed y_i with their years of formal education, x_i.

For any given years of formal education (x_i), the random variable Y_i arises from a normal distribution whose mean is $b_0 + b_1 x_i$ and whose precision is τ. You may recall that $\tau = \dfrac{1}{\sigma^2}$; low precision indicates high variance, and vice versa.

We can write this generally as:

$$Y_i \sim N(b_0 + b_1 x_i, \tau). \tag{17.15}$$

This equation indicates that the random variable Y_i is normally distributed.

- The mean of this distribution is the parameter b_0, plus a second parameter, b_1, which is multiplied by the number of years of formal education that contestant i experienced, or x_i.
- The precision of this distribution is τ.

Thus, there are three, count them, three unknown parameters to estimate in this linear model.

To summarize:

- For any given number of years of formal education, x_i, a randomly selected y_i will be plucked from a normal distribution with a mean of $b_0 + b_1 x_i$ and precision τ.
- This means that the mean of the normal distribution for those with, say, 8 years of formal education will be different than the mean of the normal distribution for those with, say, 12 years of formal education, and the level of difference depends on the size of the b_1 parameter and the number of years. For individuals with 8 years of education, the mean is $b_0 + b_1 * 8$, whereas the mean is $b_0 + b_1 * 12$ for those with 12 years of formal education.
- The precision of the normal distribution which gives rise to the data, τ, is the same regardless of the years of formal education.
- The mean, $b_0 + b_1 * x_i$, represents the model's **signal**.
- We also assume that all observations are independent.

 ?? **Is that all there is to it?**

Answer: Not yet. For our Bayesian analysis, the unknown parameters will be considered random variables that arise from some prior distribution. The prior distributions can also be thought of as the weight of belief for each alternative parameter value. It is very useful to diagram the process by which the data were generated (see Figure 17.11).

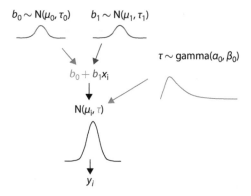

Figure 17.11

This sort of diagram was popularized by John Kruschke in his book, Doing Bayesian Data Analysis (2015). Here, a single observation from our dataset, y_i, is located at the bottom of diagram. This sample is drawn from a normal distribution whose mean is μ_i and whose precision is τ. In turn, μ_i is controlled by the parameters, b_0 and b_1, as well as x_i, the number of years of formal education completed by contestant i. The prior distributions for the three parameters are sketched as well. Here, the prior distribution for b_0 and b_1 will be a normal distribution because these parameters can take on any real value, while the prior distribution for τ will be a gamma distribution because precision cannot be negative.

Such a diagram helps us to fully specify our model. Now we are ready to begin our Bayesian analysis.

 ?? OK, where do we start?

Answer: Let's briefly remind ourselves what Bayes' Theorem is. As you've seen, a major use of Bayesian inference involves estimating parameters. You might recall from previous chapters the generic version of Bayes' Theorem for **one** continuous variable:

$$P(\theta|\text{data}) = \frac{P(\text{data}|\theta) * P(\theta)}{\int P(\text{data}|\theta) * P(\theta)d\theta}. \qquad (17.16)$$

This is the generic version of Bayes' Theorem when the posterior distribution for a single parameter, θ, given the observed data, is represented by a **pdf**.

For our Bayesian inference problem at hand (the Survivor problem), Bayes' Theorem would be:

$$P(b_0, b_1, \tau|\text{data}) = \frac{P(\text{data}|b_0, b_1, \tau) * P(b_0, b_1, \tau)}{\iiint P(\text{data}|b_0, b_1, \tau) * P(b_0, b_1, \tau)db_0, db_1, d\tau}. \qquad (17.17)$$

This is quite the mathematical challenge, but as we'll see, we can tackle the problem by considering each parameter one at a time and make use of the Bayesian conjugate solutions that we encountered in previous chapters. You may recall that the conjugate shortcuts allow us to update the prior pdf to the posterior pdf rather quickly.

Let's quickly review the steps for Bayesian analysis:

1. Identify your hypotheses. These would be the alternative hypotheses for:
   - b_0, the model intercept
   - b_1, the model slope
   - τ, the model precision
2. Express your belief that each hypothesis is true in terms of prior probabilities.
3. Gather the data.
4. Determine the likelihood of the observed data, assuming each hypothesis is true.
5. Use Bayes' Theorem to compute the posterior probabilities for each parameter of interest.

We've identified our hypotheses, so now we are ready for step 2.

 ?? And what is step 2?

Answer: Set prior distributions for each of the parameters of interest: b_0, b_1, τ. Notice that we will color-code them from this point forward so that it will be easier to trace through our calculations. Also note that τ does not have a subscript.

Let's work on the two regression coefficients, b_0 and b_1, first.

We need to express the prior distribution for the y-intercept b_0. Let's use a **normal prior distribution** that has a mean, μ_0, of 0 and a standard deviation, σ_0, of 100. This means that the variance $\sigma_0^2 = 100 * 100 = 10{,}000$. And, as we've just learned, we can also express the spread in terms of τ_0, which is $\frac{1}{\sigma_0^2}$. If $\sigma_0^2 = 10{,}000$, then $\tau_0 = \frac{1}{10{,}000} = 0.0001$. This is a vague (non-informative) prior distribution and suggests that we have quite a bit of uncertainty about the value of b_0. Recall that the parameters for this prior distribution are called hyperparameters. One more thing very important thing to note: the hyperparameters here have subscripts of 0 to remind us that we are talking about the parameter, b_0. These subscripts are critical—don't lose track of them or you may confuse τ_0 (a hyperparameter for b_0) with τ (one of the unknown parameters of interest).

So, for the parameter b_0, our prior distribution is a normal distribution with **hyperparameters** (see Figure 17.12):

- $\mu_0 = 0$
- $\sigma_0 = 100$; $\sigma_0^2 = 10000$; $\tau_0 = 0.0001$

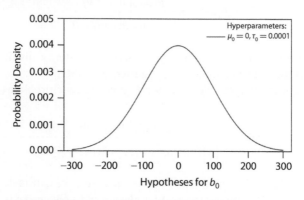

Figure 17.12 Normal prior distribution for b_0.

Next, we need a prior distribution for b_1, the slope of our regression equation. Without any previous knowledge of the effect of **Years** of formal education on **Success**, let's go with a normal distribution with a mean of 0 and a high standard deviation of 100. This means that the variance $\sigma_1^2 = 100 * 100 = 10{,}000$. It also means that τ_1, which is $\frac{1}{\sigma_1^2}$, is equal to 0.0001. Notice that the subscripts for these hyperparameters are 1, to remind us that we are talking about the parameter b_1 (see Figure 17.13):

- $\mu_1 = 0$.
- $\sigma_1 = 100$; $\sigma_1^2 = 10000$; $\tau_1 = 0.0001$.

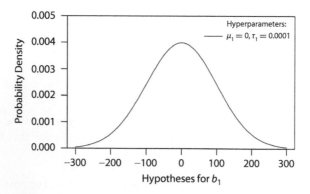

Figure 17.13 Normal prior distribution for b_1.

Finally, we need to set a prior distribution for the precision of our error terms, τ. We know that σ^2, and therefore τ, must be positive, so you may recall that a gamma distribution might be an appropriate prior distribution for τ. We will employ it here.

Remember that the gamma distribution is controlled by two parameters: α (also called the "shape" parameter) and β (also called the "rate" parameter). Let's let $\alpha_0 = 0.01$ and $\beta_0 = 0.01$, which results in a gamma distribution that is heavily weighted toward tiny numbers (i.e., low precision; see Figure 17.14).

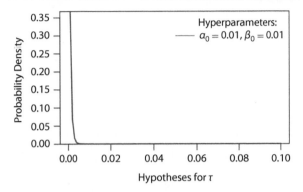

Figure 17.14 Gamma prior distribution for τ.

So, here we have our three prior distributions for our three unknown parameters, b_0, b_1, and τ, each of which is defined by their corresponding **hyperparameters** (see Figure 17.15).

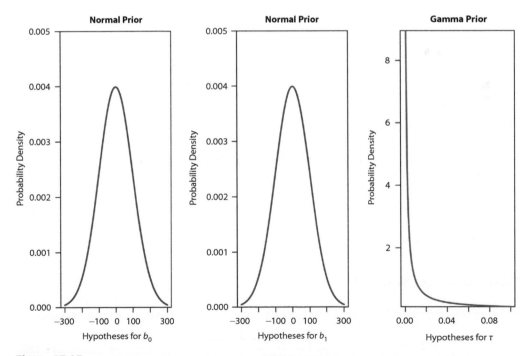

Figure 17.15

?? All of these prior distributions are vague. Couldn't we have used existing Survivor data to help set our priors?

Answer: We could have! As the analyst, you must set prior distributions for each parameter and be able to defend your choices! For instance, you could have used OLD Survivor data to generate the prior, excluding any data that will be used to update the prior.

Now that we have set our prior distributions, we are ready for step 3.

?? What is step 3?

Answer: Collect data! We already have our dataset in hand, and we are particularly interested in just two of the columns: **Success** and **Years** of formal education (see Table 17.3).

Table 17.3

ID	Success	IQ	Years in School	Grit
1	33.48	112	12	2.2
2	42.53	94	14	3.2
3	48.53	118	18	3.4
4	30.21	87	10	1.8
5	38.76	96	13	2.8
6	38.59	106	22	0.2
7	52.93	71	17	4.4
8	32.65	91	15	1.0
9	52.42	95	16	4.6
10	22.22	94	9	0.4
11	41.40	100	19	1.6
12	16.28	98	8	0.0
13	40.83	94	20	1.2
14	24.43	113	11	0.6
15	56.38	85	21	4.2

?? On to steps 4 and 5?

Answer: You got it! We now have an infinite number of hypotheses for b_0, an infinite number of hypotheses for b_1, and an infinite number of hypotheses for τ. For each and every combination of parameters, we need to determine the likelihood of the observed data, assuming each hypothesis is true, and then use Bayes' Theorem to compute the posterior densities for each combination. This will allow us to build our posterior distribution, which is a joint distribution.

?? How do we get started?

We mentioned earlier that, with three parameters to estimate, Bayes' Theorem becomes:

$$P(b_0, b_1, \tau | \text{data}) = \frac{P(\text{data}|b_0, b_1, \tau) * P(b_0, b_1, \tau)}{\iiint P(\text{data}|b_0, b_1, \tau) * P(b_0, b_1, \tau) db_0, db_1, d\tau}. \qquad (17.18)$$

The solution here is fairly intractable. Luckily, with multiple parameters to estimate, MCMC with Gibbs sampling can be used to tackle the problem. Here, we assume that all three parameters are independent of each other.

 ?? What is Gibbs sampling again?

Answer: A key feature is that Gibbs sampling allows one to evaluate multiple parameters (e.g., b_0, b_1, and τ). In this method, in each MCMC trial you focus on each parameter one at a time, **conditional** on the values of the remaining parameters for that trial. For a target parameter in a given trial, Bayes' Theorem is used to compute the parameters of the posterior distribution, and then draw a proposal from this new distribution. Then, still within the same trial, you cycle to the next parameter, which now is the target parameter. When this process is repeated across a large number of MCMC trials, the proposals are used to **estimate** the joint posterior distribution, as well as the marginal posterior distribution for each parameter. We introduced Gibbs sampling in Chapter 16—if this doesn't ring a bell, you really (truly) should review that material before proceeding!

In each trial, Gibbs sampling can make use of the Bayesian conjugate solutions to compute a target parameter's posterior distribution, conditional on the values of the remaining parameters for that trial.

In Chapter 16, we showed that conjugate solutions provide a rapid way of updating the prior distribution to the posterior distribution with some mathematical shortcuts. Be assured that Bayes' Theorem is at the heart of the calculations; however, the shortcuts allow one to move rapidly from the prior to the posterior with a few mathematical calculations that avoid the messy integral in the denominator of Bayes' Theorem.

We used these conjugates in our solving the Maple Syrup Problem with Gibbs sampling. Remember? In Figure 17.16, the data are denoted as x_i:

- In the left red panel, we are dealing with a normal distribution where the precision, τ, is **known**. We are trying to estimate the unknown mean, or μ. The prior distribution for the unknown parameter, μ, is a normal distribution with hyperparameters μ_0 and τ_0. The bottom box provides the calculations needed to find the hyperparameters for the posterior distribution (i.e., the conjugate shortcuts).
- In the right red panel, we are dealing with a normal distribution where the mean, μ, is **known**. We are trying to estimate the unknown parameter, τ, or precision. The prior distribution for the unknown parameter, τ, is a gamma distribution with hyperparameters α_0 and β_0. The bottom box provides the shortcut calculations needed to find the hyperparameters for the posterior distribution.

These are conjugate solutions for variables that are distributed as $N(\mu, \tau)$; we will need to modify them to fit within a linear regression framework. For the grit problem, our variables are distributed as $N(b_0 + b_1 x, \tau)$. Thus there are three unknown parameters. For each of these parameters, we specify a prior distribution. For each MCMC trial, we will update the prior pdf to a posterior pdf using a Bayesian conjugate solution.

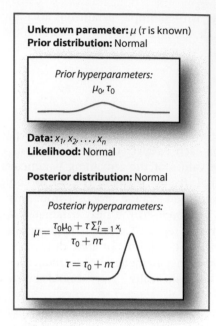

Unknown parameter: μ (τ is known)
Prior distribution: Normal

Prior hyperparameters:
$$\mu_0, \tau_0$$

Data: $x_1, x_2, \ldots, x_n$
Likelihood: Normal

Posterior distribution: Normal

Posterior hyperparameters:
$$\mu = \frac{\tau_0 \mu_0 + \tau \Sigma_{i=1}^{n} x_i}{\tau_0 + n\tau}$$
$$\tau = \tau_0 + n\tau$$

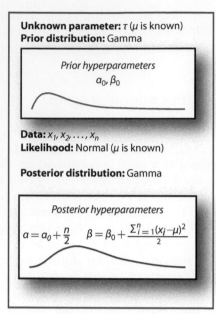

Unknown parameter: τ (μ is known)
Prior distribution: Gamma

Prior hyperparameters
$$\alpha_0, \beta_0$$

Data: $x_1, x_2, \ldots, x_n$
Likelihood: Normal (μ is known)

Posterior distribution: Gamma

Posterior hyperparameters
$$a = \alpha_0 + \frac{n}{2} \qquad \beta = \beta_0 + \frac{\Sigma_{i=1}^{n}(x_i - \mu)^2}{2}$$

Figure 17.16

 ??   ## How do we make this modification?

Answer: Before we get started, it will be very helpful to study a roadmap so we have the big picture in mind (see Figure 17.17).

Let's study this roadmap carefully. Here, you see three boxes, one for each of our unknown parameters (b_0, b_1, and τ). Notice the arrows that link the three boxes. For **each** of these parameters, we specify a prior distribution, shown in black for all three cases. For **each** MCMC trial, we will update the prior pdf to a posterior pdf using a Bayesian conjugate solution. The posterior distributions for a hypothetical MCMC trial are shown in color (blue for b_0, red for b_1, and green for τ).

The general idea is that we estimate the posterior distribution for τ in trial [t] in the green box, given $b_{0[t-1]}$ and $b_{1[t-1]}$ from a previous trial, and draw a proposal from this posterior Figure 17.7(a). This proposal, $\tau_{[t]}$, is then used in the next box (blue), which focuses on b_0 Figure 17.17(b). We estimate the posterior distribution for $b_{0[t]}$ in trial [t] in the blue box, given $\tau_{[t]}$ and $b_{1[t-1]}$, and draw a proposal from this posterior. This proposal, $b_{0[t]}$, is then used in the next box (red), which focuses on b_1 Figure 17.17(c). We then estimate the posterior distribution for $b_{1[t]}$ in trial [t] in the red box, given $\tau_{[t]}$ and $b_{0[t]}$, and draw a proposal from this posterior, which brings us back to the green box. We continue looping around the race track, where each loop is one MCMC trial.

Let's now dig deeper, starting with the green box Figure 17.17(a).

- **The green box focuses on the unknown parameter** τ. We already indicated that our prior distribution is a gamma distribution with hyperparameters $\alpha_0 = 0.01$ and

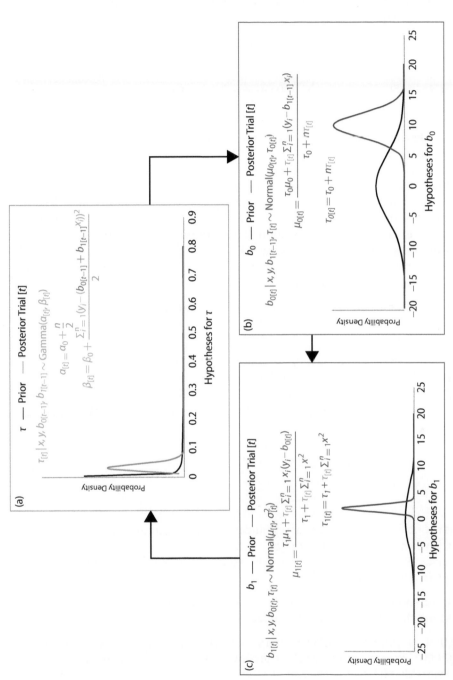

Figure 17.17

$\beta_0 = 0.01$. We will use these values in each and every trial. Now, assuming that we have b_0 and b_1 from a **previous** MCMC trial, designated as $[t - 1]$, the posterior distribution for τ for the MCMC trial $[t]$ is another gamma distribution with hyperparameters $\alpha_{[t]}$ and $\beta_{[t]}$. **Here are the updates:**

$$\alpha_{[t]} = \alpha_0 + \frac{n}{2} \tag{17.19}$$

$$\beta_{[t]} = \beta_0 + \frac{\sum_{i=1}^{n} (y_i - (b_{0[t-1]} + b_{1[t-1]} x_i))^2}{2}. \tag{17.20}$$

- Note that $\alpha_{[t]}$ depends only on the hyperparameter α_0 and sample size, n. Thus, it will be constant across all MCMC trials.
- Here, y_i is the observed number of days candidate i lasts in a game of Survivor ($i = 1$ to n). And x_i is the years of formal education for candidate i.
- Take a good, long look at the numerator for the $\beta_{[t]}$ calculation in Equation 17.20. The numerator here is formally called the Sum of Squared Errors, or SSE for short. The error for each datapoint is calculated as in Equation 17.11 and then squared. The sum of these squared error terms across the dataset is the SSE.
- Once we have calculated $\alpha_{[t]}$ and $\beta_{[t]}$, we now draw a proposal from this posterior distribution and **accept it**. This is now $\tau_{[t]}$.

Next we move to the blue box, which focuses on the unknown parameter b_0, conditional on the values $\tau_{[t]}$ and $b_{1[t-1]}$; see Figure 17.17(b).

- We indicated that our prior distribution is a normal distribution with hyperparameters $\mu_0 = 0$ and $\tau_0 = 0.0001$. Pay attention to those subscripts! We will be using these values in each and every MCMC step. Now, assuming we have our new proposal for $\tau_{[t]}$ from this trial, and $b_{1[t-1]}$ from a **previous** MCMC trial, the posterior distribution for b_0 is another normal distribution defined by parameters $\mu_{0[t]}$ and $\tau_{0[t]}$. **Here are the updates:**

$$\mu_{0[t]} = \frac{\tau_0 \mu_0 + \tau_{[t]} \sum (y_i - b_{1[t-1]} x_i)}{\tau_0 + n \tau_{[t]}}. \tag{17.21}$$

$$\tau_{0[t]} = \tau_0 + n \tau_{[t]}. \tag{17.22}$$

- Again, y_i is the observed number of days candidate i lasts in a game of Survivor ($i = 1$ to n). And x_i is the years of formal education for candidate i.
- Once we have calculated $\mu_{0[t]}$ and $\tau_{0[t]}$, we draw a proposal from this posterior distribution and **accept it**. This is now $b_{0[t]}$.

Now we move to the red box, which focuses on the unknown parameter b_1, conditional on the values $\tau_{[t]}$ and $b_{0[t]}$ (Figure 17.17 (c)).

- We know that our prior distribution is a normal distribution with hyperparameters $\mu_1 = 0$ and $\tau_1 = 0.0001$. Remember those subscripts, and note that these values will be used in each and every MCMC step! Now, assuming we have our proposal for $\tau_{[t]}$ from this trial, and $b_{0[t]}$ from this trial, the posterior distribution for b_1 is another normal distribution defined by the parameters $\mu_{1[t]}$ and $\tau_{1[t]}$. **Here are the updates:**

$$\mu_{1[t]} = \frac{\tau_1 \mu_1 + \tau_{[t]} \sum x_i(y_i - b_{0[t]})}{\tau_1 + \tau_{[t]} \sum x_i^2} \tag{17.23}$$

$$\tau_{1[t]} = \tau_1 + \tau_{[t]} \sum x_i^2. \tag{17.24}$$

- Once again, y_i is the observed number of days for candidate i lasts in a game of Survivor ($i = 1$ to n). And x_i is the years of formal education for candidate i.
- Once we have calculated $\mu_{1[t]}$ and $\tau_{1[t]}$, we now draw a proposal from this posterior distribution and **accept it**. This is now $b_{1[t]}$.

We now move back to the green box and then move to the next MCMC iteration. This process is repeated across several (thousands) of MCMC trials. In each case, we **store** our proposals for the unknown parameters, $b_{0[t]}$, $b_{1[t]}$, and $\tau_{[t]}$. For convenience, we will present this same roadmap multiple times in this chapter.

The histogram of the proposals for each parameter individually across the MCMC trials will provide the posterior marginal distributions. In addition, the joint distribution of the parameters can be estimated by summarizing the joint combinations of all three parameters.

We encourage you to study the conjugate solution updates in Appendix 4, which are firmly rooted in Bayes' Theorem. It is so easy to forget this fact! The main point is that, in each MCMC trial, you update a target parameter's prior distribution to the posterior distribution via the conjugate solutions and then draw a random proposal from the posterior.

 ?? Brilliant! Can we get started now?

Answer: Hold your horses! Now we are ready to work through our Bayesian regression analysis with MCMC using Gibbs sampling. But, to give you a general idea of where we are headed, if our MCMC consisted of only 10 trials, we would fill in Table 17.4 with the accepted proposals and then use the results to obtain the posterior distribution.

Table 17.4

Trial	b_0	b_1	τ
1	-	-	-
2	-	-	-
3	-	-	-
4	-	-	-
5	-	-	-
6	-	-	-
7	-	-	-
8	-	-	-
9	-	-	-
10	-	-	-

?? **Now can we see the Gibbs sampler in action?**

Hold onto your hat while we work through the steps!

Let's first set up a table to hold not only our parameters of interest but also some "helper" columns that we will use to store updated hyperparameters (see Table 17.5). We'll stick with just 10 trials so we can focus on the mechanics.

Table 17.5

Info		b_0 Proposal Distribution			b_1 Proposal Distribution				τ Proposal Distribution			
		Hyperparameters			Hyperparameters					Hyperparameters		
Trial [t]	n	$\mu_{0[t]}$	$\tau_{0[t]}$	$b_{0[t]}$	$\mu_{1[t]}$	$\tau_{1[t]}$	$b_{1[t]}$	$SSE_{[t]}$	$a_{[t]}$	$\beta_{[t]}$	$\tau_{[t]}$
0	15	0.000	0.0001		0.000	0.0001			0.010	0.010	

Let us walk you through this table, as you will soon learn to love it. There are five main sections:

1. In the column group called **Info**, we list the MCMC trial [t] and the sample size, n (which is 15 Survivor datapoints).
2. In the column group called **b_0 Proposal Distribution** (columns 3–5), we store information related to the unknown parameter b_0 (the model intercept). **Trial 0 holds the hyperparameters for the b_0 prior distribution**, which, if you recall, is a normal distribution with mean $\mu_0 = 0$ and precision $\tau_0 = 0.0001$. In each trial (1 through 10) (not shown yet), we will update the prior distribution to the posterior distribution for b_0. Once we have this, we will draw a proposal from this posterior distribution and will store it in the blue-shaded column.
3. In the column group called **b_1 Proposal Distribution** (columns 6–8), we store information related to the unknown parameter b_1 (the model slope). **Trial 0 holds the hyperparameters for the b_1 prior distribution**, which, if you recall, is a normal distribution with mean $\mu1 = 0$ and precision $\tau1 = 0.0001$. In each trial (1 through 10) (not shown yet), we will update the prior distribution to the posterior distribution for b_1.

Once we have this, we will draw a proposal from this posterior distribution and will store it in the red-shaded column.

4. The final column group is labeled τ **Proposal Distribution.** Here, we store information related to the unknown parameter τ (the model's precision). **Trial 0 holds the hyperparameters for the τ prior distribution**, which, if you recall, is a gamma distribution with $\alpha_0 = 0.01$ and $\beta_0 = 0.01$. In each trial (1 through 10) (not shown yet), we will update the prior distribution to the posterior distribution for τ. Part of the solution involves computing the Sums of Squares Error term, which we will store for reference. Once we have the posterior distribution, we will draw a proposal from it and will store it in the green-shaded column.

Take some time to orient yourself to this table. Make sure you understand where the prior distribution's hyperparameters are stored for each of our three unknown parameters! The hyperparameters for each of our unknown parameters are identified with trial 0. This is not conventional... we are just storing it as trial 0 to help reference the numbers in our narrative. Also make sure you grasp the idea that the proposals drawn from a posterior distribution for any given trial will be stored in one of the color-shaded cells.

To begin with, we need to "prime" the MCMC pump by entering the first values for each parameter in the chain.

These are:

- $b_0 = 6.000$
- $b_1 = 0.3000$
- $\tau = \text{NA}$

These reasonable values for b_0 and b_1 are our MCMC starting values. Given these values, you'll soon see that an initial value for τ can be determined. We can now add them to the table (see Table 17.6).

Table 17.6

Info	b_0 Proposal Distribution			b_1 Proposal Distribution			τ Proposal Distribution			
	Hyperparameters			Hyperparameters			Hyperparameters			
Trial $[t]$ n	$\mu_{0[t]}$	$\tau_{0[t]}$	$b_{0[t]}$	$\mu_{1[t]}$	$\tau_{1[t]}$	$b_{1[t]}$	$SSE_{[t]}$	$\alpha_{[t]}$	$\beta_{[t]}$	$\tau_{[t]}$
0 15	0.000	0.0001	6.000	0.000	0.0001	0.3000		0.010	0.010	NA

Now, **given the parameter coefficients of our trial 0 model**, we can calculate the SSE **for trial 0**, which is the sum of the squared error terms given the model. First, let's compute the predicted y_i for each observation, given trial 0 parameters:

$$\text{predicted } y_{i[0]} = b_{0[0]} + b_{1[0]} * x_i \tag{17.25}$$

$$\text{predicted } y_{i[0]} = 6.000 + 0.3000 * \text{years}_i \tag{17.26}$$

Next, let's calculate the error for each observation at trial 0:

$$\text{error}_{i[0]} = y_i - (b_{0[0]} + b_{1[0]} * \text{years}_i) \qquad (17.27)$$

Given this model, the predicted values, error values, and squared error values for trial 0 are shown in Table 17.7.

Table 17.7

ID	Success	Years	Predicted$_{[0]}$	Error$_{[0]}$	Error$^2{}_{[0]}$
1	33.48	12	9.600	23.880	570.254
2	42.53	14	10.200	32.330	1045.229
3	48.53	18	11.400	37.130	1378.637
4	30.21	10	9.000	21.210	449.864
5	38.76	13	9.900	28.860	832.900
6	38.59	22	12.600	25.990	675.480
7	52.93	17	11.100	41.830	1749.749
8	32.65	15	10.500	22.150	490.622
9	52.42	16	10.800	41.620	1732.224
10	22.22	9	8.700	13.520	182.790
11	41.4	19	11.700	29.700	882.090
12	16.28	8	8.400	7.880	62.094
13	40.83	20	12.000	28.830	831.169
14	24.43	11	9.300	15.130	228.917
15	56.38	21	12.300	44.080	1943.046
					13055.067

The sum of the final column is 13055.067 for trial 0, and we can enter that result into our table under the column SSE$_{[t]}$, as in Table 17.8.

Table 17.8

Info	b_0 Proposal Distribution			b_1 Proposal Distribution			τ Proposal Distribution			
	Hyperparameters			Hyperparameters				Hyperparameters		
Trial [t] n	$\mu_{0[t]}$	$\tau_{0[t]}$	$b_{0[t]}$	$\mu_{1[t]}$	$\tau_{1[t]}$	$b_{1[t]}$	SSE$_{[t]}$	$\alpha_{[t]}$	$\beta_{[t]}$	$\tau_{[t]}$
0 15	0.000	0.0001	6.000	0.000	0.0001	0.300	13055.067	0.010	0.010	NA

Do you still have your hat? Now we are ready for trial 1, which consists of several calculations. Here, $[t] = 1$ and $[t - 1] = 0$. We'll begin by focusing on the unknown model precision parameter, τ, for trial 1 (see Figure 17.18(a)).

Our prior distribution for τ is a gamma distribution with hyperparameters $\alpha_0 = 0.01$ and $\beta_0 = 0.01$. Now, assuming that we have b_0 and b_1 from a previous MCMC trial, designated with the subscript $[t-1]$ in square brackets, the posterior distribution for τ for MCMC trial $[t]$ is another gamma distribution with hyperparameters $\alpha_{[t]}$ and $\beta_{[t]}$:

$$\alpha_{[t]} = \alpha_0 + \frac{n}{2} = 0.01 + \frac{15}{2} = 7.510 \qquad (17.28)$$

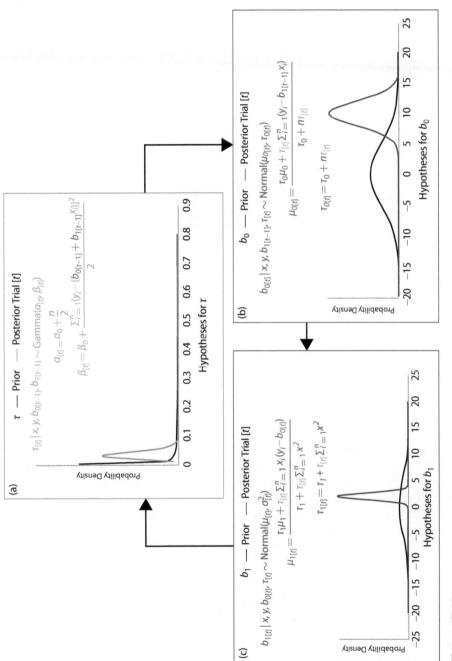

Figure 17.18

$$\beta_{[t]} = \beta_0 + \frac{\sum(y_i - b_{0[t-1]} - b_{1[t-1]}x_i)^2}{2} = \beta_0 + \frac{SSE_{[t-1]}}{2} = 0.010 + \frac{13055.067}{2} = 6527.543 \tag{17.29}$$

This is trial 1's posterior gamma pdf for the unknown parameter, τ, which is fully conditional on the parameters $b_{0[t-1]}$ and $b_{1[t-1]}$. Next, we will draw a random variable from this pdf. Let's assume its value is 0.001. **We will accept this value and store it**. Our MCMC table now looks like the one shown in Table 17.9.

Table 17.9

Info	b_0 Proposal Distribution			b_1 Proposal Distribution			τ Proposal Distribution			
	Hyperparameters			Hyperparameters			Hyperparameters			
Trial [t] n	$\mu_{0[t]}$	$\tau_{0[t]}$	$b_{0[t]}$	$\mu_{1[t]}$	$\tau_{1[t]}$	$b_{1[t]}$	$SSE_{[t]}$	$a_{[t]}$	$\beta_{[t]}$	$\tau_{[t]}$
0 15	0.000	0.0001	6.000	0.000	0.0001	0.300	13055.067	0.010	0.010	NA
1 15								7.51	6527.543	0.001

Now we are ready to focus on b_0 for trial 1 (see Figure 17.18(b)). Our prior distribution for b_0 is a normal distribution with hyperparameters $\mu_0 = 0$ and $\tau_0 = 0.0001$. Now, assuming that we have b_1 from the previous trial and τ from the current trial, the posterior distribution for b_0 for MCMC trial [t] is another normal distribution defined by parameters $\mu_{0[t]}$ and $\tau_{0[t]}$:

$$\mu_{0[t]} = \frac{\tau_0\mu_0 + \tau_{[t]}\sum(y_i - b_{1[t-1]}x_i)}{\tau_0 + n\tau_{[t]}} = \frac{0.0001 * 0 + 0.001\sum(y_i - 0.3000 * x_i)}{0.0001 + 15 * 0.001} = 33.387 \tag{17.30}$$

$$\tau_{0[t]} = \tau_0 + n\tau_{[t]} = 0.0001 + 15 * 0.001 = 0.0151 \tag{17.31}$$

This is our posterior normal pdf for the unknown parameter, b_0, for trial 1, conditional on $\tau_{[t]}$ and $b_{1[t-1]}$. Next, we will draw a random value from this pdf: a normal distribution with a mean of 33.387 and precision of 0.0151. Let's assume it is 31.888. **We will accept this value and store it**. Our MCMC table now looks like the one in Table 17.10.

Now we can move to b_1 for trial 1 (see Figure 17.18(c)). Our prior distribution for b_1 is a normal distribution with hyperparameters $\mu_1 = 0$ and $\tau_1 = 0.0001$. Now, assuming that we have b_0 from this trial as well as τ from this trial, the posterior distribution for $b_{1[t]}$ for MCMC trial [t] is another normal distribution defined by parameters $\mu_{1[t]}$ and $\tau_{1[t]}$:

$$\mu_{1[t]} = \frac{\tau_1\mu_1 + \tau_{[t]}\sum x_i(y_i - b_{0[t]})}{\tau_1 + \tau_{[t]}\sum x_i^2} = \frac{0.0001 * 0 + 0.001\sum x_i(y_i - 33.388)}{0.0001 + 0.001\sum x_i^2} = 0.538 \tag{17.32}$$

$$\tau_{1[t]} = \tau_1 + \tau_{[t]}\sum x_i^2 = 0.0001 + 0.001\sum x_i^2 = 3.6500 \tag{17.33}$$

This is our posterior normal pdf for the unknown parameter, b_1 for trial 1. It is fully conditional on the parameter values for b_0 and τ for this trial. Next, we will draw a random value from this pdf: a normal distribution with a mean of 0.538 and precision of 3.6500.

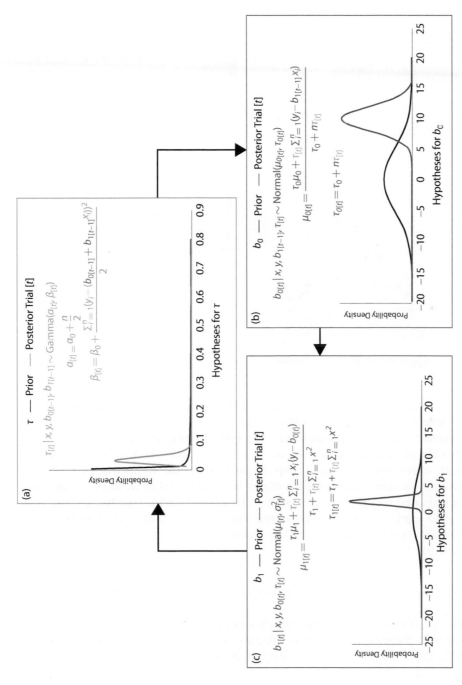

Figure 17.19

Table 17.10

Info	b_0 Proposal Distribution			b_1 Proposal Distribution			τ Proposal Distribution			
	Hyperparameters			Hyperparameters			Hyperparameters			
Trial [t] n	$\mu_{0[t]}$	$\tau_{0[t]}$	$b_{0[t]}$	$\mu_{1[t]}$	$\tau_{1[t]}$	$b_{1[t]}$	$SSE_{[t]}$	$\alpha_{[t]}$	$\beta_{[t]}$	$\tau_{[t]}$
0 15	0.000	0.0001	6.000	0.000	0.0001	0.300	13055.067	0.010	0.010	NA
1 15	33.387	0.0151	31.888					7.510	6527.543	0.001

Let's assume it is –0.18. **We will accept this value and store it**. Our MCMC table now looks like the one in Table 17.11.

Believe it or not, we've come full circle. And I still have my hat!

  Now that we have an updated b_0 and b_1, we can compute the SSE for trial 1. Then, we move to trial 2, use the SSE from trial 1 to calculate the posterior gamma distribution hyperparameters for trial 2, and then draw a random value from this distribution to give us $\tau_{[2]}$ Figure 17.19(a). Once we know $\tau_{[2]}$, we can use it, along with $b_{1[1]}$, to compute new posterior hyperparameters for the normal distribution associated with b_0 in trial 2 and then draw a random value from this distribution to give us $b_{0[2]}$ Figure 17.19(b). Finally, we can use $\tau_{[2]}$ and $b_{0[2]}$ to update the posterior hyperparameters associated with b_1 in trial 2 and then draw a random value from this distribution to give us $b_{1[2]}$ Figure 17.19(c)). And so on.

 ?? Goodness! That is a lot of subscripting to track and calculating to do!

Answer: Yes, it is! Fortunately, given today's computing power, you can program the MCMC if you are a programmer. But even if you are not, there are many software packages that will do the calculations for you provided that you supply:

1. A dataset
2. A description of your model.
3. The prior distributions for all model unknowns.
4. The likelihood function.
5. The know-how to tell the software what to do.

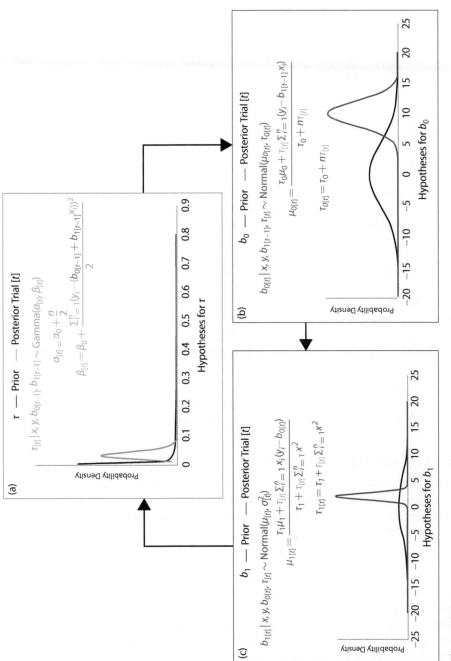

Figure 17.20

Table 17.11

Info		b_0 Proposal Distribution			b_1 Proposal Distribution			τ Proposal Distribution			
		Hyperparameters			Hyperparameters			Hyperparameters			
Trial [t]	n	$\mu_{0[t]}$	$\tau_{0[t]}$	$b_{0[t]}$	$\mu_{1[t]}$	$\tau_{1[t]}$	$b_{1[t]}$	$SSE_{[t]}$	$\alpha_{[t]}$	$\beta_{[t]}$	$\tau_{[t]}$
0	15	0.000	0.0001	6.000	0.000	0.0001	0.300	13055.067	0.010	0.010	NA
1	15	33.387	0.0151	31.888	0.538	3.6500	−0.180	3341.533	7.510	6527.543	0.001

 ?? And where does Bayes' Theorem fit in?

Answer: You know the answer to this, right? We had three chapters featuring Bayesian conjugates, and we've made use of them here (Figure 17.20). Remember, a conjugate distribution is a distribution that can be analytically updated in the presence of new information without using Bayes' Theorem directly. In other words, they may not look like Bayes' Theorem, but they are analytically equivalent.

 ?? Are the Geman brothers responsible for this approach?

Answer: The Geman brothers introduced the world to Gibbs sampling and used it for image restoration. But using Gibbs sampling with MCMC for **parameter estimation** has launched a modern Bayesian revolution, allowing even us commoners to estimate the posterior probability densities for unknown parameters in a statistical model. A seminal paper was written by Alan Gelfand and Adrian Smith.

> A. E. Gelfandand A. F. M Smith. "Sampling-based approaches to calculating marginal densities." *Journal of the American Statistical Association* 85.410 (1990): 398–409.

This article was followed by a friendly version:

> A. F. M. Smith and A. E. Gelfand. "Bayesian statistics without tears: A sampling–resampling perspective." *The American Statistician* 46.2: 84–8. Stable URL: http://www.jstor.org/stable/2684170

Their first abstract reads: "Stochastic substitution, the Gibbs sampler, and the sampling–importance–resampling algorithm can be viewed as three alternative sampling- (or Monte Carlo-) based approaches to the calculation of numerical estimates of marginal probability distributions. The three approaches will be reviewed, compared, and contrasted in relation to various joint probability structures frequently encountered in applications. In particular, the relevance of the approaches to calculating Bayesian posterior densities for a variety of structured models will be discussed and illustrated."

Their second abstract reads: "Even to the initiated, statistical calculations based on Bayes' Theorem can be daunting because of the numerical integrations required in all but the simplest applications. Moreover, from a teaching perspective, introductions to Bayesian statistics—if they are given at all—are circumscribed by these apparent calculational difficulties. Here we offer a straightforward sampling-resampling perspective on Bayesian inference, which has both pedagogic appeal and suggests easily implemented calculation strategies."

 ?? Why the emphasis on marginal densities?

Answer: Well, think about it. Each MCMC trial provides us with one sample of our unknown parameters, b_0, b_1, and τ from their posterior distribution. For any given parameter in any given trial, we **fix** or **condition** on the other two unknown parameters. Thus, for a given trial, we draw a random sample for the given parameter **conditioned** on the current values of the other two parameters. However, **across** MCMC trials, we explore the full sample space for our parameter of interest. Thus, in the end, our MCMC chains provide us with the marginal distribution of each parameter.

 ?? What do the results look like for all 10 trials?

Answer: Here they are in Table 17.12 (recall that trial 0 stores our prior distributions for each parameter).

Table 17.12

Info	b_0 Proposal Distribution			b_1 Proposal Distribution			τ Proposal Distribution			
	Hyperparameters			Hyperparameters				Hyperparameters		
Trial [t] n	$\mu_{0[t]}$	$\tau_{0[t]}$	$b_{0[t]}$	$\mu_{1[t]}$	$\tau_{1[t]}$	$b_{1[t]}$	$SSE_{[t]}$	$\alpha_{[t]}$	$\beta_{[t]}$	$\tau_{[t]}$
0	15 0.000	0.0001	6.000	0.000	0.0001	0.300	13055.067	0.010	0.010	NA
1	15 33.387	0.0151	31.888	0.538	3.6500	−0.180	3341.533	7.510	6527.543	0.001
2	15 40.719	0.0450	39.885	0.045	10.9890	0.096	1984.146	7.510	1670.776	0.003
3	15 36.639	0.1200	36.121	0.277	29.4120	0.161	1762.932	7.510	992.083	0.008
4	15 35.673	0.1650	36.395	0.260	40.0000	0.201	1744.393	7.510	881.476	0.011
5	15 35.071	0.1500	33.245	0.454	37.0370	0.393	1549.736	7.510	872.207	0.010
6	15 32.171	0.0750	31.459	0.564	18.1820	0.781	1607.288	7.510	774.878	0.005
7	15 26.377	0.1500	24.838	0.972	37.0370	0.613	1596.203	7.510	803.654	0.010
8	15 28.887	0.1050	28.780	0.729	25.6410	0.896	1399.926	7.510	798.111	0.007
9	15 24.660	0.2550	24.527	0.991	62.5000	1.018	1116.588	7.510	699.973	0.017
10	15 22.827	0.1800	18.459	1.364	43.4780	1.352	923.707	7.510	558.304	0.012

Remember that we are interested in the results associated with the colored columns. The results are MCMC draws from a posterior distribution for a given parameter, **conditional** on the values of the other parameters.

 ?? How do we know if our MCMC results are any good?

Answer: As with previous MCMC chapters, you MUST carefully consider the resulting posterior distributions and run some diagnostic tests. Setting the number of MCMC trials (usually several thousand) and discarding the initial trials (say, the first 1000 trials or more) as a burn-in are standard procedure. After that, one of the first tests usually involves analysis of the traceplots.

Let's look at our results for an MCMC analysis of 20,000 trials, with a 5000 trial burn-in and pruning every other result. To begin, let's look at the traceplots for the first 100 trials after the burn-in (see Figure 17.21).

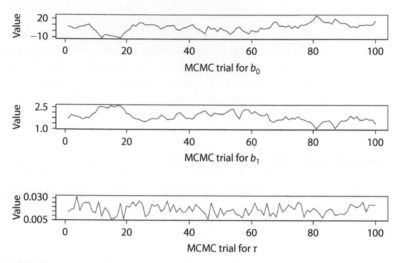

Figure 17.21

The traceplots show our 100 MCMC trials after our burn-in and pruning. There are multiple diagnostics that you can use to help determine if your MCMC is on target. Here, we are looking to see that the proposals fully explore the posterior distribution space.

We ran quite a few trials for this analysis. We needed to run this many to converge on reasonable estimates of the posterior distributions. Software programs that do MCMC often have a trick or two up their sleeves to get solutions more quickly. For example, the raw data may be **standardized** so that the MCMC chains explore the posterior distribution space more efficiently, allowing you to run fewer trials overall (Appendix 5). Definitely look into some of the more technical books on Bayesian statistical analysis on this front. We're just hoping to give you the big picture.

 ?? **And how do we summarize our results?**

Answer: There are a few ways to do this. You can summarize the posterior distributions for each of the unknown parameters as a histogram, which provides the shape of the parameter's marginal distribution. Figure 17.22 shows the histograms for our trials after the burn-in and prune. Because you may be more used to working with variance instead of precision, we'll take the inverse of our τ proposals to give us an estimate of variance, or σ^2, as well.

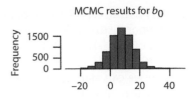

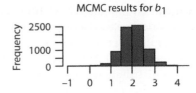

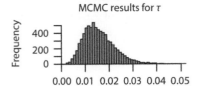

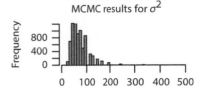

Figure 17.22

 ?? **What about the Bayesian credible intervals?**

Answer: Let's compute the 95% Bayesian credible intervals for all three parameters. Remember there are various ways to do this. In Table 17.13, we will report the 0.025 and the 0.975 quantiles for each parameter (and add in the 25th, 50th, and 75th percentiles for fun).

Table 17.13

Quantiles	b_0	b_1	τ
2.5%	−9.319	1.001	0.006
25%	2.392	1.683	0.012
50%	7.634	2.031	0.015
75%	13.028	2.372	0.020
97.5%	23.604	3.103	0.031

The credible intervals for each parameter can be found by listing the values in the first and last rows. It appears that the credible interval for the slope is positive and does not include 0, which would suggest that there is evidence that years of formal education is positively related to the number of days that a participant lasts in a game of Survivor.

There are other ways to summarize our results. In Table 17.14, we summarize the MCMC chains with a variety of common statistics, such as the minimum, maximum, mean, and so on.

Table 17.14

	Minimum	Maximum	Mean	Median	Std
b_0	−28.884	49.863	7.644	7.634	8.290
b_1	−0.539	4.337	2.031	2.031	0.532
τ	0.002	0.052	0.016	0.015	0.006
σ	19.231	500.000	73.074	66.667	35.042

Here, the mean and standard deviation outputs are very useful, especially for our parameters b_0 and b_1 since we used a normal distribution for a prior distribution. Remember that our prior distribution set the mean to 0 and a precision of 0.0001, which translates to a standard deviation of 100. Now that we've run our MCMC analysis, we can use our results to set the prior distribution for our next analysis, which we can start on right after next year's Survivor competition is complete! This is a primary benefit of the Bayesian method over other methods such as maximum likelihood and the least squares: the ability to incorporate existing knowledge into your analysis. In next year's analysis:

- The prior for b_0 would be a normal distribution with a mean of 7.644 and a precision of $\frac{1}{8.290^2}$, or $1/68.72 = 0.131$.

- The prior for b_1 would be a normal distribution with a mean of 2.031 and a precision of $\frac{1}{0.532^2}$, or $1/0.283 = 0.492$.

 ?? Can we do something similar for τ?

Answer: Yes! Remember that our prior distribution for τ was a gamma distribution with shape and rate parameters. If you recall, the mean of a gamma distribution is:

$$\mu_\tau = \frac{\alpha}{\beta} \qquad (17.34)$$

We can find parameters for our posterior distribution as follows:

$$\alpha_{posterior} = \alpha_0 + \frac{n}{2} = 0.01 + \frac{15}{2} = 7.51 \qquad (17.35)$$

So now we just need to estimate β using the mean τ from our MCMC trials:

$$\beta_{posterior} = \frac{\alpha_{posterior}}{\bar{\tau}} = \frac{7.51}{0.016} = 469.375 \qquad (17.36)$$

This can be the gamma prior for τ in our next analysis. There are other ways to estimate β too. Can you think of some other ways?

 ?? So what is our linear equation?

Answer: Excellent question. You may be **tempted** to use the mean of the posterior estimates for b_0 and b_1 to calculate a predicted y_i for a given x_i, where "prediction" is designated by the little hat:

$$\hat{y}_i = \bar{b}_0 + \bar{b}_1 * x_i \qquad (17.37)$$

$$\hat{y}_i = 7.644 + 2.031 * x_i \qquad (17.38)$$

Figure 17.23 shows our estimate of the model for our small dataset of 15 Survivor contestants. These are based on the marginal distributions for each parameter. Let's have a look.

Although the blue line might be used to predict the success of a new contestant, Thomas Bayes would probably not approve of this approach. Why? Because it is based on **point estimates** (i.e., specific hypotheses) from the marginal distribution of our parameters, rather than the full joint posterior distribution.

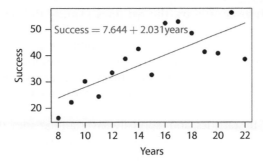

Figure 17.23

 ?? What would Bayes do?

Answer: We can't be 100% certain, but he would probably **embrace the uncertainty** that is reflected in the joint posterior distributions of the parameters! The MCMC trial results provide us with a way to capture this uncertainty. Remember, each trial in the MCMC provides a specific location in the joint distribution for the three parameters. In each trial, we have an estimate of b_0, an estimate of b_1, and an estimate of τ. With this information, we can do the following:

1. Determine the y_i for **each** level of x in **each** MCMC trial as:

$$y_{i[t]} = b_{0[t]} + b_{1[t]} x_i. \tag{17.39}$$

   Then, across MCMC trials, we can compute the 95% credible intervals for each level, which are shown in blue in Figure 17.24.
2. Determine the **posterior predictive distribution** of y_i for **each** level of x in each MCMC trial by recalling that:

$$y_{i[t]} \sim N(b_{0[t]} + b_{1[t]} x_i, \tau_{[t]}). \tag{17.40}$$

   This means that we can generate a single **random** success value for people with 8, 9, 10, ..., 22 years of formal education for **each** MCMC trial by drawing a random value from a normal distribution defined by the trial's mean, $b_{0[t]} + b_{1[t]} x_i$ and a standard deviation of $\frac{1}{\sqrt{\tau}}$. Then, we plot the credible intervals of choice of the predicted values **across trials**, as shown in red in Figure 17.24.

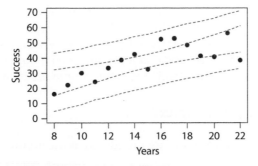

Figure 17.24

Technically, for each level of years, the red lines identify the **posterior predictive distribution**.

The Oxford Dictionary of Social Research Methods (Elliot et al., 2016) describes posterior predictive distribution this way: "A Bayesian analysis involves using the current data and [the] prior distribution to obtain a posterior distribution of parameters defining a model of interest and then using the model to determine a posterior predictive distribution for the unobserved elements of the population."

Wikipedia explains, "In statistics, and especially Bayesian statistics, the posterior predictive distribution is the distribution of unobserved observations (prediction) conditional on the observed data" (article accessed August 21, 2017).

So, if you have 18 years of formal education, the model suggests that you may last between 26.56 and 62.48 days in a game of Survivor. That answer makes use of the full joint distribution from our MCMC analysis. It includes not only uncertainty in the estimate of the mean but also uncertainty associated with "error" as modeled by the parameter τ.

 ?? Does the posterior predictive distribution help us assess model fit?

Answer: Yes, it is a good place to start. First of all, the observed data in Figure 17.24 fall within the credible intervals, which is a good sign. Goodness of fit is simply "a test of the fit of some model to a set of data" (Upton and Cook, 2014).

 ?? Do you get different results if you use a different prior?

Answer: You may! This is why the selection of the prior is so critical. That is one of the many decisions that you, the analyst, must decide!

In our analysis, we used **vague priors** so that the prior distribution played a tiny role in estimating the posterior distribution. You can verify this for yourself by looking at how the prior parameters affect the posterior parameters. For example, in updating α from α_0 to $\alpha_{posterior}$, you can see that our α_0 of 0.01 has little effect on the resulting $\alpha_{posterior}$:

$$\alpha_{posterior} = \alpha_0 + \frac{n}{2} = 0.01 + \frac{15}{2} = 7.51 \tag{17.41}$$

We could have set α_0 to something even smaller, such as 0.001, to make it even less informative:

$$\alpha_{posterior} = \alpha_0 + \frac{n}{2} = 0.001 + \frac{15}{2} = 7.501 \tag{17.42}$$

You can do the same thing with the β parameter. Additionally, for the parameters b_0 and b_1, we set the prior mean to 0, and τ to 0.0001 (standard deviation = 100). If you scan back to the conjugate solutions, you'll see that these hyperparameters have almost no bearing on the posterior parameters. The likelihood dominates (in a major way) the resulting posterior distribution. It becomes even more dominant as the sample size increases.

 ?? Can we summarize this chapter?

Answer: Let's have a look at the terms we introduced at the beginning of the chapter:

- Linear equation
- Linear model

- Sums of Squares
- Linear regression with MCMC and Gibbs sampling
- Posterior predictive distribution

We've covered a lot of ground, starting with functions, then linear equations and linear models. We then introduced linear regression and tediously worked our way through a simple analysis of a hypothetical Survivor dataset. We talked about how to present results in terms of marginal distributions, and how to create the posterior predictive distribution as a means for estimating success in Survival, based on the number of years of formal education. All of this was done within the context of the scientific method, where we noted that many statistical procedures are firmly rooted in inductive science; in doing the analysis, we have updated our knowledge about the parameters of the **years** model (b_0, b_1, and τ). However, we have other hypotheses that may explain why some people last longer in a game of Survivor than others (IQ, grit).

 ?? **What's next?**

Answer: Head on over to Chapter 18 after you've had a break. There, we will explore another hypothesis as to why some people last longer in Survivor. In particular, we conduct linear regression analysis on whether grit is a good predictor of success in a Survivor challenge. We will also take a (shallow) dive into the concepts of model fit and model selection, where we will compare the **years** model and the **grit** model.

CHAPTER 18

The Survivor Problem Continued: Introduction to Bayesian Model Selection

In this chapter, we continue our journey into the world of statistical inference, which is the process of identifying patterns from data. As in Chapter 17, we will use a statistical approach called **simple linear regression** to generate a general pattern from specific observations in a dataset. We will formalize a general pattern as a **model**, specifically, a **linear model.** If a researcher has collected the data to test a specific hypothesis, the **linear model** output can help determine whether the hypothesis is supported or not.

In Chapter 17, we analyzed a small, hypothetical dataset that is (sort-of) based on research by Angela Duckworth (University of Pennsylvania).

As a quick reminder, Duckworth and her colleagues were keenly interested in answering the following question: **"Why do some individuals accomplish more than others of equal intelligence?"**

They explain: "In addition to cognitive ability, a list of attributes of high-achieving individuals would likely include creativity, vigor, emotional intelligence, charisma, self-confidence, emotional stability, physical attractiveness, and other positive qualities. A priori, some traits seem more crucial than others for particular vocations. Extra-version may be fundamental to a career in sales, for instance, but irrelevant to a career in creative writing. However, some traits might be essential to success no matter the domain. We suggest that one personal quality is shared by the most prominent leaders in every field: grit."

 ?? Grit?

Answer: The researchers define grit as "perseverance and passion for long-term goals."

They put forth the hypothesis that "grit" is a characteristic that leads some people to accomplish more than others. The grit hypothesis is in the top of our scientific wheel in Figure 18.1.

Duckworth and colleagues predicted that a person's grit, if it can be measured, is one factor that can explain variability in accomplishment among a group of individuals. To test this hypothesis, along with competing hypotheses, they collected data on thousands of individuals and analyzed the patterns within.

You might recall that we have **completely fabricated a tiny dataset** in the spirit of the Duckworth et al. analysis so that we can illustrate a simple regression analysis with Bayesian techniques. In addition to grit, our little dataset includes not only how many

Bayesian Statistics for Beginners: A Step-by-Step Approach. Therese M. Donovan and Ruth M. Mickey,
Oxford University Press (2019). © Ruth M. Mickey 2019.
DOI: 10.1093/oso/9780198841296.001.0001

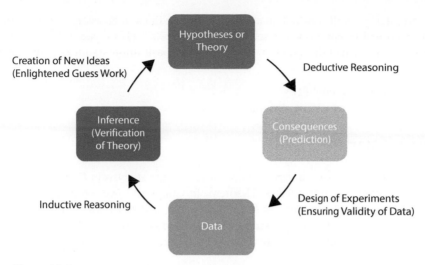

Figure 18.1

days an individual lasted on Survivor but also their IQ and how many years of formal education they have had.

Let's have a look (see Table 18.1).

Table 18.1

ID	Success	IQ	Years in School	Grit
1	33.48	112	12	2.2
2	42.53	94	14	3.2
3	48.53	118	18	3.4
4	30.21	87	10	1.8
5	38.76	96	13	2.8
6	38.59	106	22	0.2
7	52.93	71	17	4.4
8	32.65	91	15	1.0
9	52.42	95	16	4.6
10	22.22	94	9	0.4
11	41.40	100	19	1.6
12	16.28	98	8	0.0
13	40.83	94	20	1.2
14	24.43	113	11	0.6
15	56.38	85	21	4.2

Our goal for this chapter is to analyze the dataset and look for **patterns** in the data. Specifically, we will be looking to see if success in a Survivor competition can be predicted based on the following variables (known as predictor variables):

1. **Years in School**, which is the number of years that a participant attended a formal school.
2. **Grit**, measured on a scale between 0 and 5.

In Chapter 17, we stepped through a Bayesian linear regression analysis with Gibbs sampling, in which we modeled **success** as a function of **years** of formal education.

In this chapter, we will conduct an additional simple linear regression analysis. Then, we will compare the different models to see which hypothesis (**grit** vs. **years**) is best supported by the data. By the end of this chapter, you will have a basic understanding of the following concepts:

- Model assessment (model fit)
- DIC (Deviance Information Criterion)

 ?? Shall we get started?

Answer: Yes. Let's start with a very quick summary of the analysis in Chapter 17, where we estimated a linear model using Bayes' Theorem. In Chapter 17, we estimated the following model, where y_i is **Success** in Survivor (number of days), and x_i is **Years** of formal education.

For our Survivor data, we presume that each datapoint, y_i, is generated from a normal distribution whose mean is $b_0 + b_1 x_i$ and whose precision is τ. We can write:

$$Y_i \sim N(b_0 + b_1 x_i, \tau) \tag{18.1}$$

This model equation fully captures our assumed data-generating process and is an essential first step in a Bayesian statistical analysis. The observed number of days that contestant i lasts in a Survivor competion, y_i, arises from a normal distribution whose mean is $b_0 + b_1 x_i$ and precision is τ. In this problem, we seek the posterior distributions for each of the three unknown parameters.

Let's quickly review the steps for Bayesian analysis:

1. Identify your hypotheses–these would be the alternative hypotheses for b_0, and alternative hypotheses for b_1, and alternative hypotheses for τ.
2. Express your belief that each hypothesis is true in terms of prior probabilities.
3. Gather the data—we have that in hand already.
4. Determine the likelihood of the observed data, assuming each hypothesis is true.
5. Use Bayes' Theorem to compute the posterior probabilities for each parameter of interest.

Don't worry about your hat for this chapter!

 ?? OK, phew! What were the results again?

Answer: Well, we ran 20,000 MCMC trials for this analysis, and, with each trial, we store proposals from the posterior distribution. Let's have a look at the first 10 trial results **after** a burn-in of 5,000 trials, pruning every other trial (see Table 18.2).

Table 18.2

Info		b_0 Proposal Distribution		b_1 Proposal Distribution			τ Proposal Distribution		
		Hyperparameters			Hyperparameters			Hyperparameters		
Trial [t]	n	$\mu_{0[t]}$	$\tau_{0[t]}$	$b_{0[t]}$	$\mu_{1[t]}$	$\tau_{1[t]}$	$b_{1[t]}$	$a_{[t]}$	$\beta_{[t]}$	$\tau_{[t]}$
5002	15	8.090	0.210	7.364	2.047	50.000	1.903	7.510	450.301	0.014
5004	15	6.397	0.255	6.068	2.127	62.500	2.272	7.510	521.529	0.017
5006	15	6.187	0.270	3.957	2.257	66.667	2.194	7.510	405.404	0.018
5008	15	3.684	0.480	4.391	2.230	111.111	2.136	7.510	407.457	0.032
5010	15	4.118	0.240	7.126	2.062	58.824	2.047	7.510	405.054	0.016
5012	15	8.752	0.360	7.515	2.038	90.909	1.876	7.510	430.834	0.024
5014	15	7.357	0.375	9.027	1.945	90.909	2.043	7.510	398.051	0.025
5016	15	9.697	0.375	10.839	1.833	90.909	1.981	7.510	407.293	0.025
5018	15	7.941	0.225	4.232	2.240	55.556	2.347	7.510	428.011	0.015
5020	15	1.928	0.195	0.197	2.489	47.619	2.490	7.510	462.432	0.013

You might recall that the three colored columns give the **proposals** that have been drawn from the joint posterior distribution for each of our unknown parameters. And you might further recall that we generated random values for each level of **years** in **each** MCMC trial, given the proposals for that trial. This allowed us to create the posterior predictive distribution for each level of years (see Figure 18.2).

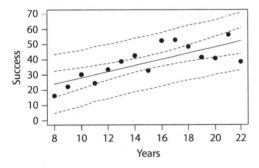

Figure 18.2

 ?? **Are there other ways to assess fit besides the posterior predictive distributions?**

Answer: Yes. This brings us to the concept of model fit and metrics that will help us to quantify how well a model fits to the data.

 ?? **Model fit?**

The Oxford Reference tells us that "model fit measures the degree to which a model is in agreement with the empirical data." In Chapter 17, we computed the Sums of Squares Error

(SSE), or residual sum of squares, in each trial of the MCMC. You may recall that SSE is computed in each trial as:

$$SSE = \sum_{i=1}^{n} (y_i - (b_0 + b_1 x_i))^2 \tag{18.2}$$

Here, we measure the difference between an observed datapoint (y_i), and what is predicted from the model ($b_0 + b_1 x_i$), and square the result. The sum of these squared deviations is the SSE. This metric is an intuitive measure of fit: the smaller the SSE, the less deviation there is between the observed data and that which is predicted by the model.

In this chapter, we will focus on computing the likelihood $\mathcal{L}$, the log likelihood $\ln\mathcal{L}$, and the −2 log likelihood −2 $\ln\mathcal{L}$ for each trial as a way of assessing the fit of a model. Unlike SSE, the higher a model's likelihood, the more closely the observed data match up with that which is predicted by the model. Let's have a look (see Table 18.3).

Table 18.3

	Proposals			Model Fit		
Trial [t]	$b_{0[t]}$	$b_{1[t]}$	$\tau_{[t]}$	$\mathcal{L}$	$\ln\mathcal{L}$	$-2\ln\mathcal{L}$
5004	6.068	2.272	0.017	3.30e−23	−51.76486	103.5297
5006	3.957	2.194	0.018	5.07e−23	−51.33613	102.6723
5008	4.391	2.136	0.032	9.38e−24	−53.02337	106.0467
5010	7.126	2.047	0.016	6.09e−23	−51.15252	102.3050
5012	7.515	1.876	0.024	1.70e−23	−52.42597	104.8519
5014	9.027	2.043	0.025	3.12e−23	−51.82250	103.6450
5016	10.839	1.981	0.025	1.60e−23	−52.48851	104.9770
5018	4.232	2.347	0.015	3.69e−23	−51.65249	103.3050
5020	0.197	2.490	0.013	2.76e−23	−51.94293	103.8859
5022	−5.374	2.976	0.026	1.23e−24	−55.05540	110.1108

The likelihood of the model **for each trial**, given the trial's parameters, can be calculated with the the normal pdf. Recall how the normal pdf is often displayed:

$$f(x) = \frac{1}{\sqrt{2\pi}\sigma} e^{-(x-\mu)^2/(2\sigma^2)}. \tag{18.3}$$

We use the normal distribution because we assumed that this was the distribution that generated our data. Remember? We worked with τ instead of σ^2, so our normal pdf was rewritten:

$$f(x) = \frac{\sqrt{\tau}}{\sqrt{2\pi}} e^{\frac{-\tau}{2}(x-\mu)^2}. \tag{18.4}$$

Look more closely for the portion of the function that hints vaguely of our residual term, "something minus something else, squared":

$$f(x) = \frac{\sqrt{\tau}}{\sqrt{2\pi}} e^{\frac{-\tau}{2}(x-\mu)^2}. \tag{18.5}$$

See it? If we let y_i represent an observed variable (standing in for x), and $b_0 + b_1 x_i$ stand in for μ, the likelihood of observing **a specific datapoint**, given the trial's parameters is:

$$\mathcal{L}(y_i; b_0, b_1, \tau) = \frac{\sqrt{\tau}}{\sqrt{2\pi}} e^{\frac{-\tau}{2}(y_i - (b_0 + b_1 x_i))^2}. \tag{18.6}$$

Now, if we plug in the trial's values for b_0, b_1, and τ, we can estimate the likelihood of observing **each** datapoint in our dataset. To get the likelihood of observing the **full dataset** in a given trial, the equation is:

$$\mathcal{L}(y_1, y_2, \ldots, y_n; b_0, b_1, \tau) = \prod_{i=1}^{n} \frac{\sqrt{\tau}}{\sqrt{2\pi}} e^{\frac{-\tau}{2}(y_i - (b_0 + b_1 x_i))^2}. \tag{18.7}$$

This says: the likelihood is equal to the product of n terms (one for each observation), in which the likelihood of each observation is obtained from the normal pdf. We then multiply the individual likelihoods together to get the model's likelihood. As such, we assume that the likelihoods among individual contestants are **independent** of each other (which brings us back to the laws of probability introduced in our first chapters).

Take a good look at the likelihood values in Table 18.3; they are truly tiny numbers. The column labeled lnL is just the natural log of the likelihood values. Notice that the smaller the likelihood, the smaller the log likelihood.

You might recall that the laws of logarithms let us compute the log likelihood for each trial, in which we **add** the log likelihoods together instead of multiplying them:

$$\mathit{ln}\Big(\mathcal{L}(y_1, y_2, \ldots, y_n; b_0, b_1, \tau)\Big) = \sum_{i=1}^{n} \mathit{ln}\left(\frac{\sqrt{\tau}}{\sqrt{2\pi}} e^{\frac{-\tau}{2}(y_i - (b_0 + b_1 x_i))^2}\right). \tag{18.8}$$

The final gray-shaded column in Table 18.3 is –2 times the log likelihood of the dataset. Because we multiply by a negative number, the smaller the likelihood or log likelihood, the larger the $-2\mathit{lnL}$.

These three columns provide us with information about model fit and can be used to help us compare two or more models. This brings us to our second model, the grit model.

 ?? ## OK, shall we evaluate the grit model?

Answer: OK, then! Let's now estimate a model that relates success in Survivor as a function of grit.

Here, we will estimate the following model, where y_i is **Success** in Survivor (number of days) and x_i is **Grit**:

$$y_i = b_0 + b_1 * x_i + \epsilon_i. \tag{18.9}$$

Our dataset for this problem is highlighted in Table 18.4 in gray.

As a preliminary, essential step, we need to specify the process by which we assume the data have been generated. Again, presume that each datapoint, y_i, is generated from a normal distribution whose mean is $b_0 + b_1 x_i$ and whose precision is τ. We can write:

$$Y_i \sim N(b_0 + b_1 x_i, \tau) \tag{18.10}$$

$$Y_i \sim N(b_0 + b_1 * \mathrm{grit}_i, \tau). \tag{18.11}$$

The observed number of days that contestant i lasts in a Survivor competion, y_i, arises from this normal distribution; the parameters have distributions associated with them (see Figure 17.11).

Table 18.4

ID	Success	IQ	Years in School	Grit
1	33.48	112	12	2.2
2	42.53	94	14	3.2
3	48.53	118	18	3.4
4	30.21	87	10	1.8
5	38.76	96	13	2.8
6	38.59	106	22	0.2
7	52.93	71	17	4.4
8	32.65	91	15	1.0
9	52.42	95	16	4.6
10	22.22	94	9	0.4
11	41.40	100	19	1.6
12	16.28	98	8	0.0
13	40.83	94	20	1.2
14	24.43	113	11	0.6
15	56.38	85	21	4.2

Let's quickly review the steps for Bayesian analysis for this model:

1. Identify your hypotheses—these would be the alternative hypotheses for b_0, and alternative hypotheses for b_1, and τ.
2. Express your belief that each hypothesis is true in terms of prior probabilities. Here, we'll use the same prior distributions that we used with our first analysis. **However, this is not required!** There are three parameters to estimate, each with a prior distribution defined by hyperparameters:
   - b_0's prior distribution is a normal distribution with hyperparameters $\mu_0 = 0$ and $\tau_0 = 0.0001$.
   - b_1's prior distribution is a normal distribution with hyperparameters $\mu_1 = 0$ and $\tau_1 = 0.0001$.
   - τ's prior distribution is a gamma distribution with hyperparameters $\alpha_0 = 0.01$ and $\beta_0 = 0.01$.
3. Gather the data—we have that in hand already, shown in gray in Table 18.4. Although the data could be standardized for analysis (see Appendix 5), we will use the raw data.
4. Determine the likelihood of the observed data, assuming each hypothesis is true.
5. Use Bayes' Theorem to compute the posterior probabilities for each parameter of interest.

We won't go into the details for this analysis, but essentially we do everything we did before in Chapter 17 except that we are evaluating success as a function of grit. Let's have a peak at a summary of the MCMC trials for the **grit model** (see Table 18.5).

Table 18.5

	Minimum	Maximum	Mean	Median	Std
b_0	10.456	41.55	24.586	24.598	3.065
b_1	−0.007	12.053	6.418	6.416	1.178
τ	0.004	0.08	0.026	0.024	0.01

Remember that these statistics summarize the posterior distributions of each unknown parameter; they are marginal distributions. We can also find and plot the posterior predictive intervals as we did in Chapter 17 (see Figure 18.3).

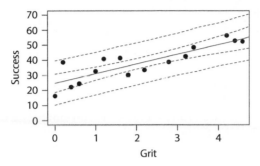

Figure 18.3

So, with this model, if you have grit score of 3, the 95% posterior predictive interval suggests that you may last between 29.99 and 57.76 days in a game of Survivor. The tighter interval indicates that you, the researcher, have more confidence in this prediction.

This is a good time to remind you that the dataset was fabricated so that grit was a strong predictor of success, which has nothing to do with the Duckworth et al. dataset. However, in our own experience, rolling up your sleeves and tackling problems with fierce determination is a skill worth building!

 ?? ## So which model is the better model?

Answer: This brings us to the topic of model selection, which we will briefly introduce and then point you to the experts. Upton and Cook (2014) define model selection as "a procedure for choosing between competing models that is based on balancing model complexity against the quality of that model's fit to the given data."

The first step is to identify a model set—or the statistical models that will be pitted against each other. Here, we treat each model as a hypothesis (see Figure 18.4).

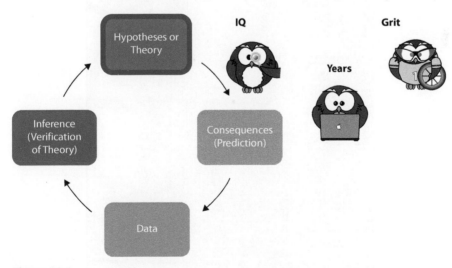

Figure 18.4

So far, we have considered two alternative models which make up our model set. We designate the model set as:

$$M = \{M_1, M_2\} \tag{18.12}$$

- M_1 = the years model
- M_2 = the grit model

But we could have considered more hypotheses about what drives success. Duckworth and colleagues, for example, analyzed levels of education, age, sex, conscientiousness, neuroticism, agreeableness, extraversion, and openness to experience (among other variables).

In their book Model Selection and Multimodel Inference: A Practical Information–Theoretic Approach, Ken Burnham and David Anderson emphasize that the models within a model set should be based on sound scientific principles (Burham and Anderson, 2004). In other words, models represent scientific hypotheses and are intentionally evaluated.

 ?? **Then what?**

Answer: Once the set of candidate models has been chosen, we need to balance model complexity against the quality of that model's fit. The opposite of complexity is simplicity, so we are aiming for a model that is simple but also fits the observed data well.

 ?? **Simplicity?**

Answer: This notion is based on Occam's razor (see Figure 18.5). Encyclopedia Britannica explains: "Occam's razor, also spelled Ockham's razor, also called law of economy or law of parsimony, [is a] principle stated by the Scholastic philosopher William of Ockham (1285–1347/49)

Figure 18.5 William of Ockham.

that *pluralitas non est ponenda sine necessitate*, 'plurality should not be posited without necessity.' The principle gives precedence to simplicity: of two competing theories, the simpler explanation of an entity is to be preferred."

The idea here is that "parameters" are akin to assumptions. For models with the **same** fit (e.g., the same model likelihood), we should select the model with the fewest parameters.

A quick example is in order. Let's consider a new, hypothetical dataset and fit a few models to the data (see Figure 18.6). The blue line in Figure 18.6 is the equation of a straight line, so two parameters are estimated. The black line includes 1 additional term (a second order polynomial), and the red line includes 2 additional terms (a third order polynomial).

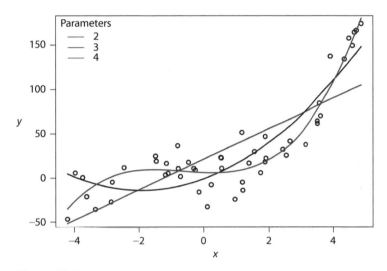

Figure 18.6

Which model best fits the data: red, black, or blue? Well, there is a distinct "downward dip" in the data, which the blue line misses. The black line captures the dip but misses the fact that the data are relatively constant for low values of x. The red line seems to capture the data best, but it does so by estimating an additional parameter.

 ?? So why not go with the red line?

Answer: Well, it is always easy to add more parameters to explain the data. In a famous quote, John von Neumann said "With four parameters I can fit an elephant, and with five I can make him wiggle his trunk." Apparently it's true!

J. Mayer, K. Khairy, and J. Howard. "Drawing an elephant with four complex parameters." *American Journal of Physics* 78.6 (2010): 648–9. DOI:10.1119/1.3254017.
J. Wei. "Least square fitting of an elephant." *Chemtech* 5.2 (1975): 128–9.

You can see this in action at http://demonstrations.wolfram.com/FittingAnElephant/. What von Neumann meant was that, with enough parameters, we can find a function that will really capture every datapoint of interest.

?? What's wrong with that?

Answer: Well, in statistics we develop models from a **sample** of data and are trying to make inferences to a broader **population**. In other words, we don't want to draw conclusions about success for only the 15 Survivor contestants that make up our dataset; we want to draw conclusions about ALL participants of Survivor. If you use a lot of parameters to explain the data in hand (the sample), you may have captured your particular dataset but completely miss the mark for the population as a whole! This is known as "overfitting."

'Tis a gift to be simple.

?? OK, so how do we quantify this trade-off between "fit" and "simplicity" for a given model?

Answer: We need some measure of fit for each model, and some measure of simplicity for each model. These two measures will be combined to give us some quantitative metric for each model.

?? How is this metric computed?

Answer: There are several alternative methods, and the one we will discuss is called DIC, or Deviance Information Criterion. It requires that you specify the following information about each model and the dataset:

- n = the sample size of the dataset
- k = the number of parameters that are estimated (model simplicity)
- model fit:
  - $\mathcal{L}$ = the model's likelihood
  - $ln\mathcal{L}$ = the model's log likelihood (natural log, or ln)
  - $-2\,ln\mathcal{L} = -2 *$ the log likelihood

We previously showed that these "fit" statistics can be measured for **each** MCMC trial (Equations 18.7 and 18.8). In addition to summarizing the proposals for each parameter, we can summarize these statistics across trials for our models.

Table 18.6 shows the results for our **years** model.

Table 18.6

	Minimum	Maximum	Mean	Median	Std
$\mathcal{L}$	9.72e−29	6.74e−23	2.16e−23	1.78e−23	1.73e−23
$ln\mathcal{L}$	−64.50106	−51.05081	−52.75127	−52.38553	1.41026
$-2\,ln\mathcal{L}$	102.10162	129.00213	105.50253	104.77106	2.82051

And Table 18.7 shows the summary for our **grit** model.

Now that we have these summary statistics in place, we can compute the **Deviance Information Criterion**, or DIC for short.

Table 18.7

	Minimum	Maximum	Mean	Median	Std
$\mathcal{L}$	1.01e−26	2.03e−21	6.55e−22	5.44e−22	5.19e−22
$\ln\mathcal{L}$	−59.85392	−47.64556	−49.33342	−48.96228	1.39335
−2 $\ln\mathcal{L}$	95.29112	119.70785	98.66683	97.92457	2.7867

- The SAS Institute defines deviance information criterion as: "The deviance information criterion (DIC) is a model assessment tool...The DIC uses the posterior densities, which means that it takes the prior information into account. Calculation of the DIC in MCMC is trivial [easy]... A smaller DIC indicates a better fit to the data set." This method was developed by Spiegelhalter et al. in 2002.

> D. J. Spiegelhalter, N. G. Best, B. P. Carlin, et al. "Bayesian measures of model complexity and fit." *Journal of the Royal Statistical Society: Series B (Statistical Methodology)* 64.4 (2002): 583–639. DOI: 10.1111/1467-9868.00353. http://doi.org/10.1111%2F1467-9868.00353.

$$DIC = \bar{D} + p_D \tag{18.13}$$

The term D, generally speaking, is the **deviance** of a model and is −2 times the log likelihood function. This is a standard term for this metric. Notice the bar over the D in the DIC calculation. This signifies that we take the mean of the −2$ln\mathcal{L}$ across MCMC trials, which is provided in the last row of each model's summary table (Tables 18.6 and 18.7).

- The term p_D is a measure of simplicity, where p_D is called the **effective number of parameters** and is calculated as $p_D = \bar{D} - \hat{D}$. To calculate $\hat{D}$, you plug in the average estimates for the model parameters across MCMC trials and then calculate the −2 $ln\mathcal{L}$ associated with this model. In the words of the WinBUGS helpfile, "p_D is the posterior mean of the deviance minus the deviance of the posterior means."

?? Do these criteria somehow help us to compare models?

Answer: You got it. The idea is that models with smaller DIC should be preferred to models with larger DIC. Let's start with the **years** model (see Table 18.8).

Table 18.8

	Minimum	Maximum	Mean	Median	Std
b_0	−28.884	49.863	7.644	7.634	8.29
b_1	−0.539	4.337	2.031	2.031	0.532
τ	0.002	0.052	0.016	0.015	0.006
−2 $ln\ \mathcal{L}$	102.102	129.002	105.503	104.771	2.821

Remember that DIC is computed as:

$$DIC = \bar{D} + p_D \tag{18.14}$$

where

$$p_D = \bar{D} - \hat{D}. \tag{18.15}$$

For the **years** model, $\bar{D}$ is the average of $-2 \ln \mathcal{L}$ across MCMC trials, or 105.503:

$$\bar{D} = 105.503 \qquad (18.16)$$

The term $p_D = \bar{D} - \hat{D}$ requires that we compute $\hat{D}$, which is the deviance of the model when the average parameter estimates across trials are used. We can calculate this by plugging in the MCMC average values for each parameter estimate into the normal pdf and calculating the -2 log likelihood for this model.

Here, that would be:

$$\hat{D} = -2 * \ln\left(\mathcal{L}(\text{data}; \bar{b}_0, \bar{b}_1, \bar{\tau})\right) = -2 * \sum_{i=1}^{n} \ln\left(\frac{\sqrt{\bar{\tau}}}{\sqrt{2\pi}} e^{-\frac{\bar{\tau}}{2}(y_i - (\bar{b}_0 + \bar{b}_1 * x_i))^2}\right) \qquad (18.17)$$

$$\hat{D} = -2 * \ln\left(\mathcal{L}(\text{data}; \bar{b}_0, \bar{b}_1, \bar{\tau})\right) = -2 * \sum_{i=1}^{n} \ln\left(\frac{\sqrt{0.016}}{\sqrt{2\pi}} e^{-\frac{0.016}{2}(y_i - (7.644 + 2.031 * x_i))^2}\right) = 102.284. \qquad (18.18)$$

Thus, the DIC score for the **years** model is:

$$DIC = \bar{D} + p_D = 105.503 + (105.503 - 102.284) = 108.721. \qquad (18.19)$$

Now for the **grit** model. Here, once again, are our summarized MCMC results (see Table 18.9).

Table 18.9

	Minimum	Maximum	Mean	Median	Std
b_0	10.456	41.55	24.586	24.598	3.065
b_1	−0.007	12.053	6.418	6.416	1.178
τ	0.004	0.08	0.026	0.024	0.01
$-2 \ln \mathcal{L}$	95.291	119.708	98.667	97.925	2.787

For the **grit** model which we just ran, $\bar{D}$ is the average of $-2\ln\mathcal{L}$ across MCMC trials, or 98.667. The term $p_D = \bar{D} - \hat{D}$ for this model is 98.667 minus the model's $-2\ln\mathcal{L}$ when we plug in the average values for each parameter estimate into the normal pdf.

Here, that would be:

$$\hat{D} = -2 * \ln\left(\mathcal{L}(\text{data}; \bar{b}_0, \bar{b}_1, \bar{\tau})\right) = -2 * \sum_{i=1}^{n} \ln\left(\frac{\sqrt{\bar{\tau}}}{\sqrt{2\pi}} e^{-\frac{\bar{\tau}}{2}(y_i - (\bar{b}_0 + \bar{b}_1 * x_i))^2}\right) \qquad (18.20)$$

$$\hat{D} = -2 * \ln\left(\mathcal{L}(\text{data}; \bar{b}_0, \bar{b}_1, \bar{\tau})\right) = -2* * \sum_{i=1}^{n} \ln\left(\frac{\sqrt{0.026}}{\sqrt{2\pi}} e^{-\frac{0.026}{2}(y_i - (24.586 + 6.418 * x_i))^2}\right) = 95.42. \qquad (18.21)$$

The final DIC result for the **grit** model is:

$$DIC = \bar{D} + p_D = 98.667 + (98.667 - 95.42) = 101.914. \qquad (18.22)$$

So, the **grit** model has a DIC value of 101.914, while the **years** model has a DIC value of 108.721, a difference of 6.807 units.

- *Years$_{DIC}$* = 108.721
- *Grit$_{DIC}$* = 101.914

?? How do you use these two results to draw conclusions?

Answer: Well, you're looking for the model with the smallest DIC value, which in this case is the **grit** model.

The authors of WinBUGS provide these thoughts on DIC in their user guide: "It is difficult to say what would constitute an *important* difference in DIC. Very roughly, differences of more than 10 might definitely rule out the model with the higher DIC, differences between 5 and 10 are substantial. But if the difference in DIC is, say, less than 5, and the models make very different inferences, then it could be misleading just to report the model with the lowest DIC."

Essentially, we have two hypotheses related to success: grit vs. years of education. Our DIC results suggest that the grit model is more strongly supported.

Technically, we should have standardized our dataset as discussed in Chapter 17. Keep in mind that the aim throughout this book has been to provide a big-picture overview of Bayesian topics, and you will certainly want to dig deeper.

?? Can we summarize the chapter so far?

Answer: Sure. Model selection is directly related to the scientific race track. Alternative models represent alternative hypotheses that explain some phenomena. Here, grit versus years of formal education are competing explanations (hypotheses) for why some people last longer in a game of Survivor. After collecting data, we used MCMC approaches to develop two linear regression models, one for each hypothesis. Each model had prior distributions for each of the model's parameters, and the MCMC analysis provided updated posterior distributions for these parameters, as well as a joint posterior distribution. With the results of the two MCMC analyses in hand, we used a particular model selection metric, Deviance Information Criterion (DIC), to help identify which hypothesis (model) is most strongly supported by the observed data, while accounting for simplicity (see Figure 18.7).

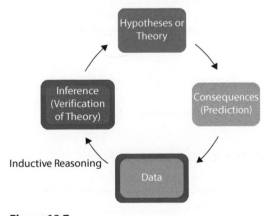

Figure 18.7

?? **Since each model is a hypothesis regarding success, can't we use Bayes' Theorem in some way and update our beliefs that each model is true?**

Answer: Ahhh, yes.

Here is the discrete version of Bayes' Theorem that we have come to love:

$$\Pr(H_i \mid \text{data}) = \frac{\Pr(\text{data} \mid H_i) * \Pr(H_i)}{\sum_{j=1}^{n} \Pr(\text{data} \mid H_j) * \Pr(H_j)}. \tag{18.23}$$

The prior probability for each discrete hypothesis is shown in blue, and the likelihood of observing the data under each hypothesis is shown in red. Bayes' Theorem will return the posterior probability of hypothesis *i* given the data.

If alternative **models** can be thought of as alternative hypotheses, we can make use of the discrete version of Bayes' Theorem to update the prior probability that each model (M_i) is true. In other words:

$$\Pr(M_i \mid \text{data}) = \frac{P(\text{data} \mid M_i) * \Pr(M_i)}{\sum_{j=1}^{K} P(\text{data} \mid M_j) * \Pr(M_j)}. \tag{18.24}$$

In this case, we have a set of *K* models, written as $M = \{M_1, M_2, \ldots, M_k\}$, under consideration. Each model is provided a prior probability of being the correct model, and the prior probabilities must sum to 1. The prior probabilities are assigned **before** you do your data analysis, and you should make use of any previous knowledge that you have.

The red portion of the equation is the probability of the data given a specific model. This is the marginal distribution of the data. Some books refer to this as the marginal data distribution for the model (e.g., Hobbs and Hooten, 2015). This can be calculated in general terms as:

$$P(\text{data}|M_i) = \int P(\text{data}|\theta, M_i) P(\theta) d\theta. \tag{18.25}$$

Here, θ is a vector of parameters to be estimated for model M_i.

 ?? **This looks familiar. Have I seen this before?**

Answer: Yes! You've seen this many, many times in this book. This is what is typically in the denominator of Bayes' Theorem.

$$P(\theta|\text{data}) = \frac{P(\text{data}|\theta) * P(\theta)}{\int P(\text{data}|\theta) * P(\theta) d\theta}. \tag{18.26}$$

This is the generic version of Bayes' Theorem when the posterior distribution for a single parameter, given the observed data, is represented by a **pdf**.

For our specific linear regression model, we wrote this as:

$$P(b_0, b_1, \sigma^2|\text{data}) = \frac{P(\text{data}|b_0, b_1, \sigma^2) * P(b_0, b_1, \sigma^2)}{\int \int \int P(\text{data}|b_0, b_1, \tau) * P(b_0, b_1, \tau) db_0, db_1, d\tau}. \tag{18.27}$$

 ?? **Well, how do we calculate the marginal distribution of the data for a given model?**

Answer: This is beyond the scope of our book. We've just touched the surface. We haven't even scratched it yet!

 ?? **How do I get below the surface?**

Answer: As we've mentioned, there's no shortage of help. Here are a few books and articles that grace our shelves:

- A. Gelman, J. B. Carlin, H. S. Stern, and D. B. Rubin. Bayesian Data Analysis. Chapman & Hall, 2004.
- N. T. Hobbs and M. B. Hooten. Bayesian Models: A Statistical Primer for Ecologists. Princeton University Press, 2015.
- M. B. Hooten and N. T. Hobbs. "A guide to Bayesian model selection for ecologists." Ecological Monographs 85.1 (2015): 3–28.
- J. V. Stone. Bayes' Rule: a Tutorial Introduction to Bayesian Analysis. Sebtel Press, 2014.

For programming:

- J. Kruschke. Doing Bayesian Data Analysis: A Tutorial with R, JAGS, and Stan. Elsevier, 2015..
- J. Albert. Bayesian Computation with R. Springer New York, 2009.

For ecologists in particular:

- J. A. Royle and M. Kery. Applied Hierarchical Modeling in Ecology. Elsevier, 2016.
- B. Bolker. Ecological Models and Data in R. Princeton University Press, 2008.
- M. A. McCarthy. Bayesian Methods for Ecology. Cambridge University Press, 2007.
- M. Kéry. Introduction to WinBUGS for Ecologists. Elsevier, 2010.
- W. Link and R. Barker. Bayesian Inference. Elsevier, 2010.

?? **What's next?**

CHAPTER 19

The Lorax Problem: Introduction to Bayesian Networks

Welcome to our chapter on Bayesian networks, a very useful technique that can be used to answer a variety of questions and aid decision making. Conditional and marginal probabilities rule the day in a Bayesian network.

By the end of this chapter, you will have a firm understanding of the following concepts:

- Bayesian network
- Directed acyclic graph
- Parent or root node
- Child node
- Influence diagram
- Conditional probability table (CPT)
- Chain rule for joint probability

To set the stage for this chapter, we'll begin with a story called The Lorax, one of Dr. Seuss's personal favorites. You may have read this story or even watched the film.

Pour yourself a cold glass of milk and grab a cookie or two, as we'll take a moment to read the plot line from Wikipedia:

A boy living in a polluted area visits a strange isolated man called the Once-ler in the Street of the Lifted Lorax. The boy pays the Once-ler fifteen cents, a nail, and the shell of a great-great-great grandfather snail to hear the legend of how the Lorax was lifted away.

The Once-ler tells the boy of his arrival in a beautiful valley containing a forest of Truffula trees and a range of animals.

The Once-ler, having long searched for such a tree as the Truffula, chops one down and uses its wool-like foliage to knit a Thneed, an impossibly versatile garment. The Lorax, who "speaks for the trees" as they have no tongues, emerges from the stump of the Truffula and voices his disapproval both of the sacrifice of the tree and of the Thneed itself. However, the first other person to happen by purchases the Thneed for $3.98, so the Once-ler is encouraged and starts a business making and selling Thneeds.

The Once-ler's small shop soon grows into a factory. The Once-ler's relatives all come to work for him and new vehicles and equipment are brought in to log the Truffula forest and ship out Thneeds.

The Lorax appears again to report that the small bear-like Bar-ba-loots, who eat Truffula fruits, are short of food and must be sent away to find more. The Lorax later returns to complain that the factory has polluted the air and the water, forcing the Swomee-Swans and Humming-Fish to migrate as well.

Bayesian Statistics for Beginners: A Step-by-Step Approach. Therese M. Donovan and Ruth M. Mickey,
Oxford University Press (2019). © Ruth M. Mickey 2019.
DOI: 10.1093/oso/ 9780198841296.001.0001

The Once-ler is unrepentant and defiantly tells the Lorax that he will keep on "biggering" his business, but at that moment one of his machines fells the very last Truffula tree. Without raw materials, the factory shuts down and the Once-ler's relatives leave.

What a bleak, bleak story! In the end, everybody loses.

 How could this tragedy have been averted?

Answer: Wouldn't it be nice to understand how Thneed production, Truffula trees, Bar-ba-loots, Swomee-Swans and Humming-Fish influence each other? If the Once-ler had known about **Bayesian networks**, he could have used them as an aid to his business so that the trees were harvested in a sustainable manner.

 Sustainable manner?

Answer: Encyclopedia Britannica says that "sustainability is presented as an alternative to short-term, myopic, and wasteful behaviour."

Of course, if the Once-ler was interested in such an approach, he wouldn't advocate squandering the resource (Truffula trees) that fuels his business, and the name "Once-ler" wouldn't really apply!

 What is a Bayesian belief network?

Answer: Wikipedia tells us that "a Bayesian network, Bayes network, belief network, Bayes (ian) model or probabilistic directed acyclic graphical model is a probabilistic graphical model (a type of statistical model) that represents a set of random variables and their conditional dependencies via a directed acyclic graph (DAG)" (article accessed August 21, 2017).

 Aack! Can we see an example?

Answer: That's what we're here for. Let's take a quick look at the example shown on the Wikipedia page (see Figure 19.1).

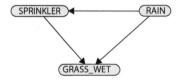

Figure 19.1

This example is also provided by the makers of a software program called Netica, which enables analysis of Bayesian networks. This particular diagram is a **model** of why the grass could be wet or dry, or why the sprinkler may be on or off.

Here, we see three variables of interest, each shown by an oval. These variables are often called "nodes."

You've probably noticed that some variables (nodes) have arrows leading out of them, and some have arrows leading into them. The RAIN node has arrows leading out of it, while SPRINKLER node has arrows leading into it and out of it, and GRASS WET node has arrows leading into it. The nodes and arrows collectively make up a model of how the system works, loosely corresponding to cause and effect. Rain affects whether the grass is wet, and it also affects whether the sprinkler is turned on. The sprinkler also affects whether the grass is wet. If the grass is wet, it could be due to the sprinkler, the rain, or both.

The Netica tutorial indicates that "the direction of the link arrows roughly corresponds to causality." The node that is the origin of an arrow is sometimes called the "parent node" or a "root node," and the node to which an arrow leads is sometimes called the "child node." In our example, Rain is a parent node, Sprinkler is both a parent and a child node, and Grass Wet is a child node. The diagram in Figure 19.1 is called an influence diagram because it specifies how variables influence one another.

Now let's look at this same network as depicted in the Wikipedia example, where we see a table associated with each node (see Figure 19.2).

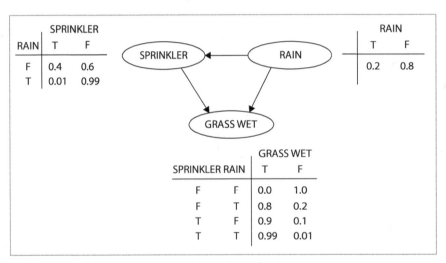

Figure 19.2 A simple Bayesian network with conditional probability tables.

1. The SPRINKLER has two states: it can be on (T) or off (F).
2. RAIN has two states: it can be raining (T) or not (F).
3. The GRASS WET variable has two states: it can be wet (T) or dry (F).

The definition of a **state** here is identical to the definition of an **event** introduced in Chapter 1. There, we quoted from the Oxford Dictionary of Statistics (Upton and Cook, 2014):

"An event is a particular collection of outcomes, and is a subset of the sample space." For example, when a die is thrown and the score observed, the sample space is {1, 2, 3, 4, 5, 6}, and a possible event is the score is even, that is, {2, 4, 6}. If all the possible outcomes are equally likely, then the probability of an event A is given by

$$\Pr(A) = \frac{\text{Number of events in subset of sample space corresponding to } A}{\text{Number of events in sample space}}. \qquad (19.1)$$

In a network, we take a given variable, like RAIN, and write out the probability of observing each state.

- Let's let S represent the sprinkler is on (SPRINKLER = T) and $\sim S$ represent the sprinkler is off (SPRINKLER = F).
- Let's let R represent it is raining (RAIN = T) and $\sim R$ represent not raining (RAIN = F)
- Let's let G represent the grass is wet (GRASS WET = T) and $\sim G$ represent the grass is dry (GRASS WET = F).

Then we can express the probability that it is raining as:

$$Pr(R) = 0.2 \tag{19.2}$$

The probability that it is not raining is:

$$Pr(\sim R) = 0.8. \tag{19.3}$$

See Figure 19.3.

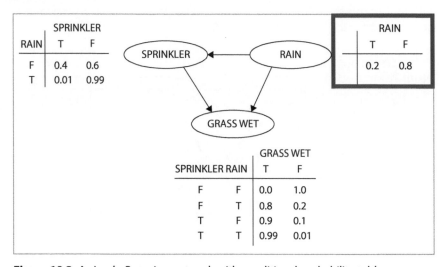

Figure 19.3 A simple Bayesian network with conditional probability tables.

These probabilities for RAIN are **marginal probabilities**: they are the sum over all states for any other variable that may influence whether it rains or not. We will pick up on this topic soon.

 ?? So, is this a probabilistic model?

Answer: Yes! It is a network **model** that consists of three variables, and each variable state has an associated probability.

We could generalize the problem like this, which is an example of a directed acyclic graph (DAG; see Figure 19.4).

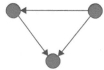

Figure 19.4 Directed acyclic graph.

?? Directed acyclic graph?

Answer: The Oxford Dictionary of Social Research Methods (Elliot et al., 2016) states that "a directed acyclic graph (DAG) is a directed graph that contains no cycles. Any two nodes can be joined or not (by an edge) and edges can either directed or undirected."

Our graph in Figure 19.4 has three nodes (points or vertices) and three directed arrows (edges) that connect them. The arrows specify directionality. The graph is designed so that there is no way to start at any one point and return to it later (cycling).

The general idea of directed acyclic graphs is not as complicated as it seems. Have you ever worked in a spreadsheet? A spreadsheet can be modeled as a DAG, where entries in one cell are used in other cells. If you've used a spreadsheet, have you ever generated a circular error? This occurs when the rules of DAGs are broken. For instance, if the formula in cell D3 specifies adding the values stored in cell D1, D2, and D3, an error will result (see Figure 19.5).

Sum	▼	÷	X	√	f_x	=D1 + D2 + D3	

	A	B	C	D	E
1				3481	
2				4129	
3				=D1 + D2 + D3	
4					
5					

Figure 19.5

This equation in cell D3 can be graphed as in Figure 19.6.

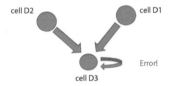

Figure 19.6

Since cell D3 feeds back or cycles back into itself, this is not a DAG and an error is generated. Similarly, a Bayesian network is a DAG, and it will not work if it does not follow DAG rules.

 ?? OK. Got it! Who created the sprinkler–rain–grass network?

Answer: We're not sure! The main point is that you take a problem of interest, identify the relevant variables, and then draw arrows to show how they relate to one another. Notice that whoever created this network did not indicate that rain depends on the sprinkler event. You can probably see why the arrows are necessary in our network. Does it make any sense that the probability of rain is influenced by whether the sprinkler is on or off? But notice that the network creator also did not let the state of the grass influence the state of the sprinkler. If you are a homeowner, you might be tempted to turn on the sprinkler if the grass is too dry. However, that would create a double-headed arrow, and the graph would not be a DAG.

?? All right then. What about the SPRINKLER node of our diagram?

See Figure 19.7.

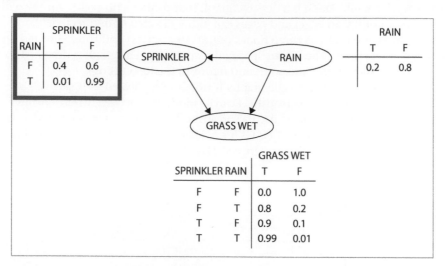

Figure 19.7 A simple Bayesian network with conditional probability tables.

Answer: The arrow from the RAIN node to the SPRINKLER node indicates that the SPRINKLER node is *conditional*, that is, it depends on the state of the RAIN node. This harkens back to ideas presented in Chapter 2 when we introduced the terms **joint probability**, **marginal probability**, and **conditional probability**.

It will be well worth our time to review these terms. Let's suppose we collect a dataset on sprinkler use and rain across 100,000 days. Here's the dataset, shown in the dark turquoise (see Table 19.1).

Table 19.1

	SPRINKLER		
	On	Off	Sum
RAIN			
No	32,000	48,000	80,000
Yes	200	19,800	20,000
Sum →	32,200	67,800	100,000

The rows indicate whether it rained on a given day or not, and the columns indicate whether the sprinkler was on or off. The number 32,000 in the upper left cell indicates that we observed 32,000 days in which it did not rain and the sprinkler was on. The number 48000 in the upper right cell indicates that we observed 48,000 days in which it did not rain and the sprinkler was off. And so on.

The margins of the table give the sums of the rows and columns. In this instance, we observed 80,000 days of no rain and 20,000 days with rain. In addition, we observed 32,200 days when the sprinkler was on and 67,800 days in which the sprinkler was off (light turquoise).

These raw numbers can be turned into probabilities by dividing the entire table by the number of samples, or 100,000 (see Table 19.2).

Table 19.2

	SPRINKLER		
	On	Off	Sum
RAIN			
No	0.32	0.48	0.8
Yes	0.002	0.198	0.2
Sum →	**0.322**	**0.678**	**1**

This table, sometimes called a **conjoint table**, provides us with the **joint** and **marginal** probabilities of interest. A **joint probability**, shaded in dark turquoise, is the probability of two things happening together, such as the probability that it rained AND that the sprinkler was on. The answer is 0.002. This can be written:

$$\Pr(R \cap S) = 0.002. \tag{19.4}$$

When you hear the word **joint**, you should think of the word **AND**.

In addition to the joint probabilities, the table also provides the **marginal** probabilities, which look at the probability of rain or no rain (aggregating over the state of the sprinkler) and the probability of sprinkler on versus sprinkler off (aggregating over the state of the rain). The word **marginal** is defined in the dictionary as "pertaining to the margins; or situated on the border or edge." In calculating a marginal probability, the variables that are aggregated are said to be "marginalized out."

In Table 19.2, the marginal probabilities are just the totals for one characteristic of interest accumulated over the other characteristics that might be listed in the table.

The marginal probabilities associated with RAIN in column 3 are not shaded. Here, as we've seen earlier, the marginal probability of no rain = 0.8, and the marginal probability of rain is 0.2. The marginal probabilities associated with SPRINKLER in row 3 of our table are shaded light turquoise. Here, the marginal probability that the SPRINKLER is on is 0.322, while the marginal probability that the SPRINKLER is off = 0.678. In these calculations, we have "marginalized out" the variable, RAIN. These can be written:

$$\Pr(S) = 0.322 \tag{19.5}$$

$$\Pr(\sim S) = 0.678. \tag{19.6}$$

 ?? **Are there any other probabilities stored in this table?**

Answer: One more type of probability can be **calculated** from this table, and that is **conditional probability**. You may recall that conditional probability measures the probability of an event given that another event has occurred. Conditional probability is written as:

- $\Pr(S \mid R)$, which is read, "the probability that the sprinkler is on, given rain."
- $\Pr(S \mid \sim R)$, which is read, "the probability that the sprinkler is on, given no rain."

- Pr($R \mid \sim S$), which is read, "the probability that it is raining, given the sprinkler is off."
- Pr($R \mid S$), which is read, "the probability that it is raining, given the sprinkler is on" (article accessed August 21, 2017).

 ?? ## Remind me again ... how do you calculate conditional probability?

Answer: You "zoom" to the marginal probability for the conditioning event of interest and then ask "what portion of the total marginal probability is made up of the other characteristic of interest?"

For example, the conditional probability that the sprinkler is on, given that it is raining, is computed as:

$$\Pr(S \mid R) = \frac{\Pr(S \cap R)}{\Pr(R)} = \frac{0.002}{0.2} = 0.01. \tag{19.7}$$

The conditional probability that the sprinkler is off, given that it is raining, is computed as:

$$\Pr(\sim S \mid R) = \frac{0.198}{0.2} = 0.99. \tag{19.8}$$

The conditional probability that the sprinkler is on, given that is does not rain, is computed as:

$$\Pr(S \mid \sim R) = \frac{0.32}{0.8} = 0.4. \tag{19.9}$$

The conditional probability that the sprinkler is off, given that it does not rain, is computed as:

$$\Pr(\sim S \mid \sim R) = \frac{0.48}{0.8} = 0.6. \tag{19.10}$$

Now let's have another look at the Wikipedia network, paying close attention to the SPRINKLER section (see Figure 19.8). Do the numbers look familiar?

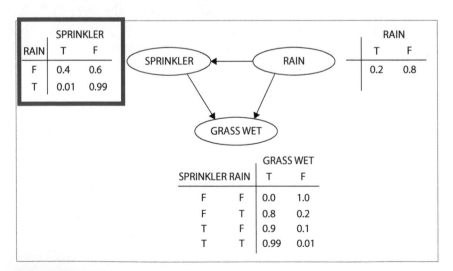

Figure 19.8 A simple Bayesian network with conditional probability tables.

?? Um...the tables consist of conditional probabilities?

Answer: You got it! In addition to the influence diagram, a Bayesian network requires tables that hold **conditional probabilities**. These tables are called, not surprisingly, **conditional probability tables**, or **CPT's** for short.

Now we can make some important, general comments about a Bayesian network:

- Nodes with no arrows leading to them have tables that provide **marginal probabilities**.
- Nodes with arrows leading to them have tables that provide **conditional probabilities**.

 ?? What do you notice about the setup of the SPRINKLER table?

Answer: The states of the sprinkler make up the columns of this table, while the states of the variable that influences it are provided as rows. Because sprinkler state (T or F) depends on the rain state (T or F), the table is a two-by-two table that is filled with conditional probabilities.

Note that the sum of each row equals 1 because we completely accounted for both of the sprinkler events for a specific rainfall condition.

 ?? And what about the GRASS WET variable?

Answer: Once again, notice that the columns of the table GRASS WET provide the two possible states for the node of interest. Also notice that the combinations of the variables that influence it are provided in the rows of the table.

What should the rows consist of? Well, the state for GRASS WET (T or F) is conditional upon both the state of the sprinkler and the state of rain. The arrows in the network tell us that this is the case. Since SPRINKLER has two states, and RAIN has two states, there are four possible combinations of SPRINKLER and RAIN that make up the rows in the GRASS WET table (see Figure 19.9).

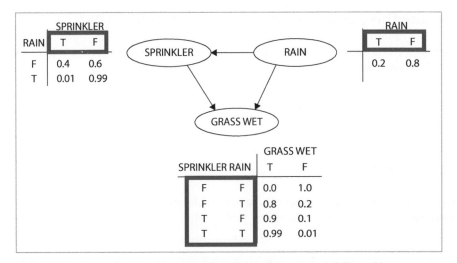

Figure 19.9 A simple Bayesian network with conditional probability tables.

Once again, it should be clear that the numbers in this table are **conditional probabilities** – each row sums to 1.0.

 ?? How do we calculate the values in this table?

Answer: Well, we started off with the dataset shown in Table 19.3, which was very appropriate for computing the SPRINKLER conditional probabilities.

Table 19.3

	SPRINKLER		
	On	Off	Sum
RAIN			
No	32,000	48,000	80,000
Yes	200	19,800	20,000
Sum →	32,200	67,800	100,000

This dataset contains **no information** on grass conditions. To get the conditional probabilities in the GRASS WET table, we need a more refined dataset.

 ?? What if you don't have any data for the necessary calculations?

Answer: The conditional probabilities associated with any of the tables in the network can come from data, but they can also come from your best guess. Or they could come from a combination of data and expert guesses. In the words of Bayesian belief network guru Bruce Marcot, "That's cool!"

We were not provided with a refined dataset that included the state of the grass. As a result, we will assume that the conditional probabilities were provided by experts, perhaps a friendly neighborhood lawn company.

 ?? Can we use this particular network to answer questions?

Answer: Once our network is fully specified as in Figure 19.9, we can use it to answer all kinds of questions, and Bayes' Theorem plays a central role. For instance:

- If the grass is wet, what are the chances it was caused by rain?
- If the chance of rain increases, how does that affect the amount of time I'll need to spend watering the lawn?

 ?? How do we start?

Answer: Since both questions above include the state of the grass, we will need to create a **conjoint table** based on the conditional probabilities in the GRASS WET node, and ANY parent nodes that affect it.

> To use the network, the relevant underlying conjoint tables must be computed.

Our table will end up looking something like that shown in Table 19.4; we will fill in the turquoise cells shortly.

Table 19.4

	SPRINKLER	RAIN	GRASS WET T	GRASS WET F
1	False	False		
2	False	True		
3	True	False		
4	True	True		

Notice that the structure of this table mirrors the structure of the conditional probability table . . . the main difference is that it will hold **joint probabilities** in the dark turquoise cells instead of conditional probabilities. As you'll see, the sum of ALL turquoise cells will be 1.0.

We can calculate each joint probability easily by making use of the **chain rule in probability**, which is also called the **general product rule** and is not to be confused with the chain rule in calculus. The chain rule is just an extension—a repeated application—of the joint probability formula given in Chapter 2. Wikipedia tells us: "The rule is useful in the study of Bayesian networks, which describe a probability distribution in terms of conditional probabilities."

Wikipedia continues with an example of four events, where the chain rule produces this product of conditional probabilities to compute a joint probability of 4 events:

$$\Pr(A_4 \cap A_3 \cap A_2 \cap A_1) = \Pr(A_4 \,|\, A_3 \cap A_2 \cap A_1) * \Pr(A_3 \,|\, A_2 \cap A_1) * \Pr(A_2 \,|\, A_1) * \Pr(A_1).$$
(19.11)

Hopefully, the pattern is apparent. Let's use the chain rule to compute the joint probability that the grass is wet, the sprinkler is on, and it is raining (see Figure 19.10). We can write this as $\Pr(G, R, S)$, and it can be calculated as:

$$\Pr(G \cap R \cap S) = \Pr(G \,|\, S \cap R) * \Pr(S \,|\, R) * \Pr(R) = 0.99 * 0.01 * 0.2 = 0.00198.$$
(19.12)

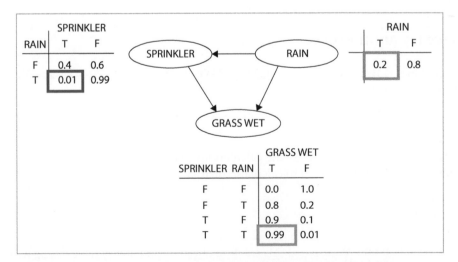

Figure 19.10 A simple Bayesian network with conditional probability tables.

Notice the calculation simply multiplies a series of conditional probabilities that lead back to the root node. This should make sense, knowing which nodes influence the others. Three terms will be multiplied:

- $\Pr(G \mid S \cap R) = 0.99$
- $\Pr(S \mid R) = 0.01$
- $\Pr(R) = 0.2$

The answer is 0.00198. We can add this result to our joint probability table (see Table 19.5).

Table 19.5

			GRASS WET	
SPRINKLER		RAIN	T	F
1	False	False		
2	False	True		
3	True	False		
4	True	True	0.00198	

 ?? Can you tell me more about the chain rule?

Answer: Indeed. Formally, the chain rule is:

$$\Pr\left(\bigcap_{k=1}^{n} A_k\right) = \prod_{k=1}^{n} \left(A_k \mid \bigcap_{j=1}^{k-1} A_j\right). \qquad (19.13)$$

In a nutshell, this equation basically says that the joint probability of n events called $A_1, A_2, A_3, \ldots, A_n$ occurring is the product of $n - 1$ conditional probability terms and one marginal probability term. The equation looks scarier than its implementation, don't you agree? In our example, there are $n = 3$ events, and the joint probability was calculated as the product of 2 conditional probabilities—$\Pr(G \mid S \cap R)$ and $\Pr(S \mid R)$—and one marginal probability—$\Pr(R)$.

 ?? Can we try another one?

Answer: Sure! How about the upper right cell? We can write this as $\Pr(\sim G, \sim R, \sim S)$, and it can be calculated as:

$$\Pr(\sim G \cap \sim R \cap \sim S) = \Pr(\sim G \mid \sim S \cap \sim R) * \Pr(\sim S \mid \sim R) * \Pr(\sim R) = 1.0 * 0.6 * 0.8 = 0.48000. \qquad (19.14)$$

Let's use the Bayes' net to visualize the calculations (see Figure 19.11).
    Three terms will be multiplied to calculate the result:

- $\Pr(\sim G \mid \sim S \cap \sim R) = 1$
- $\Pr(\sim S \mid \sim R) = 0.6$
- $\Pr(\sim R) = 0.8$

The answer is 0.48000. We can add this result to our joint probability table, along with rest of the joint probability calculations. **Crank out your pencil and double check our**

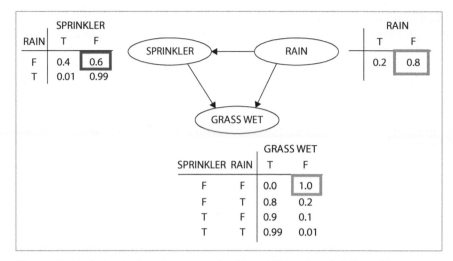

Figure 19.11 A simple Bayesian network with conditional probability tables.

results before continuing! Notice that we've also added in the marginal calculations and that the sum of the full table is 1.0 (see Table 19.6).

Table 19.6

		GRASS WET		
SPRINKLER	RAIN	T	F	Sum
1 False	False	0.00000	0.48000	0.480
2 False	True	0.15840	0.03960	0.198
3 True	False	0.28800	0.03200	0.320
4 True	True	0.00198	0.00002	0.002
Sum →		0.44838	0.55162	1.000

The chain rule rocks! Because the network is composed of conditional probabilities (except for root nodes, which are marginal probabilities), we can use it to quickly compute the joint probabilities which can be used to answer all kinds of questions.

 ?? **Why is this important?**

Answer: Because a Bayesian network simplifies the number of calculations dramatically by computing only those joint probabilities that are required. Think about it: if we have a network of one node that consists of two states (rain or no rain), we need to estimate two parameters: $\Pr(R)$ and $\Pr(\sim R)$ (or only one if you can get the second by subtraction). Let's add a new node: the sprinkler can be on or off. That results in four joint probability estimates. Let's add yet another new node: the lawn can be wet or dry. That results in 8 joint probabilities to estimate. Suppose we have one more node: the hose is working (yes or no). That brings us to 16 joint parameters to estimate.

Do you see a pattern here? 2, 4, 8, 16, 32, . . .

The network size grows geometrically, with each new node doubling the number of parameters to estimate.

If we had 100 nodes, each with just two options, that results in 2^{100} parameters! That's 1.26765e+30. That is one large number! That is just too many parameters for any computer program to handle.

Our friends at Netica suggest that a better way is needed, and that "Bayesian nets are one such way. Because a Bayes net only relates nodes that are probabilistically related by some sort of causal dependency, an enormous saving of computation can result. There is no need to store all possible configurations of states, all possible worlds, if you will. All that is needed to store and work with is all possible combinations of states between sets of related parent and child nodes (families of nodes, if you will). This makes for a great saving of table space and computation."

 ?? **How do we actually use the Bayes' network for addressing practical problems?**

Answer: Let's find out. Let's answer the following question:

> Given the lawn is wet, what are the chances it was caused by rain?

We'll solve this **four** different ways. First, we'll use the conjoint tables directly. Second, we'll use Bayes' Theorem. Third, we'll use a Bayesian inference approach. Finally, we'll use a Bayesian network software program that will provide the answer (all four will give the same result).

First, let's use our new conjoint table. The probability that it is raining given that the grass is wet is:

$$\Pr(R \mid G) = \frac{0.15840 + 0.00198}{0.44838} = 0.3576877. \tag{19.15}$$

Did you see how we used the conjoint table (Table 19.6) in these calculations? We were told the grass is wet, which occurs 44.838% of the time. Two entries in our conjoint table make up the numerator (first column, second and fourth row of the table).

 ?? **And where does Bayes' Theorem come into play?**

Answer: In our second approach, we'll solve this problem with Bayes' Theorem. In Section 1, we learned that the following joint probability of A and B occurring can be expressed as:

$$\Pr(A \cap B) = \Pr(A \mid B) * \Pr(B). \tag{19.16}$$

In words: **the joint probability of A and B is the product of the conditional probability of A, given B and the marginal probability of B.** Look familiar? This is a mini-chain rule problem!

Similarly, the joint probability of A and B occurring can be expressed as:

$$\Pr(B \cap A) = \Pr(B \mid A) * \Pr(A). \tag{19.17}$$

Therefore:

$$\Pr(A \mid B) * \Pr(B) = \Pr(B \mid A) * \Pr(A). \tag{19.18}$$

And dividing both sides by Pr(B) gives us **Bayes' Theorem** (at least one version of it!):

$$Pr(A \mid B) = \frac{Pr(B \mid A) * Pr(A)}{Pr(B)}. \tag{19.19}$$

We can use Bayes' Theorem to calculate the probability that it is raining, given the grass is wet as:

$$Pr(R \mid G) = \frac{Pr(G \mid R) * Pr(R)}{Pr(G)}. \tag{19.20}$$

$$\frac{\frac{0.15840 + 0.00198}{(0.198 + 0.002)} * (0.198 + 0.002)}{0.44838} = 0.3576877. \tag{19.21}$$

Same answer as before, right? The first method took fewer steps because we ignored the terms in the numerator that cancel out.

 ?? And how does Bayesian inference fit into this network?

Answer: Our third approach will place this same problem within a Bayesian inference context. Think back to the Author Problem in Chapter 5 (you know the one ... who penned the unsigned Federalist paper: Alexander Hamilton or James Madison). In that chapter, we noted that Bayes' Theorem expresses how a subjective degree of belief should rationally change to account for evidence.

To use Bayes' Theorem for scientific inference, it's useful (actually, essential) to replace the marginal denominator Pr(B) as the sum of the joint probabilities that make it up:

$$Pr(A \mid B) = \frac{Pr(B \mid A) * Pr(A)}{Pr(A \cap B) + Pr(\sim A \cap B)}. \tag{19.22}$$

Now, let's replace the **joint probabilities** with their conditional probability equivalents:

$$Pr(A \mid B) = \frac{Pr(B \mid A) * Pr(A)}{Pr(B \mid A) * Pr(A) + Pr(B \mid \sim A) * Pr(\sim A)}. \tag{19.23}$$

This version of Bayes' Theorem will work well for a Bayesian inference problem, where we have multiple, competing hypotheses. For a given hypothesis, Bayes' Theorem can be used to update the prior probability of a hypothesis to a posterior probability of a hypothesis, given new data. Do you remember the terms?

Let's use this framework and our Bayes network to address our problem: The grass is wet ... what is the likely cause? Instead of hypotheses A and $\sim A$ as shown in Figure 19.12, the conjoint table shows that we have four hypotheses:

1. It's not raining and the sprinkler is off.
2. It is raining and sprinkler is off.
3. It's not raining and the sprinkler is on.
4. It is raining and the sprinkler is on.

Here, instead of determining the probabilities for each of these four hypotheses, we'll focus on obtaining the marginal probability that the wet grass was caused by rain versus the marginal probability that it was caused by the sprinkler. Thus, the "rain" hypothesis, R, will

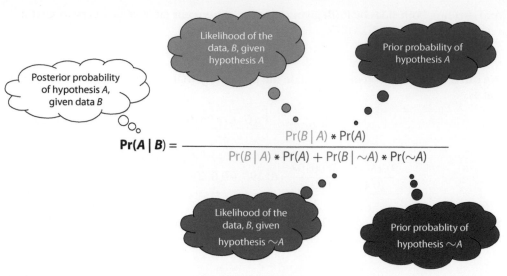

Figure 19.12

capture hypotheses 2 and 4 above, and the "sprinkler" hypothesis, S, will capture hypotheses 3 and 4 above.

The posterior probability of the rain hypothesis, given the grass is wet is:

$$Pr(R \mid G) = \frac{Pr(G \mid R) * Pr(R)}{Pr(G \mid R) * Pr(R) + Pr(G \mid {\sim}R) * Pr({\sim}R)} = 0.35768. \tag{19.24}$$

This is the same result as for our first two examples.

 ?? What about the hypothesis that the grass is wet because the sprinkler is on?

Answer: This is an alternative hypothesis for why the grass is wet. The posterior probability of the sprinkler hypothesis, given the grass is wet is:

$$Pr(S \mid G) = \frac{Pr(G \mid S) * Pr(S)}{Pr(G \mid S) * Pr(S) + Pr(G \mid {\sim}S) * Pr({\sim}S)} \tag{19.25}$$

Again, the conjoint table can provide this answer quickly:

$$Pr(S \mid G) = \frac{0.28800 + 0.00198}{0.44838} = 0.6467282. \tag{19.26}$$

> Note, however, that these two posterior probabilities do not sum to 1!

That is, $0.3576877 + 0.6467282 = 1.0044159$. The reason? It's because rain and sprinkler are NOT mutually exclusive. You can see this in the influence the diagram, and also see in the conjoint table, and by looking at hypothesis 4 above. Look for the entry in the conjoint table where the grass is wet, it is raining, and the sprinkler is on. The probability for that joint entry is 0.00198. If we restrict ourselves to the grass being wet, the probability that it is wet due to both factors is:

$$\frac{0.00198}{0.44838} = 0.0044159. \tag{19.27}$$

This tiny number in the conjoint table (0.00198) is included in both the calculation of $\Pr(S \mid G)$ and $\Pr(R \mid G)$, and thus we've double-dipped! In other words, the conjoint table shows us clearly that there are four hypotheses when the grass is wet, but we have collapsed these into two hypotheses.

Given our results, if we had to choose which hypothesis is most supported, we would go with the sprinkler hypothesis. This is the maximum *a posteriori* probability, or the updated posterior probability with the most support.

Bayes' Theorem and Bayesian inference are at the heart of the Bayes network, but you can use simple, direct calculations to generate your results. That is, it can be easy to miss Bayes' Theorem in a Bayesian network, even though it is lurking in every node.

 ?? **What is the fourth approach?**

Answer: To answer this question, we are going to use Netica, a popular computer program used for Bayesian network analysis. With Netica, you create a DAG, and each node is associated with either a conditional probability table (CPT) or a marginal probability table. Once these tables are populated, Netica *displays* the marginal probabilities for each node as horizontal bars (see Figure 19.13).

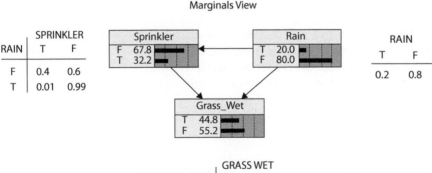

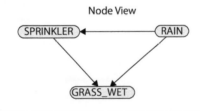

Figure 19.13

In the top portion of this figure, the node view is shown. In the bottom portion, the yellow boxes display the marginal probabilities associated with each node. This is what Netica displays. The bars inside these boxes are called "belief bars." They are calculated from entries that the Netica user (you!) specifies in the conditional or marginal probability tables. For example, the belief bars associated with the rain node show the probability of rain is 20%, and the probability of no rain is 80%. The table to the right of this box shows the entries made by the Netica user, and this is a root node. Root nodes (or parent nodes) are nodes that have arrows coming FROM them but nothing leading TO them.

The belief bars associated with the sprinkler node indicate that the chance that the sprinkler is on is 32.2%, and the chance that the sprinkler is off is 67.8%. These values are computed from the CPT entries made by the Netica user for this node. For example, the probability that the sprinkler is on is computed as:

$$\Pr(S) = \Pr(S \cap R) + \Pr(S \cap \sim R) = \Pr(S \mid R) * \Pr(R) + \Pr(S \mid \sim R) * \Pr(\sim R) \tag{19.28}$$

$$\Pr(S) = 0.01 * 0.2 + 0.4 * 0.8 = 0.322 \tag{19.29}$$

In Netica, this result is multiplied by 100 to give 32.2% chance. Compare this result with Table 19.2.

Once the network is fully specified, you can use it as means of computing posterior probabilities *after* you have observed a state. For example, if we observe that the grass is wet, we "click" on this belief bar, and Netica updates the entire network, as shown in Figure 19.14.

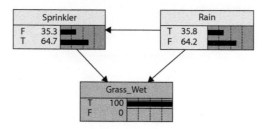

Figure 19.14

The gray box indicates that we have observed wet grass. The two arrows into this box indicate we have two hypotheses: it's raining or the sprinkler is on. The probability that the sprinkler is on is 64.7%, while the probability that it is raining is 35.8%. We calculated these by hand in our previous three examples. Notice once again that these don't add to 100% because of the double-dipping issue! In short, before we make an observation, the marginals act like a prior probability for a hypothesis. After we make an observation, the marginals are updated (and are interpreted as posteriors).

 ?? OK, I think I've got it. Can we try setting up the Once-ler's network?

Answer: We'll guide you through the development of the Once-ler's Bayesian network.

The Once-ler could build a Bayesian network to gain a holistic perspective on how his business affects the state of his raw materials (Truffula Trees) as well as other resources that may be affected by business activities. We start by drawing an influence diagram highlighting the key variables of interest and how they relate to each other. In Figure 19.15, we've

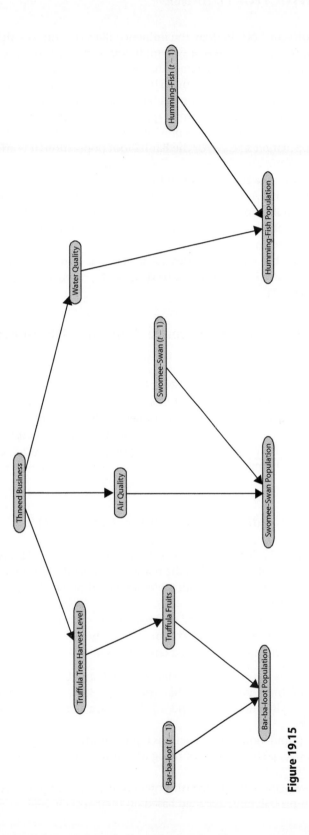

Figure 19.15

used the Netica software both to draw the influence diagram and to assign probabilities to states. Remember, this is a **model** of a system. If you were to watch or read The Lorax and create an influence diagram, your diagram may look different than ours!

Here, at the top of the diagram is our Thneed business, which will have two states: sustainable and unsustainable. The Thneed business directly influences the Truffula Tree harvest level, the air quality, and the water quality. The air and water quality variables, in turn, along with population size in the previous year ($t - 1$), affect the Swomee-Swan and Humming-Fish population size. As for the Bar-ba-loot population, it is affected by the state of the previous year's population size as well as the Truffula fruit production, which in turn is affected by Truffula tree harvest probability. For our three wild species, the reason we include the population size in the previous year ($t - 1$) should make some sense: if the population size is small, there will be fewer breeders present to produce offspring. Thus, even if Truffula fruits abound, the Bar-ba-loot population size may still be small if there are just a few animals to begin with.

Look closely at this diagram and identify the root nodes. There are four of them: Thneed Business, Bar-ba-loot ($t - 1$), Swomee-Swan ($t - 1$), and Humming-Fish ($t - 1$).

 ?? ## Is there a way to create an influence diagram without drawing it like you did?

Answer: There is! Often, a Bayesian network is specified by an expert and is then used to perform inference. We're not experts, but this is the approach we just used for the Lorax.

In other cases, humans may not be able to define the network because it may be too complex. However, machine learning algorithms can be used on a complex dataset to infer the influence diagram from an existing dataset. In other words, feed a complex dataset to a computer, and, with the right algorithm, a graph structure of a Bayesian network can be found. Double cool!

 ?? ## What about the conditional probability tables (CPT's)?

Answer: Right you are! A Bayesian network consists of both the DAG and the underlying conditional probability tables as well as the marginal probabilities for the root nodes. Let's add those now, keeping in mind that all tables are based on our limited expertise! If only Dr. Seuss were still alive!

Let's start by identifying the states and **marginal** probabilities for the states in each root node. These are nodes that have arrows coming FROM them but nothing leading TO them, and are identified in Figure 19.16 with red check marks.

- The Thneed business has two states: Sustainable (0.3) and Unsustainable (0.7). These are marginal probabilities that suggest that there is a 30% chance that the business is operating in a sustainable manner, and a 70% chance that it is operating in an unsustainable manner.
- The Bar-ba-loot population in year $t - 1$ has 0 probability of being extinct, a 20% chance of being low in size, a 60% chance of being medium in size, and a 20% chance of being high in size.
- The Swomee-Swan and Humming-Fish population sizes are scored with the same procedure, but with probabilities assigned based on their specific biology.

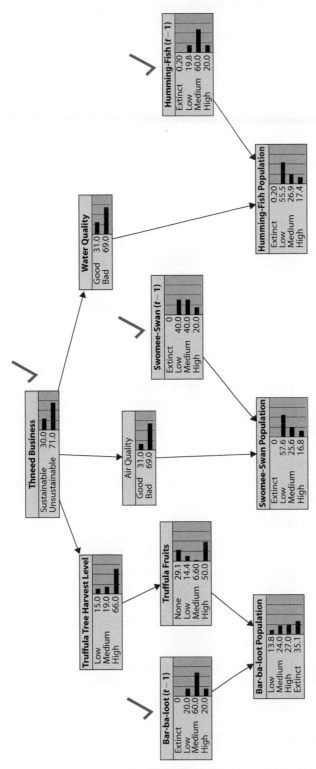

Figure 19.16

- Notice that the Netica diagram displays percentages rather than probabilities. So, a probability of 0.60 is displayed as 60.0 in the diagram.
- Notice that **all** tables in Netica show **marginal probabilities**, even for those nodes that have an arrow leading into them. We'll come back to this shortly.

 ?? Where are the CPT's?

Answer: The Netica user must provide them in order to show the belief bars in Figure 19.16. Table 19.7 shows the CPT that we entered in Netica for the Truffula harvest level.

Table 19.7

	Harvest Level			
	Low	Medium	High	Sum
Sustainable	0.5	0.4	0.1	1
Unsustainable	0	0.1	0.9	1

Here, you can see that the columns are the three states of harvest level (low, medium, and high). The rows are the two states of the parent node (sustainable and unsustainable). When the Thneed business is sustainable, there is a 50% chance that the Truffula harvest will be low, a 40% chance that the Truffula harvest will be medium, and a 10% chance that the Truffula harvest will be at a high level. **Although these numbers can be estimated from real data, we made these numbers up!** When the Thneed business is operating in an unsustainable manner, there is a 90% chance that the harvest level will be high, and no chance at all that the harvest level will be low. Notice once again that the row sums are 1.0.

Given these conditional probabilities, it is straightforward to understand why Netica's marginal probabilities for Truffula harvest probability are 0.15, 0.19, and 0.66 for low, medium, and high harvest, respectively—just use the **chain rule** to convert the CPT to joint probabilities. Remember that our conjoint table will have the same structure as the underlying CPT table; the difference is that it will hold **joint probabilities** instead of **conditional probabilities** (see Table 19.8).

Table 19.8

	Harvest Level			
	Low	Medium	High	Sum
Sustainable	0.15	0.12	0.03	0.3
Unsustainable	0	0.07	0.63	0.7
Sum	**0.15**	**0.19**	**0.66**	**1**

Compare the numbers in the final row of this table with the Netica Truffula Tree Harvest Level node (Figure 19.16). They match. In summary, the CPT tables are there, but they don't appear in the Netica diagram; the marginal probabilities for each table are displayed instead.

 ?? Can we look at one more CPT?

Answer: Of course. Let's look at the Water Quality node, which has two states. Netica tells us that the marginal probability that water quality is good is 0.31, and the marginal probability that the water quality is bad is 0.69. These numbers are calculated from the underlying CPT, which is shown in Table 19.9.

Table 19.9

	Water Quality		Sum
	Good	*Bad*	
Sustainable	0.8	0.2	1
Unsustainable	0.1	0.9	1

Now let's use the chain rule and convert these conditional probabilities to joint probabilities (see Table 19.10).

Table 19.10

	Water Quality		Sum
	Good	Bad	
Sustainable	0.24	0.06	0.3
Unsustainable	0.07	0.63	0.7
Sum	0.31	0.69	1

Once again, verify that the last row of the conjoint table, which gives the marginal probability of Good or Bad water quality, matches with the Netica diagram (Figure 19.16).

 ?? OK, now what?

Answer: Now that our network diagram, marginal probabilities, and CPT's are defined, we can use this network to address many questions. **The key is to enter observations in this network.** Your observations for a given variable are your data, or evidence. In a nutshell, you identify which state you have actually observed. Then, given your observations about a variable, the entire network will update using the rules of probability.

For instance, suppose we observe that the water quality is **bad**. In Netica, the updated network looks like the one shown in Figure 19.17.

Notice that the marginal probability of bad water quality is now 1.00 (or 100, in percentages). Take a look at the right side of the diagram. We see that last year's Humming-Fish state probabilities are unchanged, but that the current Humming-Fish population's state probabilities have all been updated because the population is influenced by water quality. Originally, the percentages associated with extinct, low, medium, and high population size were 0.2, 55.5, 26.9, and 17.4, respectively. After observing that the state of water quality is bad, these state probabilities have been updated to 0.2, 67.8, 22.0, and 9.98, respectively.

Cool! . . . Well, maybe not for the Humming-Fish!

In addition, we can see the probability that the Thneed business is run in an unsustainable manner jumped from 0.70 (70%) to 0.913 (91.3%).

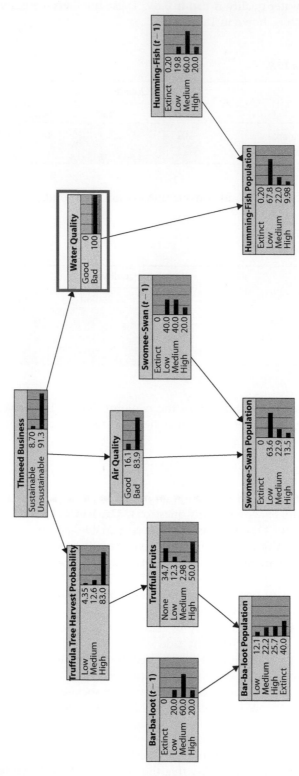

Figure 19.17

?? **The arrows do not show that water quality influences the Thneed business, so what is going on here?**

Answer: Well, for this example, the CPT table is fixed. Here it is again, with our observed water quality information highlighted in turquoise (see Table 19.11).

Table 19.11

	Water Quality		
	Good	Bad	Sum
Sustainable	0.8	0.2	1
Unsustainable	0.1	0.9	1

If we **observe** that water quality is bad, then the marginal table that feeds into it must be adjusted accordingly. How is this adjusted? **Bayes' Theorem, of course!** We use Bayes' Theorem to compute the posterior probability that the Thneed business is operating sustainably, as well as the posterior probability that the business is operating unsustainably.

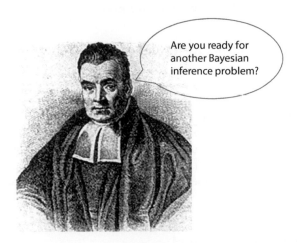

Are you ready for another Bayesian inference problem?

Here's the Theorem:

$$\Pr(A\,|\,B) = \frac{\Pr(B\,|\,A)*\Pr(A)}{\Pr(B\,|\,A)*\Pr(A)+\Pr(B\,|\,\sim A)*\Pr(\sim A)} \tag{19.30}$$

Let's let S be the prior probability of running a sustainable business, $\sim S$ be the prior probability of running an unsustainable business, and bad represent our observed water quality data. Then we can calculate the posterior probability that the business is operating sustainably as:

$$\Pr(S\,|\,\text{bad}) = \frac{\Pr(\text{bad}\,|\,S)*\Pr(S)}{\Pr(\text{bad}\,|\,S)*\Pr(S)+\Pr(\text{bad}\,|\,\sim S)*\Pr(\sim S)} \tag{19.31}$$

$$\Pr(S\,|\,\text{bad}) = \frac{0.2*0.3}{0.2*0.3+0.9*0.7} = \frac{0.06}{0.69} = 0.087 \tag{19.32}$$

And although we can calculate the posterior probability that the business is operating unsustainably by subtraction, let's use Bayes' Theorem again:

$$\Pr(\sim S \,|\, \text{bad}) = \frac{\Pr(\text{bad} \,|\, \sim S) * \Pr(\sim S)}{\Pr(\text{bad} \,|\, \sim S) * \Pr(\sim S) + \Pr(\text{bad} \,|\, S) * \Pr(S)} \tag{19.33}$$

$$\Pr(\sim S \,|\, \text{bad}) = \frac{0.9 * 0.7}{0.9 * 0.7 + 0.2 * 0.3} = \frac{0.63}{0.69} = 0.913 \tag{19.34}$$

These updated posterior probabilities are displayed in Thneed business node (Figure 19.17). Way cool!

What if you have more observations?

Answer: Let's suppose you also observe the Swomee-Swan population is low. Again, this represents our observed data (or evidence), and the network is automatically updated (see Figure 19.18).

Of course, we haven't shown you the CPT's we entered for the Swomee-Swans, nor the CPT's for air quality, but you can see the entire network has been updated. And we hope you get the idea of how powerful Bayesian networks can be.

Who came up with the idea of Bayesian networks?

Answer: The term "Bayesian networks" was coined by Judea Pearl in 1985. Encyclopedia Brittanica can help us here with some background:

"Pearl introduced the messiness of real life to artificial intelligence. Previous work in the field had a foundation in Boolean algebra, where statements were either true or false. Pearl created the Bayesian network, which used graph theory (and often, but not always, Bayesian statistics) to allow machines to make plausible hypotheses when given uncertain or fragmentary information. He described this work in his book Probabilistic Reasoning in Intelligent Systems.

Can we summarize this chapter?

Answer: We've seen that a Bayesian network is a probabilistic graphical model (a type of statistical model) that represents a set of random variables and their conditional dependencies via a directed acyclic graph (DAG). Each node in the graph is a variable that has alternative states, and each state occurs with some probability. The variables themselves are linked with arrows, and the direction of the link roughly corresponds to causality. Bayes' Theorem is at the heart of these connections; it can be used to estimate the probability of observing a state *conditional* on the state of the linked variables, and it can be used to *update* the probabilities once a particular state has been observed.

Bayesian networks have been used for:

- diagnosis
- prediction
- classification
- decision-making

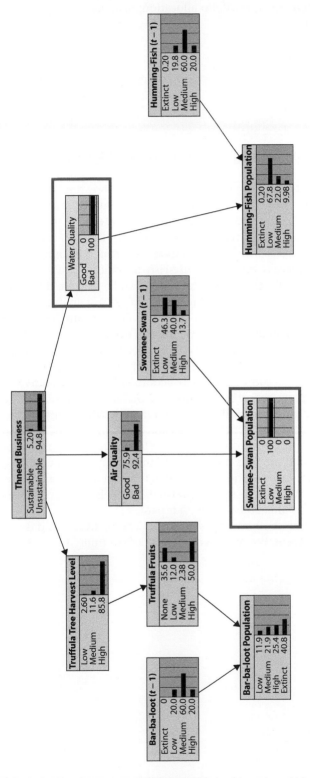

Figure 19.18

- financial risk management, portfolio allocation, insurance
- modeling ecosystems
- sensor fusion
- monitoring and alerting

McCann et al. (2006) provide an overview of the approach and its uses in ecology and natural resource management. We've just scratched the surface. Who knows? Maybe the next version of this book will shed light on some of these examples.

 ?? **I noticed decision-making is on the list. Can we see an example of this?**

Answer: At the end of the story, the Once-ler gives the boy the last Truffula seed and tells him "You're in charge of the last Truffula seed, and Truffula trees are what everyone needs. Plant a new Trufulla. Treat it with care. Give it clean water, give it fresh air. Grow a forest, protect it from axes that hack, then the Lorax and all of his friends may come back."

For the Lorax system, it sounds like the old Once-ler now has some interest in the state of Bar-ba-loots, Swomee-Swans, and Humming-Fish. Specifically, he is interested in anything and everything except the non-extinct state. How can he manage this? Well, the arrows in the influence diagram show that his business choice (sustainable versus unsustainable) directly influences Truffula Tree Harvest Level, Air Quality, and Water Quality. Thus, we can convert the Thneed Business node to a decision node and then look at updated results.

There's one more concept, though, with respect to decision-making, and this is the concept of utility. Nicholson (2014) defines utility as "a measure of the total perceived value resulting from an outcome or course of action. This may be negative, for example if an oil exploration company undertakes a survey showing that no oil can be extracted, the costs of undertaking the survey will be reflected in a negative utility for that outcome."

If we were to use our Bayes' net for decision-making, and the wild species were our primary concern, we would need to assign a utility score for each state of each variable. We could then see what happens if we plug in "sustainable" for the Thneed Business node and see how our wildlife populations will fare in terms of overall utility. By comparing the utilities associated with the two separate business models, we have some information in hand that may help us ultimately choose our path.

Chapter 20 will show you how Bayes' Theorem can be used to aid decision-making via the use of **decision trees**. We will cover the concept of utility in much more detail there!

CHAPTER 20

The Once-ler Problem: Introduction to Decision Trees

Welcome to our chapter on decision trees, a very useful technique that can be used to answer a variety of questions and assist in making decisions. We'll assume that you've read our chapter on Bayesian networks. If so, you already know that Bayes' nets can be an aid in decision-making. Decision trees are very closely related to Bayes' networks, except that they take the shape of a tree rather than a net.

By the end of this chapter, you will have a firm understanding of the following concepts:

• Decision tree
• Decision node
• Chance node
• Payoff
• Utility
• Linked decisions

To set the stage for this chapter, we'll quickly revisit a story called The Lorax that we introduced in Chapter 19.

Here's the plot line from Wikipedia:

A boy living in a polluted area visits a strange isolated man called the Once-ler in the Street of the Lifted Lorax. The boy pays the Once-ler fifteen cents, a nail, and the shell of a great-great-great grandfather snail to hear the legend of how the Lorax was lifted away.

The Once-ler tells the boy of his arrival in a beautiful valley containing a forest of Truffula trees and a range of animals. The Once-ler, having long searched for such a tree as the Truffula, chops one down and uses its wool-like foliage to knit a Thneed, an impossibly versatile garment. The Lorax, who "speaks for the trees" as they have no tongues, emerges from the stump of the Truffula and voices his disapproval both of the sacrifice of the tree and of the Thneed itself. However, the first other person to happen by purchases the Thneed for $3.98, so the Once-ler is encouraged and starts a business making and selling Thneeds. The Once-ler's small shop soon grows into a factory. The Once-ler's relatives all come to work for him and new vehicles and equipment are brought in to log the Truffula forest and ship out Thneeds.

The Lorax appears again to report that the small bear-like Bar-ba-loots, who eat Truffula fruits, are short of food and must be sent away to find more. The Lorax later returns to complain that the factory has polluted the air and the water, forcing the Swomee-Swans and Humming-Fish to migrate as well.

Bayesian Statistics for Beginners: A Step-by-Step Approach. Therese M. Donovan and Ruth M. Mickey, Oxford University Press (2019). © Ruth M. Mickey 2019.
DOI: 10.1093/oso/9780198841296.001.0001

The Once-ler is unrepentant and defiantly tells the Lorax that he will keep on "biggering" his business, but at that moment one of his machines fells the very last Truffula tree. Without raw materials, the factory shuts down and the Once-ler's relatives leave.

At the end of the story, the Once-ler regrets his business practices and gives the boy the last Truffula seed in hopes that the environment can be restored.

 ?? How could this tragedy have been averted?

Answer: Well, the Once-ler could have used a **decision tree** to aid his business so that the resource that fuels his business, the Truffula tree, persists through time, together with the wild species.

 ?? Decision tree?

Answer: The Oxford Dictionary of Statistics describes a decision tree as "a graphical representation of the alternatives in a decision-making problem." Wikipedia tells us: "In decision analysis a decision tree and the closely related influence diagram are used as a visual and analytical decision support tool, where the expected values (or expected utility) of competing alternatives are calculated."

As you'll see, a decision tree is a very handy tool. As a special bonus, all-time great New York Yankee baseball player Yogi Berra will serve as our guide through this chapter.

 ?? Why is it called a tree?

Answer: Think about a tree for a moment. A tree emerges from the root, and as it grows, it develops branches. Branches continue to fork and split until, ultimately, you end up at a terminal branch, or the tip (see Figure 20.1).

With this in mind, we can now label the different parts of our decision tree. Decision trees consists of three types of nodes, as illustrated in Figure 20.1.

1. Assume you start at the root and advance to the first main branches. Here, you face a decision and outline the alternative choices that can be selected. Our tree shows five branches (alternatives) you can take. This junction in the tree is called a **decision node**. In decision trees, decision nodes are often represented by squares. Here, it is shown in yellow.

Figure 20.1

2. Assume you select the fifth path on the tree's right and come to another fork. This fork could be another decision node, or it may be a **chance node**, where the next path is associated with a probability, not a decision. With chance nodes, some branches may be more likely than others, or they may all be equally likely. Here, the chance node is shown as a green circle with three branches emerging from this node.
3. Assume the middle branch is manifested. Again, you come to forks which may be either decision nodes or chance nodes. Ultimately, you end at a tip, such as the tip crowned by the blue triangle. Tips are often referred to as **end nodes**, where some reward or payoff awaits the decision-maker.

Let's now create such a tree for our story. The Once-ler arrives in a beautiful valley and considers setting up a business selling Thneeds. He makes his decision rapidly, perhaps using the decision tree depicted in Figure 20.2. Decision trees are often depicted "sideways," as shown.

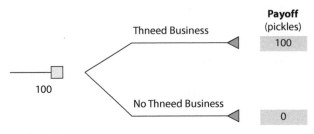

Figure 20.2

Here, the decision is whether to start a Thneed business or not. The yellow square is the root of this tree and is a decision node, and the tree has two branches which represent **decision alternatives**. The tips of the tree are depicted with end nodes that are represented by blue triangles, where we see the payoff for each alternative. In the spirit of Dr. Seuss, we'll measure the payoff in something silly, like pickles. If a business is not launched, the Once-ler will gain 0 pickles. If he launches the business, he will gain 100 pickles.

This is a pretty straightforward decision, right? He looked at the pickle payoff and made his choice. **Look for the number 100 near the yellow decision box, which indicates that the business route is the optimal choice with a payoff of 100.**

?? **Is that all there is to it?**

Answer: Heavens, no! Suppose the Once-ler had considered whether the business, once launched, might not succeed. We can add the probability that the business will succeed as a **chance node** to our diagram, which is shown by the green circle (see Figure 20.3).

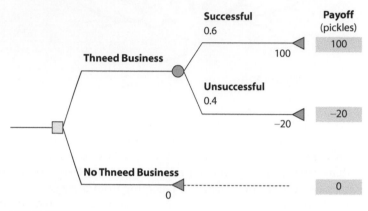

Figure 20.3

Here, let's assume that the probability that the business will be successful is 0.6, and the probability that it is unsuccessful is 0.4. **Notice that these sum to 1.0.** The addition of a chance node to the decision tree is a way in which we can incorporate uncertainty into the decision process.

 ?? **What kind of probabilities do these represent?**

Your answer here!
Answer: These are **conditional probabilities**. The probability that the business will be successful, *given* that the business was launched, is 0.6, and the probability that the business will be unsuccessful, *given* that the business was launched, is 0.4.

Now there are three tree tips, or outcomes, associated with this decision process. If the business is launched and is successful, the payoff is 100 pickles. If the business is launched but is unsuccessful, it will cost the Once-ler 20 pickles. And if the business is not launched, the payoff in pickles is 0.

 ?? Now what's the best decision?

Answer: Well, the decision consists of two alternatives, but we have three tips. If the business is not started, the payoff is 0. If it is started, the payoff may be 100 if it is successful or –20 if it is not. We need the payoff associated with the business **alternative** as a whole, however, and not the different tips.

The payoff is a function of the variable, "business," which has two states (successful and unsuccessful; see Table 20.1).

Table 20.1

Business	Probability	Payoff
Successful	0.6	100
Not Successful	0.4	−20

To calculate the expected payoff (expected number of pickles) for the business alternative, we take the weighted average of the payoffs:

$$\text{Expected Payoff for Business Alternative} = (0.6 * 100) + (0.4 * -20) = 52 \qquad (20.1)$$

Now let's add this information to our tree (see Figure 20.4).

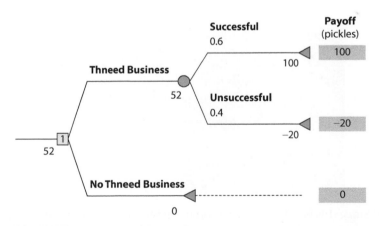

Figure 20.4

You can see the number 52 displayed below the yellow decision node in the tree in Figure 20.4. It suggests that the decision to go into the Thneed business has an expected payoff of 52, while the decision to not go into the Thneed business has an expected payoff of 0. Look for the number 52 displayed by the yellow box (our decision node), and the number 1 inside this box. This indicates that the top branch is the optimal decision, which yields an expected payoff of 52 pickles.

 ?? Does this answer tell the Once-ler what he should do?

Answer: Ha ha! Good one! The Once-ler will make the decision, not the decision tree. A decision tree is just a tool. It provides the expected monetary value for this decision problem. The Concise Oxford Dictionary of Mathematics (Clapham and Nicholson, 2014) tells us, "Where the payoff to the player in a game is dependent on chance outcomes, the expected monetary value is the average gain the player would achieve per game in the long run." It is calculated as the sum of each payoff multiplied by the probability of that payoff. In our case, the "game" corresponds to whether or not the business is successful, and the expected payoff is measured in pickles.

Technically, the expected value of a function $f(x)$ of the discrete random variable X is written:

$$E[f(X)] = \sum Pr(x)f(x) \tag{20.2}$$

In our case, X is the variable "business," x is the value that the variable X can take on (e.g., 1 = successful and 0 = unsuccessful), $f(x)$ is the payoff (100, –20), and $Pr(x)$ is the probability of x (0.6, 0.4).

The tree calculations suggest that if the Once-ler stumbled into a forest of Truffula trees thousands of times, in the long run he would have gained 52 pickles, on average.

But the Once-ler can't play this experiment thousands of times. He has one shot. The tree calculations provide him with guidance, but, ultimately, the choice is his! Moreover, he would not actually make 52 pickles, the average payoff for going into business. If he goes into business, he can only make 100 pickles or lose 20 pickles.

So true, Yogi! Since this is a one-time decision (a decision that won't be repeated thousands of times), the Once-ler could replace the **payoff** (in cold, hard pickles) with **utility scores**.

 ?? Utility?

Clapham and Nicholson (2014) define utility as "a measure of the total perceived value resulting from an outcome or course of action. This may be negative, for example if an oil exploration company undertakes a survey showing that no oil can be extracted, the costs of undertaking the survey will be reflected in a negative utility for that outcome."

Let's suppose the Once-ler's new tree looked like the one in Figure 20.5.

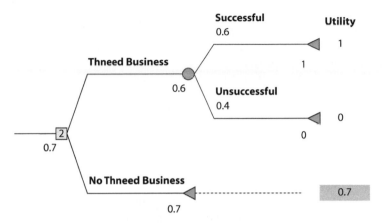

Figure 20.5

The only difference between this tree and the previous examples is that we have replaced pickles with a utility score. In this case, we've set the tip with the highest pickle gain to a utility of 1.0. And we set the tip with the lowest pickle gain (a loss) to 0. We also arbitrarily entered 0.7 as the utility associated with the No Thneed tip.

- The expected utility for the Thneed Business option is now

$$(0.6 * 1) + (0.4 * 0) = 0.6 \qquad (20.3)$$

- The expected utility for the No Thneed Business option is 0.7.
- Given these values, the better choice is to not go into business. You can see that the second option is displayed in the yellow box, with its expected value of 0.7 posted beneath it.

 ?? **How is utility determined?**

In decision-making circles, utility scores represent the **values** of the decision-maker. These scores, which generally represent some sort of satisfaction, can be **elicited** from decision-makers when they don't have a good idea how to assign numbers to each tip.

In the words of Decision Analysis for Management Judgement authors Paul Goodwin and George Wright (2014), "The most commonly used methods involve offering the decision maker a series of choices between receiving given sums of money [pickles] for certain or entering hypothetical lotteries....which will result in either the best outcome on the tree... or the worst...with specified probabilities."

We are trying to find out what the Once-ler's utility score is for the No Thneed Business tip, where the original payoff was zero pickles. Here's how this dialogue might go:

You: *"Hey, Once-ler! Choose between these two options. Option 1 is that I give you zero pickles; Option 2 is that we flip a coin where the probability of heads is 0.9. If it comes up heads, I will give you 100 pickles, but if it comes up tails, you give me 20 pickles."*

Once-ler: *"I'll take Option 2."*

You: *"Ok, then. Choose between these two options. Option 1 is that I give you zero pickles; Option 2 is that we flip a coin where the probability of heads is 0.8. If it comes up heads, I will give you 100 pickles, but if it comes up tails, you give me 20 pickles."*

Once-ler: *"I'll take Option 2."*

You: *"Ok, let's keep going. Choose between these two options. Option 1 is that I give you zero pickles; Option 2 is that we flip a coin where the probability of heads is 0.7. If it comes up heads, I will give you 100 pickles, but if it comes up tails, you give me 20 pickles."*

Once-ler: *"I'll take the zero pickles!"*

Now that the Once-ler has switched his choice, we can calculate the utility for this tip:

$$u(0 \text{ pickles}) = (0.7 * 1) + (0.3 * 0) = 0.7 \tag{20.4}$$

This equation basically says that the Once-ler's decision to accept zero pickles is the same as 70% of the tip with the highest utility, and 30% of the tip with the lowest utility. Of course, the process can be more finely tuned than we've shown here.

Notice that the Once-ler's answers are his own…your answers and ours may differ! This process of elicitation is intended to help determine the utility of each outcome that may not easily be scored.

 ?? ## Do people really use utility in practice in decision-making?

In theory, there is no difference between theory and practice. In practice, there is.

Goodwin and Wright (2014) tell us that utility theory is designed to simply aid decision-making, and if a decision-maker wants to ignore its indications, that prerogative is available. They continue, "In important problems which do involve a high level of uncertainty and risk, we do feel that utility has a valuable role to play as long as the decision maker is familiar with the concept of probability and has the time and patience to devote the necessary effort and thought to the questions required by the elicitation procedure. In these circumstances the derivation of utilities may lead to valuable insights into the decision problem."

 ?? ## What about the Bar-ba-loots, Swammy-Swans, and Humming-Fish?

Answer: Right-o. If the Once-ler considered the fates of these species in his decision, his tree would need some attention. The reason? Now he has more **objectives** to consider in his decision besides maximizing pickle profits. In addition to pickles, suppose that he is also concerned with keeping wildlife on part of the landscape. Let's suppose that the Once-ler sketches out the decision tree shown in Figure 20.6.

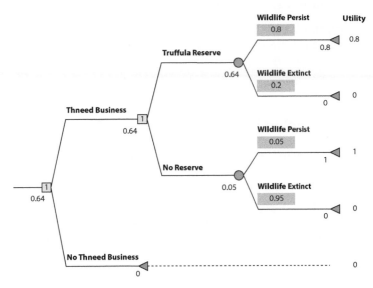

Figure 20.6

Here, the first decision node is whether to go into business or not. If he goes into business, he is faced with another decision: should he establish a wildlife reserve or not?

If the reserve is established, the chance that wildlife will persist is 0.8, and the chance that they go extinct is 0.2. If he does not establish the reserve, the chance that wildlife will persist is 0.05, and the chance they will go extinct is 0.95.

In this tree, utilities are used instead of pickles, and represent the **Once-ler's** satisfaction of each tip. We set a utility of zero for cases where wildlife goes extinct and for the case where a business is not established. We established a utility of 1 when a reserve is NOT established but the wildlife persists. This is because, in the Once-ler's opinion, profits will be maximized by not setting aside the reserve, so this is his best-case scenario. If a reserve is established and wildlife persists, the Once-ler's utility is 0.8.

With the utilities and the conditional probabilities shown in green in Figure 20.6, we can calculate the expected utility for *both* decisions. This is done by "rolling back the tree" **from the tips to the root**:

- The expected utility for setting up a reserve is:

$$(0.8 * 0.8) + (0.2 * 0) = 0.64 \tag{20.5}$$

- The expected utility for not setting up a reserve is:

$$(0.05 * 1) + (0.95 * 0) = 0.05 \tag{20.6}$$

- Thus, if the Once-ler follows the maximum utility for the reserve decision, he would elect to set up the reserve. This is shown in the yellow box associated with the reserve node (as is the maximum utility associated with this decision, 0.64). This is the expected utility for the Thneed business decision.

- Now we can move left in the tree and consider the decision about whether to go into business or not. The expected utility of not going into business is 0.
- Thus, if the Once-ler follows the maximum utility for the business decision, he would elect to go into business because 0.64 > 0.

Of course, if the utilities are changed, or if the conditional probabilities associated with chance nodes change, then the results may differ! A very useful exercise here is to twiddle with the utilities and conditional probabilities to find out what it takes to **change the decision**.

 ?? **Is that how Bayes' Theorem is used in decision trees?**

Answer: Yes. Bayes' Theorem is an essential tool when new information becomes available and the conditional probabilities associated with any chance node are updated.

To show you how this works, let's now consider a new decision tree in which the payoff is in pickles (see Figure 20.7).

In this new example, there are up to three decisions to be made. The first decision is whether to launch a business or not, where no launch results in a payoff of zero pickles.

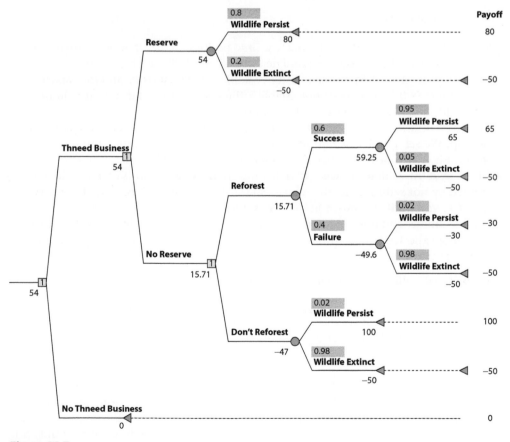

Figure 20.7

However, if the business is launched, the next decision is whether to set up a wildlife reserve or not. If a reserve is established, the probability that the wildlife species will persist is 0.8. If a reserve is not established, yet a third decision faces the Once-ler: should he reforest the Truffulas or not? That is, should he actively plant trees to replace those removed? Lack of reforestation results in a 2% chance that wildlife will persist. Alternatively, the reforestation effort has a 60% chance of succeeding, and a 40% chance of failure. If it is successful, the probability that the wildlife species will persist is 0.95. If it fails, the probability of persistence is 0.02. For each of the possible branches associated with the business start, there is a chance node regarding the persistence of the wild species, resulting in nine possible paths. Let's hope you can see that the tree structure must adequately describe the decision problem(s)...sometimes, this is tricky!

Now, let's review the payoffs. In this problem, if the wildlife species go locally extinct, the Once-ler must pay a regulatory fine of 50 pickles. The other payoffs result from a combination of the pickle gain and pickle expenses associated with each path.

To determine the expected payoffs, we again roll back the tree from right to left.

- The expected payoff for the Reserve alternative is 54, indicated underneath the top green chance node:

$$(0.8 * 80) + (0.2 * -50) = 54 \tag{20.7}$$

- The expected payoff of the No Reserve alternative requires more calculations. We must calculate the expected values for the Reforest alternative and the Don't Reforest alternative.
  - The expected payoff for the Reforest alternative if it is successful is 59.25, shown under the green node second from the top:

$$(0.95 * 65) + (0.05 * -50) = 59.25 \tag{20.8}$$

  - The expected payoff for the Reforest alternative, if it is unsuccessful, is −49.6:

$$(0.02 * -30) + (0.98 * -50) = -49.6 \tag{20.9}$$

  - The expected payoff for the Reforest alternative is 15.71:

$$(0.6 * 59.25) + (0.4 * -49.6) = 15.71 \tag{20.10}$$

- The expected payoff for the Don't Reforest alternative is –47:

$$(0.02 * 100) + (0.98 * -50) = -47 \tag{20.11}$$

- The expected payoff for the No Reserve alternative is the maximum of 15.71 and –47, or 15.71:

$$\max(15.71, -47) = 15.71 \tag{20.12}$$

- The expected payoff for the Thneed Business alternative is the maximum of 54 and 15.71, or 54:

$$\max(54, 15.71) = 54 \tag{20.13}$$

Thus, under the current tree setup, the long-term expected payoff can be maximized at a value of 54 pickles, which the Once-ler can attain by going into business and setting up a reserve.

 ?? ### Did you forget Bayes' Theorem?

Answer: We're getting there! Bayes' Theorem can be used to update the conditional probabilities as **new** information becomes available.

Are you ready for another Bayesian inference problem?

Let's assume that you obtain new information from 10 Truffula reforestation efforts, and 9 of them were successful. This represents our data.

Here's the Theorem:

$$\Pr(A|B) = \frac{\Pr(B|A) * \Pr(A)}{\Pr(B|A) * \Pr(A) + \Pr(B|\sim A) * \Pr(\sim A)} \tag{20.14}$$

In the decision tree, we pointed out that the probabilities associated with chance nodes are **conditional probabilities**. With new information in hand, we now consider various **hypotheses**. Our prior probability for the hypothesis that the reforestation effort would be successful is 0.6, and the prior probability for the hypothesis that reforestation will fail is

0.4. With these new data, we can use Bayes' Theorem to compute the posterior probability of successful reforestation, S, as:

$$\Pr(S|\text{data}) = \frac{\Pr(\text{data}|S) * \Pr(S)}{\Pr(\text{data}|S) * \Pr(S) + \Pr(\text{data}|{\sim}S) * \Pr({\sim}S)} \quad (20.15)$$

We hope you can recall the critical Bayesian inference terms (see Figure 20.8)!

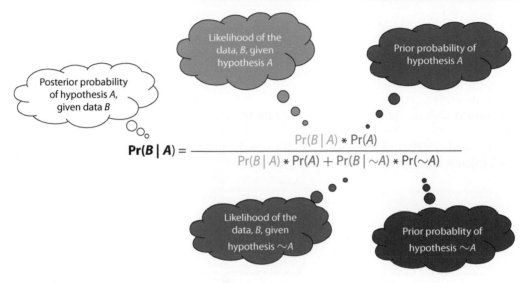

Figure 20.8

Let's run through our Bayesian analysis steps:

1. Identify the hypotheses. Easy! These are:
    - Reforestation will be successful.
    - Reforestation will be a failure.
2. Set the prior probabilities for each hypothesis. Easy again! These are given in our tree:
    - The prior probability that reforestation will be successful is 0.6
    - The prior probability that it will be a failure is 0.4.
3. Collect data!
    - Nine of 10 Truffula reforestation efforts in other locations have been successful.
4. Determine the likelihood of the observed data, assuming each hypothesis is true. The binomial probability mass function can serve us here:
    - The likelihood of observing 9 successes (y) out of 10 trials (n), given the probability of success (p) = 0.6, is:

$$f(y; n, p) = \binom{n}{y} p^y (1-p)^{(n-y)} \quad (20.16)$$

$$\mathcal{L}(y = 9;\ n = 10,\ p = 0.6) = f(y; n, p) = \binom{10}{9} 0.6^9 (1-0.6)^{(10-9)} = 0.0403 \quad (20.17)$$

- The likelihood of observing 9 successes out of 10 trials, given the probability of success = 0.4, is:

$$f(y|n,p) = \binom{n}{y} p^y (1-p)^{(n-y)} \tag{20.18}$$

$$f(y|n,p) = \binom{10}{9} 0.4^9 (1-0.4)^{(10-9)} = 0.00157 \tag{20.19}$$

5. Use Bayes' Theorem to compute the posterior probability that reforestation is successful, given the data:

$$\Pr(S|\text{data}) = \frac{\Pr(\text{data}|S) * \Pr(S)}{\Pr(\text{data}|S) * \Pr(S) + \Pr(\text{data}|\sim S) * \Pr(\sim S)} \tag{20.20}$$

$$\Pr(S|\text{data}) = \frac{0.0403 * 0.6}{0.0403 * 0.6 + 0.00157 * 0.4} = 0.97469 \tag{20.21}$$

This means that our posterior probability that the reforestation will fail is 1– 0.97469 = 0.02531.

These updated posterior probabilities can now be inserted into our tree, and the entire tree's calculations will be updated (see Figure 20.9).

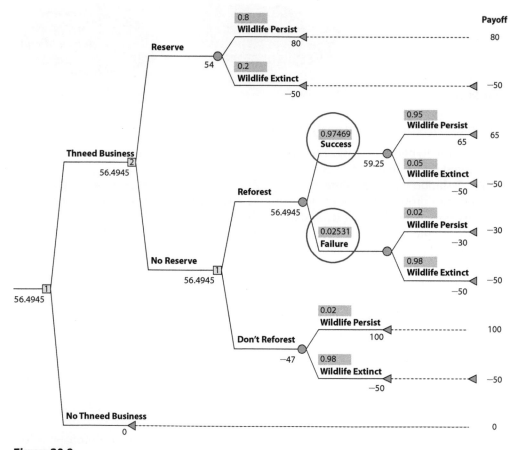

Figure 20.9

With this new, updated information, the decision outcome differs. The maximized expectation is that the Once-ler will get a payoff of ~56.5 if he goes into business and, instead of setting up a wildlife reserve, he plants Truffula trees after harvest.

 ?? **Who can we credit for developing the decision tree analysis?**

Answer: Many people have been involved, but two key players are Howard Raiffa and Robert Schlaifer, who co-authored the book Applied Statistical Decision Theory. We met these two gentlemen in Chapter 10, remember? Author Sharon Bertsch McGrayne devotes a full chapter to them in her book, The Theory That Would Not Die: How Bayes' Rule Cracked the Enigma Code, Hunted Down Russian Submarines, and Emerged Triumphant from Two Centuries of Controversy.

- Howard Raiffa is known as Mr. Decision Tree. McGrayne recounts the following exchange: "I began using decision tree diagrams that depicted the sequential nature of decision problems faced by business managers. Should I, as decision maker, act now or wait to collect further marketing information (by sampling or further engineering)?...I never made any claim to being the inventor of the decision tree but...I became known as Mr. Decision Tree." McGrayne continues, "As a pioneer in decision analysis, [Raiffa] was one of four organizers of the Kennedy School of Government at Harvard; the founder and director of a joint East–West think tank to reduce Cold War tensions long before perestroika; a founder of Harvard Law School's widely replicated role-playing course in negotiations; and scientific advisor to McGeorge Bundy, the national security assistant under Presidents Kennedy and Johnson. Raiffa also supervised more than 90 Ph.D. dissertations at Harvard in business and economics and wrote 11 books...As a Bayesian, Raiffa would cast a long shadow." Howard Raiffa died in 2016.
- Robert Schlaifer was trained as a classical Greek scholar but was a pioneer of Bayesian decision theory. How many statisticians do you know who could nimbly author "Greek theories of slavery from Homer to Aristotle" (Schlaifer, 1936)? Schlaifer was highly regarded by his peers. In the dedication of the book Introduction to Statistical Decision Theory (Pratt et al., 1995), Howard Raiffa and John Pratt said this of Schlaifer, their deceased co-author (1915–1994): "An original, deep, creative, indefatigable, persistent, versatile, demanding, sometimes irascible scholar, who always was an inspiration to us both."

 ?? **Can we summarize this chapter?**

Answer: We've seen that a decision tree is a graphical representation of the alternatives in a decision-making problem where the expected values (or expected utility) of competing alternatives are calculated. The tree itself consists of decision nodes, chance nodes, and end nodes, which provide an outcome. In the decision tree, we pointed out the probabilities associated with chance nodes are conditional probabilities, which Bayes' Theorem can be used to estimate or update.

There are many excellent books and articles on the topic of decision-making, in case you are interested. A very approachable and practical guide is the following:

J. S. Hammond, R. L. Keeney, and H. Raiffa. Smart Choices: A Practical Guide to Making Better Decisions. Harvard Business Review Press, 1999.

For any kind of managerial decision-making, Goodwin and Wright (2009) is a classic reference. For natural resource management in particular, see Conroy and Peterson (2013) and Gregory et al. (2012). See also Williams and Hooten (2016) for a more technical link between statistical analysis and decision analysis.

 ?? One more question…Did Yogi Berra really deliver the quotes in this chapter?

The Beta-Binomial Conjugate Solution

Here, we aim to show that the Bayesian posterior can be easily obtained for problems in which the prior distribution is a beta distribution and the observed data are binomial in flavor.

The prior distribution

A prior distribution is set for an unknown parameter, p, which is a beta distribution. We're interested in a beta distribution because our parameter of interest, p, can range between 0 and 1. The beta distribution is defined on the interval 0 to 1. The prior distribution has two hyperparameters, α_0 and β_0.

Our prior distribution, $P(p)$, is a beta distribution, whose probability density function looks like this:

$$P(p) = f(p; \alpha_0, \beta_0) = \frac{1}{B(\alpha_0, \beta_0)} p^{\alpha_0 - 1}(1-p)^{\beta_0 - 1} \quad 0 < p < 1 \tag{A1.1}$$

The beta function, B, is a normalization constant to ensure that the density integrates to 1. That is, different combinations of α_0 and β_0 yield a curve with a different shape, and the beta function ensures no matter what the shape, the area under the curve is 1.

An example of a beta probability distribution is shown in Figure A1.1.

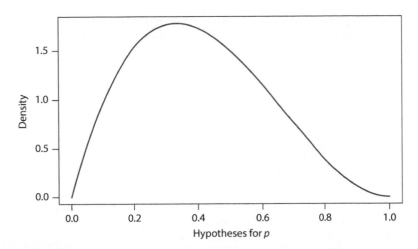

Figure A1.1

The x-axis for the beta distribution ranges between 0 and 1. The y-axis gives the **probability density**. You can think of the densities associated with each value of p as the "weight" for each hypothesis. The distribution above would put greater weight for p's between, say, 0.2 and 0.6 than other values.

The observed data

The data in this problem are binomial data, and the likelihood of observing the data can be obtained with the binomial probability mass function:

$$\mathcal{L}(y; n, p) = \binom{n}{y} p^y (1-p)^{(n-y)} \tag{A1.2}$$

The number of trials is denoted n. The number of observed successes is denoted as y. The probability of success is denoted as p.

As an example, the binomial distribution for $n = 3$ and $p = 0.7$ is shown in Figure A1.2.

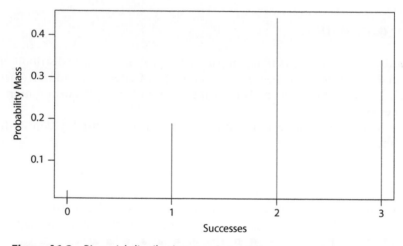

Figure A1.2   Binomial distribution: $n = 3$, $p = 0.7$

With 3 trials, and p (the probability of success) being 0.7, the binomial probability of 2 successes is 0.441). We show this only to illustrate the binomial distribution; we don't know what p is.

Bayes' Theorem

Remember that Bayes' Theorem is:

$$P(p \mid \text{data}) = \frac{P(\text{data} \mid p) * P(p)}{\displaystyle\int_0^1 P(\text{data} \mid p) * P(p)\,dp} \tag{A1.3}$$

This is the version of Bayes' Theorem when the posterior distribution for a single parameter, given the observed data, is represented by a pdf. This is designated $P(p \mid \text{data})$. This is the left side of the equation. On the right side of the equation, the numerator multiplies the prior probability **density** of p, which is written $P(p)$, by the likelihood of observing the data under a given hypothesis for p, which is written $P(\text{data}\,|\,p)$. In the denominator, we see the same terms, but this time we also see a few more symbols. The symbol $\int$ means "integrate," which roughly means "sum up all the pieces" for each tiny change in p, which is written dp. In other words, the denominator accounts for the prior density $*$ likelihood for all possible hypotheses for p, and sums them.

We'll now add color-coding to Bayes' Theorem to help track the terms:

$$P(p\,|\,\text{data}) = \frac{P(\text{data}\,|\,p) * P(p)}{\int_0^1 P(\text{data}\,|\,p) * P(p)\,dp} \tag{A1.4}$$

In this equation, $P(p\,|\,\text{data})$ and the p's in the numerator refer to a specific value of p. In the denominator, we are integrating over all possible values that p can assume (0 to 1). To avoid confusion in the derivation, we replace p with u in the denominator, where u is a variable that can take on these possible values.

$$P(p\,|\,\text{data}) = \frac{P(\text{data}\,|\,p) * P(p)}{\int_0^1 P(\text{data}\,|\,u) * P(u)\,du} \tag{A1.5}$$

For a problem in which the prior distribution is a beta distribution with hyperparameters α_0 and β_0, and in which the collected data are binomial, the posterior distribution is a beta distribution with hyperparameters:

$$\alpha_{\text{posterior}} = \alpha_0 + y \tag{A1.6}$$

$$\beta_{\text{posterior}} = \beta_0 + n - y \tag{A1.7}$$

Our goal now is to show how Bayes' Theorem generates these results.

The conjugate proof

Applying Bayes' Theorem to our problem, the prior is:

$$P(p) = f(p; \alpha_0, \beta_0) = \frac{1}{B(\alpha_0, \beta_0)} p^{\alpha_0-1}(1-p)^{\beta_0-1} \quad 0 < p < 1 \tag{A1.8}$$

And the likelihood is:

$$P(\text{data}\,|\,p) = \mathcal{L}(y; n, p) = f(y; n, p) = \binom{n}{y} p^y (1-p)^{(n-y)} \tag{A1.9}$$

Replacing the terms with their functional equivalents:

$$P(p\,|\,\text{data}) = \frac{\binom{n}{y} p^y (1-p)^{(n-y)} * \frac{1}{B(\alpha_0, \beta_0)} p^{\alpha_0-1}(1-p)^{\beta_0-1}}{\int_0^1 \binom{n}{y} u^y (1-u)^{(n-y)} * \frac{1}{B(\alpha_0, \beta_0)} u^{\alpha_0-1}(1-u)^{\beta_0-1}\,du} \tag{A1.10}$$

A couple of terms here are constants, and cancel out, which reduces the equation to:

$$P(p\,|\,\text{data}) = \frac{p^y (1-p)^{(n-y)} * p^{\alpha_0-1}(1-p)^{\beta_0-1}}{\int_0^1 u^y (1-u)^{(n-y)} * u^{\alpha_0-1}(1-u)^{\beta_0-1}\,du} \tag{A1.11}$$

The next step involves combining terms in both the numerator and denominator, where we lose our color-coding:

$$P(p \mid \text{data}) = \frac{p^{\alpha_0+y-1}(1-p)^{(\beta_0+n-y-1)}}{\displaystyle\int_0^1 u^{\alpha_0+y-1}(1-u)^{(\beta_0+n-y-1)}\,du} \tag{A1.12}$$

However, the integral (the area under the curve) in Equation A1.12 is not equal to 1.

To ensure the denominator integrates to 1.0, we bring in the **beta function**, which in our case has the parameters $\alpha_0 + y$ and $\beta_0 + n - y$. The inverse of this is:

$$\frac{1}{B(\alpha_0 + y, \beta_0 + n - y)} \tag{A1.13}$$

We now multiply both the numerator and denominator by 1 over the beta function:

$$P(p \mid \text{data}) = \frac{\frac{1}{B(\alpha_0+y,\beta_0+n-y)} * p^{\alpha_0+y-1}(1-p)^{(\beta_0+n-y-1)}}{\frac{1}{B(\alpha_0+y,\beta_0+n-y)} * \displaystyle\int_0^1 u^{\alpha_0+y-1}(1-u)^{(\beta_0+n-y-1)}\,du} \tag{A1.14}$$

Now, the denominator is equal to 1. This leaves us with just the numerator, which is our posterior distribution!

$$P(p \mid \text{data}) = \frac{1}{B(\alpha_0 + y, \beta_0 + n - y)} p^{\alpha_0+y-1}(1-p)^{(\beta_0+n-y-1)} \tag{A1.15}$$

Look familiar? Remember the form of our prior distribution (a beta distribution in Equation A1.1):

$$P(p) = \frac{1}{B(\alpha_0, \beta_0)} p^{\alpha_0-1}(1-p)^{\beta_0-1} \quad 0 < p < 1 \tag{A1.16}$$

If Equation A1.1 (or A1.16) is our prior distribution, the posterior distribution (Equation A1.15) is another beta distribution, with updated hyperparameters.

To help see this, let's bring back our color-coding (blue for prior, red for data). The alpha hyperparameter of the posterior distribution is:

$$\alpha_{\text{posterior}} = \alpha_0 + y \tag{A1.17}$$

where α_0 is from the prior distribution. The beta hyperparameter of the posterior distribution is:

$$\beta_{\text{posterior}} = \beta_0 + n - y \tag{A1.18}$$

where β_0 is from the prior distribution.

The posterior distribution is:

$$P(p \mid \text{data}) = \frac{1}{B(\alpha_0 + y, \beta_0 + n-y)} p^{(\alpha_0+y-1)}(1-p)^{(\beta_0+n-y-1)} \quad 0 < p < 1 \tag{A1.19}$$

APPENDIX 2

The Gamma-Poisson Conjugate Solution

Here, we aim to show that the Bayesian posterior can be easily obtained for problems in which the prior distribution is a gamma distribution and the observed data are Poisson in flavor.

The prior distribution

A prior distribution is set for an unknown parameter, λ, which can range from 0 to infinity. The probability density function for our gamma distribution (our prior distribution) is:

$$P(\lambda) = g(\lambda; \alpha_0, \beta_0) = \frac{\beta_0{}^{\alpha_0} \lambda^{\alpha_0-1} e^{-\beta_0 \lambda}}{\Gamma(\alpha_0)} \quad 0 < \lambda < \infty \qquad (A2.1)$$

The shape and location of our prior gamma distribution are controlled by two hyperparameters: the **shape** parameter is called alpha, α_0, and the **rate** parameter is called beta, β_0. The parameter values must be positive real numbers. Note: There are other forms of gamma distributions too, but the formulation above is standard for Bayesian problems.

Let's take a look at a few examples with the shape and rate parameterization in Figure A2.1.

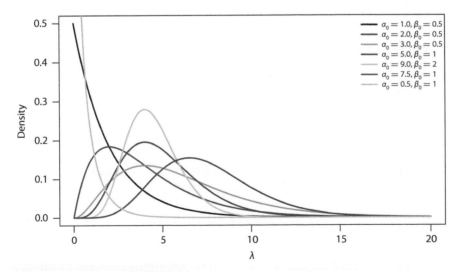

Figure A2.1 Gamma distributions.

Note that we have drawn these distributions as continuous functions (curves)—highlighting that the gamma distribution is a **probability density function**—there are an **infinite** number of hypotheses for λ.

The observed data

The data in this problem are **Poisson** data:

$$x_1, x_2, x_3, \ldots, x_n \tag{A2.2}$$

That is, they are assumed to have been generated from a Poisson distribution.

The Poisson **pmf** has just **one** parameter that controls both its shape and location. This parameter is called "lambda," and is written λ. Lambda represents the mean number of occurrences in a specified time period, such as the mean number of births in a year, the mean number of accidents in a factory in a year, the mean number of shark attacks in a year. So λ must be > 0 or the event will never occur. The units can also be volume or area instead of time; regardless, the unit must be specified in order to fully describe a Poisson distribution. By the way, λ is also the variance of the distribution.

An example of Poisson distribution where $\lambda = 2.1$ is shown in Figure A2.2.

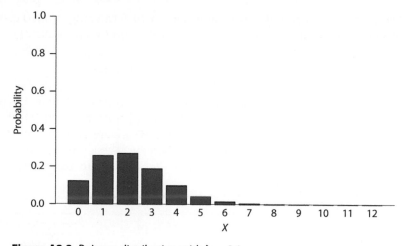

Figure A2.2 Poisson distribution with $\lambda = 2.1$.

Notice that the y-axis is labeled "Probability," and that the sum of the bars must equal 1.0. Now take a look at the x-axis; it consists of non-negative integers only. The graph is depicted in bars to indicate there are only discrete outcomes; that is, that there is no way to observe, say, $x = 2.5$. That is, the outcome x is an integer, but the average rate, λ, can be any positive number such as 2.1.

The likelihood of observing a single observation can be obtained with the Poisson probability mass function, which is written:

$$\mathcal{L}(x; \lambda) = \frac{\lambda^x e^{-\lambda}}{x!} \tag{A2.3}$$

where λ is the mean number of successes in a given period of time, x is the number of successes we are interested in, and e is the natural logarithm constant (approximately 2.718), also called Euler's number.

When there is more than one observation of x, the likelihood is written:

$$\mathcal{L}(x_1 \ldots x_n; \lambda) = \prod_{i=1}^{n} \frac{\lambda^{x_i} e^{-\lambda}}{x_i!} \tag{A2.4}$$

This can also be expressed as:

$$\mathcal{L}(x_1 \ldots x_n; \lambda) = \frac{\lambda^{\Sigma x_i} e^{-n\lambda}}{\prod x_i!} \tag{A2.5}$$

Bayes' Theorem

Remember that Bayes' Theorem is:

$$P(\lambda \mid \text{data}) = \frac{P(\text{data} \mid \lambda) * P(\lambda)}{\int_0^\infty P(\text{data} \mid \lambda) * P(\lambda) d\lambda} \tag{A2.6}$$

This is the version of Bayes' Theorem when the posterior distribution for a single parameter, given the observed data, is represented by a pdf. This is designated $P(\lambda \mid \text{data})$. This is the left side of the equation. On the right side of the equation, the numerator multiplies the prior probability **density** of λ, which is written $P(\lambda)$, by the likelihood of observing the data under a given hypothesis for λ, which is written $P(\text{data} \mid \lambda)$. In the denominator, we see the same terms, but this time we also see a few more symbols. The symbol $\int$ means "integrate," which roughly means "sum up all the pieces" for each tiny change in λ, which is written $d\lambda$. In other words, the denominator accounts for the prior density * likelihood for all possible hypotheses for λ, and sums them.

In this equation, $P(\lambda \mid \text{data})$ and the λs in the numerator refer to a specific value of λ. In the denominator, we are integrating over all possible values that λ can assume (0 to ∞). To avoid confusion in the derivation, we replace λ with u in the denominator, where u is a variable that can take on these possible values:

$$P(\lambda \mid \text{data}) = \frac{P(\text{data} \mid \lambda) * P(\lambda)}{\int_0^\infty P(\text{data} \mid u) * P(u) du} \tag{A2.7}$$

For a problem in which the prior distribution is a gamma distribution with hyperparameters α_0 and β_0, and in which the collected data are Poisson, the posterior distribution is a gamma distribution with hyperparameters:

$$\alpha_{\text{posterior}} = \alpha_0 + \sum_{i=1}^{n} x_i \tag{A2.8}$$

$$\beta_{\text{posterior}} = \beta_0 + n \tag{A2.9}$$

Our goal now is to show how Bayes' Theorem generates these results.

Conjugate proof

Now, assuming we have a prior distribution for an unknown parameter, λ, and Poisson-distributed data in hand, we can use Bayes' Theorem to update the prior distribution to the posterior distribution.

The prior pdf for our unknown parameter, λ is a gamma distribution:

$$P(\lambda) = g(\lambda; \alpha_0, \beta_0) = \frac{\beta_0^{\alpha_0} \lambda^{\alpha_0 - 1} e^{-\beta_0 \lambda}}{\Gamma(\alpha_0)} \quad 0 < \lambda < \infty \tag{A2.10}$$

The likelihood of observing the data is calculated using the Poisson pmf:

$$P(\text{data}|\lambda) = \mathcal{L}(x_1 \ldots x_n; \lambda) = \prod_{i=1}^{n} \frac{\lambda^{x_i} e^{-\lambda}}{x_i!} = \frac{\lambda^{\Sigma x_i} e^{-n\lambda}}{\prod_{i=1}^{n} x_i!} \tag{A2.11}$$

Notice that the right side is expressed in two equivalent ways. Now we can replace these terms with their functional equivalents:

$$P(\lambda | \text{data}) = \frac{\frac{\lambda^{\Sigma x_i} e^{-n\lambda}}{\prod x_i!} * \frac{\beta_0^{\alpha_0} \lambda^{\alpha_0 - 1} e^{-\beta_0 \lambda}}{\Gamma(\alpha_0)}}{\int_0^\infty \frac{u^{\Sigma x_i} e^{-nu}}{\prod x_i!} * \frac{\beta_0^{\alpha_0} u^{\alpha_0 - 1} e^{-\beta_0 u}}{\Gamma(\alpha_0)} du} \tag{A2.12}$$

Here, we replaced the prior (in blue) with the gamma pdf, and we replace the likelihood (in red) with the Poisson pmf.

Several terms here cancel out, namely $\prod x_i!$, $\Gamma(\alpha_0)$, and $\beta_0^{\alpha_0}$. In the numerator, you can also combine the terms involving λ:

$$\lambda^{\Sigma x_i} * \lambda^{\alpha_0 - 1} = \lambda^{\alpha_0 + \Sigma x_i - 1} \tag{A2.13}$$

And you can combine the e terms as well:

$$e^{-n\lambda} * e^{-\beta_0 \lambda} = e^{-\lambda(\beta_0 + n)} \tag{A2.14}$$

In the denominator, you can similarly combine terms involving u and e. These steps muddle our color-coding and reduce our equation to:

$$P(\lambda | \text{data}) = \frac{\lambda^{\alpha_0 + \Sigma x_i - 1} e^{-\lambda(\beta_0 + n)}}{\int_0^\infty u^{\alpha_0 + \Sigma x_i - 1} e^{-u(\beta_0 + n)} du} \tag{A2.15}$$

The denominator in Equation A2.15 does not integrate to 1, so we will need to address this. The next step involves multiplying both the numerator and denominator by a common term:

$$\frac{(\beta_0 + n)^{\alpha_0 + \Sigma x_i}}{\Gamma(\alpha_0 + \Sigma x_i)} \tag{A2.16}$$

This results in the following:

$$P(\lambda | \text{data}) = \frac{\frac{(\beta_0 + n)^{\alpha_0 + \Sigma x_i}}{\Gamma(\alpha_0 + \Sigma x_i)} * \lambda^{\alpha_0 + \Sigma x_i - 1} e^{-\lambda(\beta_0 + n)}}{\frac{(\beta_0 + n)^{\alpha_0 + \Sigma x_i}}{\Gamma(\alpha_0 + \Sigma x_i)} * \int_0^\infty u^{\alpha_0 + \Sigma x_i - 1} e^{-u(\beta_0 + n)} du} \tag{A2.17}$$

The denominator is now equal to 1. We're left with the numerator:

$$P(\lambda | \text{data}) = \frac{(\beta_0 + n)^{\alpha_0 + \Sigma x_i} \lambda^{\alpha_0 + \Sigma x_i - 1} e^{-\lambda(\beta_0 + n)}}{\Gamma(\alpha_0 + \Sigma x_i)} \tag{A2.18}$$

This is our posterior gamma distribution! Look familiar? Let's compare this distribution with our prior gamma distribution, which has the form:

$$P(\lambda) = \frac{\beta_0{}^{\alpha_0} \lambda^{\alpha_0-1} e^{-\beta_0\lambda}}{\Gamma(\alpha_0)} \tag{A2.19}$$

Can you see the correspondence? Here is our posterior:

$$P(\lambda \mid \text{data}) = \frac{(\beta_0 + n)^{\alpha_0 + \Sigma x_i} \lambda^{\alpha_0 + \Sigma x_i - 1} e^{-\lambda(\beta_0 + n)}}{\Gamma(\alpha_0 + \Sigma x_i)} \tag{A2.20}$$

Thus, our posterior distribution is a gamma pdf. We learned about the conjugate solution in our shark attack chapter (Chapter 11), where we had a prior distribution with hyperparameters α_0 *and* β_0. The posterior parameters are then:

$$\alpha_{0\,\text{posterior}} = \alpha_0 + \sum_{i=1}^{n} x_i \tag{A2.21}$$

and

$$\beta_{0\,\text{posterior}} = \beta_0 + n \tag{A2.22}$$

This is the gamma-Poisson conjugate solution.

The Normal-Normal Conjugate Solution

Here, we aim to show that the Bayesian posterior can be easily obtained for problems in which the prior distribution is a normal distribution and the observed data are normal in flavor.

Remember that the normal distribution is defined by two parameters: μ and some measure of spread. You may also recall that the spread of a normal distribution can be expressed in several ways:

- Standard deviation $= \sigma$.
- Variance $= \sigma^2$.
- Precision $= \tau = \frac{1}{\sigma^2}$.

The so-called normal-normal conjugate that we used in Chapter 12 requires that the measure of spread is **known**; the goal is to estimate the **unknown** parameter, μ. Here, we will assume that τ is known, such as the green ribbon shown in Figure A3.1, and the goal is to estimate the unknown parameter, μ, conditional on τ.

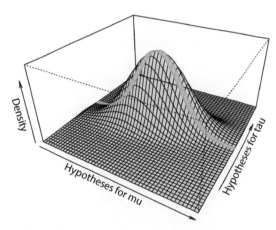

Figure A3.1

We can use Bayesian methods to update our beliefs associated with values of the unknown parameter, μ. First, we set a prior distribution for μ, then we collect data, and finally we use the conjugate solution to update the prior distribution to the posterior distribution for μ, our parameter of interest.

The prior distribution

To begin, we set a prior distribution for the unknown parameter, μ, which provides weights of belief in alternative values of μ. The prior distribution is, perhaps confusingly, a normal distribution defined by hyperparameters μ_0 and τ_0. For example, in Figure A3.2 we have set more "weight" on the belief that $\mu = 10.00$ as opposed to, say, $\mu = 5$ (or any other value for μ).

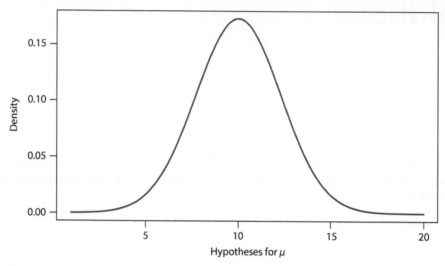

Figure A3.2 Pdf of a normal distribution

You may recall that the pdf for a normal distribution in terms of μ and σ is expressed as:

$$P(x) = f(x; \mu, \sigma) = \frac{1}{\sqrt{2\pi}\sigma} e^{-(x-\mu)^2/(2\sigma^2)}. \tag{A3.1}$$

Our prior distribution for the unknown parameter, μ, is a normal distribution with hyperparameters μ_0 and σ_0^2. Here, though, we express the parameters in terms of μ_0 and τ_0, where

$$\tau_0 = \frac{1}{\sigma_0^2} \tag{A3.2}$$

We can now express the prior distribution for the unknown parameter, μ as:

$$P(\mu) = f(\mu; \mu_0, \tau_0) = \frac{\sqrt{\tau_0}}{\sqrt{2\pi}} e^{-\frac{1}{2}\tau_0(\mu-\mu_0)^2}. \tag{A3.3}$$

Because $\sqrt{\tau_0} = \tau_0^{\frac{1}{2}}$ and $\frac{1}{\sqrt{2\pi}} = (2\pi)^{-\frac{1}{2}}$, this can be re-expressed as:

$$P(\mu) = (2\pi)^{-\frac{1}{2}} \tau_0^{\frac{1}{2}} e^{-\frac{1}{2}\tau_0(\mu-\mu_0)^2}. \tag{A3.4}$$

The observed data

The data in this problem are **normally distributed** data:

$$x_1, x_2, x_3, \ldots, x_n. \tag{A3.5}$$

That is, they are assumed to have been generated from a normal distribution.

The likelihood of observing the data, $P(\text{data} \mid \mu)$, can be obtained with the normal probability density function, which for one datapoint is written:

$$\mathcal{L}(x; \mu, \sigma) = \frac{1}{\sqrt{2\pi}\sigma} e^{-(x-\mu)^2/(2\sigma^2)}. \tag{A3.6}$$

With more than one datapoint, the likelihood is written:

$$\mathcal{L}(x_1, \ldots, x_n; \mu, \sigma) = \prod_{i=1}^{n} \frac{1}{\sqrt{2\pi}\sigma} e^{-(x_i-\mu)^2/(2\sigma^2)} \tag{A3.7}$$

We can rearrange this equation as:

$$\mathcal{L}(x_1, \ldots, x_n; \mu, \sigma) = (2\pi)^{-\frac{n}{2}}\sigma^{-n} e^{-\frac{1}{2}\Sigma(x_i-\mu)^2/\sigma^2}. \tag{A3.8}$$

Now, let's write this equation in terms of τ instead of σ^2. The likelihood can be written:

$$P(\text{data} \mid \mu) = \mathcal{L}(x_1, \ldots, x_n; \mu, \tau) = (2\pi)^{-\frac{n}{2}}\tau^{\frac{n}{2}} e^{-\frac{1}{2}\tau\Sigma(x_i-\mu)^2}. \tag{A3.9}$$

For convenience, we will refer to this likelihood below as $P(\text{data} \mid \mu)$. This is the likelihood of observing the data, given a normal distribution. Remember that the mean, μ, is unknown, and the precision, τ (with no subscript), is known!

Bayes' Theorem

For this example, Bayes' Theorem looks like this:

$$P(\mu \mid \text{data}) = \frac{P(\text{data} \mid \mu) * P(\mu)}{\int_{-\infty}^{\infty} P(\text{data} \mid \mu) * P(\mu) d\mu}. \tag{A3.10}$$

The posterior distribution of the unknown parameter, μ, is a normal distribution with hyperparameters:

$$\mu_{\text{posterior}} = \frac{\tau_0 \mu_0 + \tau \sum_{i=1}^{n} x_i}{(\tau_0 + n * \tau)} \tag{A3.11}$$

and

$$\tau_{\text{posterior}} = \tau_0 + n * \tau. \tag{A3.12}$$

Our goal here is to show how Bayes' Theorem generates these results.

Conjugate proof

We start with Bayes' Theorem:

$$P(\mu \mid \text{data}) = \frac{P(\text{data} \mid \mu) * P(\mu)}{\int_{-\infty}^{\infty} P(\text{data} \mid \mu) * P(\mu) d\mu}. \tag{A3.13}$$

In the equation, $P(\mu \mid \text{data})$ is referring to a specific value of μ. In the denominator, we are integrating over all possible values that μ can assume ($-\infty$ to $+\infty$). To avoid confusion in the derivation, we let u be a variable that takes on these possible values:

$$P(\mu \mid \text{data}) = \frac{P(\text{data} \mid \mu) * P(\mu)}{\int_{-\infty}^{\infty} P(\text{data} \mid u) * P(u)du}. \tag{A3.14}$$

Now we replace the prior and likelihood with the equations above:

$$P(\mu|\text{data}) = \frac{(2\pi)^{-\frac{n}{2}}\tau^{\frac{n}{2}}e^{-\frac{1}{2}\tau\Sigma(x_i-\mu)^2} * (2\pi)^{-\frac{1}{2}}\tau_0^{\frac{1}{2}}e^{-\frac{\tau_0}{2}(\mu-\mu_0)^2}}{\int_{\infty}^{-\infty} (2\pi)^{-\frac{n}{2}}\tau^{\frac{n}{2}}e^{-\frac{1}{2}\tau\Sigma(x_i-u)^2} * (2\pi)^{-\frac{1}{2}}\tau_0^{\frac{1}{2}}e^{-\frac{\tau_0}{2}(u-\mu_0)^2}\,du}. \tag{A3.15}$$

First, let's focus on the numerator. We can combine terms and multiply the prior density by the likelihood. In doing so, we lose our color-coding:

$$P(\mu) * P(\text{data} \mid \mu) = (2\pi)^{-\frac{n+1}{2}}\tau_0^{\frac{1}{2}}\tau^{\frac{n}{2}}e^{-\frac{1}{2}\tau_0(\mu-\mu_0)^2-\frac{1}{2}\tau\Sigma(x_i-\mu)^2}. \tag{A3.16}$$

Now let's expand the terms in the exponent (notice that only the exponent term contains μ, our unknown parameter):

$$P(\mu) * P(\text{data} \mid \mu) = (2\pi)^{-\frac{n+1}{2}}\tau_0^{\frac{1}{2}}\tau^{\frac{n}{2}}e^{-\frac{1}{2}\tau_0[\mu^2-2\mu\mu_0+\mu_0^2]-\frac{1}{2}\tau[\Sigma x_i^2-2\mu\Sigma x_i+n\mu^2]}. \tag{A3.17}$$

Now we can combine terms involving μ^2 and μ in the exponent:

$$P(\mu) * P(\text{data} \mid \mu) = (2\pi)^{-\frac{n+1}{2}}\tau_0^{\frac{1}{2}}\tau^{\frac{n}{2}}e^{-\frac{1}{2}(\tau_0+n\tau)\mu^2-\frac{1}{2}(-2\tau_0\mu_0-2\tau\Sigma x_i)\mu-\frac{1}{2}\tau_0\mu_0^2-\frac{1}{2}\tau\Sigma x_i^2}. \tag{A3.18}$$

At this point, we will add and subtract the same term to the exponent, which is the same as adding 0. This will make a future step easier. The term we will add and subtract is:

$$\frac{-\frac{1}{2}(\tau_0\mu_0 + \tau\Sigma x_i)^2}{\tau_0 + n\tau}. \tag{A3.19}$$

This gives us:

$$(2\pi)^{-\frac{n+1}{2}}\tau_0^{\frac{1}{2}}\tau^{\frac{n}{2}}e^{-\frac{1}{2}(\tau_0+n\tau)\mu^2-\frac{1}{2}(-2\tau_0\mu_0-2\tau\Sigma x_i)\mu+\frac{-\frac{1}{2}(\tau_0\mu_0+\tau\Sigma x_i)^2}{\tau_0+n\tau}-\frac{-\frac{1}{2}(\tau_0\mu_0+\tau\Sigma x_i)^2}{\tau_0+n\tau}-\frac{1}{2}\tau_0\mu_0^2-\frac{1}{2}\tau\Sigma x_i^2}. \tag{A3.20}$$

The first three terms in the exponent can be re-expressed as:

$$-\frac{1}{2}(\tau_0 + n\tau)\left[\mu-\frac{(\tau_0\mu_0 + \tau\Sigma x_i)}{\tau_0 + n\tau}\right]^2. \tag{A3.21}$$

We have now completed the square. The last three terms of the exponent don't depend on μ, our parameter of interest. Let's simplify them as C:

$$C = \frac{\frac{1}{2}(\tau_0\mu_0 + \tau\Sigma x_i)^2}{\tau_0 + n\tau} - \frac{1}{2}\tau_0\mu_0^2 - \frac{1}{2}\tau\Sigma x_i^2. \tag{A3.22}$$

Remember that we have been focusing on the numerator of Bayes' Theorem only. We can carry out these same steps for the denominator, but we won't show them because they are essentially identical.

Now we are ready to plug these new equations into Bayes' Theorem:

$$P(\mu \mid \text{data}) = \frac{P(\text{data} \mid \mu) * P(\mu)}{\int_{-\infty}^{\infty} P(\text{data} \mid u) * P(u)du} \tag{A3.23}$$

$$P(\mu \mid \text{data}) = \frac{(2\pi)^{-\frac{n+1}{2}}\tau_0^{\frac{1}{2}}\tau^{\frac{n}{2}}e^{-\frac{1}{2}(\tau_0+n\tau)\left[\mu-\frac{(\tau_0\mu_0+\tau\Sigma x_i)}{\tau_0+n\tau}\right]^2}+C}{\int_{-\infty}^{\infty}(2\pi)^{-\frac{n+1}{2}}\tau_0^{\frac{1}{2}}\tau^{\frac{n}{2}}e^{-\frac{1}{2}(\tau_0+n\tau)\left[u-\frac{(\tau_0\mu_0+\tau\Sigma x_i)}{\tau_0+n\tau}\right]^2}+C\,du}. \tag{A3.24}$$

Various terms cancel out from the numerator and denominator, so we are left with:

$$P(\mu \mid \text{data}) = \frac{e^{-\frac{1}{2}(\tau_0+n\tau)\left[\mu-\frac{(\tau_0\mu_0+\tau\Sigma x_i)}{\tau_0+n\tau}\right]^2}}{\int_{-\infty}^{\infty}e^{-\frac{1}{2}(\tau_0+n\tau)\left[u-\frac{(\tau_0\mu_0+\tau\Sigma x_i)}{\tau_0+n\tau}\right]^2}\,du}. \tag{A3.25}$$

Next, we multiply the numerator and denominator by a term highlighted in blue, which is the same thing as multiplying by 1:

$$(2\pi)^{-\frac{1}{2}}(\tau_0+n\tau)^{\frac{1}{2}}. \tag{A3.26}$$

So we have:

$$P(\mu \mid \text{data}) = \frac{(2\pi)^{-\frac{1}{2}}(\tau_0+n\tau)^{\frac{1}{2}}e^{-\frac{1}{2}(\tau_0+n\tau)\left[\mu-\frac{(\tau_0\mu_0+\tau\Sigma x_i)}{\tau_0+n\tau}\right]^2}}{\int_{-\infty}^{\infty}(2\pi)^{-\frac{1}{2}}(\tau_0+n\tau)^{\frac{1}{2}}e^{-\frac{1}{2}(\tau_0+n\tau)\left[u-\frac{(\tau_0\mu_0+\tau\Sigma x_i)}{\tau_0+n\tau}\right]^2}\,du} \tag{A3.27}$$

Now, the denominator is the integral of a pdf and thus is equal to 1. So, now we can ignore the denominator.

Still with us? Let's compare the prior distribution and the posterior distribution. The prior distribution has the following form with the mean in red and the precision in blue:

$$P(\mu) = (2\pi)^{-\frac{1}{2}}\tau_0^{\frac{1}{2}}e^{-\frac{1}{2}\tau_0(\mu-\mu_0)^2} \tag{A3.28}$$

We now know that the posterior distribution is written with the following form, with the mean in red and the precision in blue:

$$P(\mu) = (2\pi)^{-\frac{1}{2}}(\tau_0+n\tau)^{\frac{1}{2}}e^{-\frac{1}{2}(\tau_0+n\tau)\left(\mu-\frac{\tau_0\mu_0+\tau\Sigma x_i}{\tau_0+n\tau}\right)^2} \tag{A3.29}$$

These are both normal distributions. The posterior hyperparameters are:

$$\mu_{\text{posterior}} = \frac{\tau_0\mu_0+\tau\Sigma x_i}{\tau_0+n\tau} \tag{A3.30}$$

$$\tau_{\text{posterior}} = \tau_0+n\tau \tag{A3.31}$$

Compare Equation A3.30 with Equation A3.11, and Equation A3.31 with Equation A3.12. Brilliant!

APPENDIX 4

Conjugate Solutions for Simple Linear Regression

In Chapter 17 we introduced MCMC as a method for fitting a simple linear regression model. In simple linear regression, there is a dependent variable, Y, and an independent or predictor variable, X. It is "simple" because there is just one predictor variable.

Here is the statistical model we wish to build:

$$y_i = b_0 + b_1 x_i + \epsilon_i \quad i = 1, \ldots, n \tag{A4.1}$$

We presume that each variable, Y_i, is follows a normal distribution whose mean, μ_i, is equal to $b_0 + b_1 x_i$ and whose standard deviation is σ. We can write:

$$Y_i \sim \mathrm{N}(b_0 + b_1 x_i, \sigma^2) \tag{A4.2}$$

For a given datapoint x_i, y_i, arises from a normal distribution:

- The mean of this distribution is the parameter, b_0, plus a second parameter, b_1, multiplied by the level of the predictor variable, x_i.
- The standard deviation of this distribution is σ. The precision of this distribution is τ, which is $\frac{1}{\sigma^2}$.

Thus, there are three unknown parameters to estimate in this linear model. In a Bayesian analysis, a prior distribution for each parameter is specified. Then, MCMC approaches with Gibbs sampling can be used to estimate the posterior distributions of b_0, b_1, and τ. The problem is captured in the model diagram in Figure A4.1.

The likelihood of the data

The data in this problem are **normally distributed** data:

$$y_1, y_2, y_3, \ldots, y_n. \tag{A4.3}$$

That is, they are assumed to have been generated from normal distributions.

The likelihood of observing the data, $P(\text{data} \mid \mu, \sigma^2)$, can be obtained with the normal probability density function, which for one datapoint, y_i, is written:

$$f(y_i; \mu_i, \sigma^2) = \frac{1}{\sqrt{2\pi}\sigma} e^{-(y_i - \mu_i)^2/(2\sigma^2)}. \tag{A4.4}$$

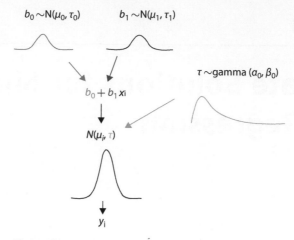

Figure A4.1

With more than one datapoint, the likelihood is written:

$$f(y_1; \mu_1, \sigma^2) \ldots f(y_n; \mu_n, \sigma^2) = \prod_{i=1}^{n} \frac{1}{\sqrt{2\pi}\sigma} e^{-(y_i - \mu_i)^2/(2\sigma^2)} \tag{A4.5}$$

We can rearrange this equation to remove the denominator:

$$f(y_1; \mu_1, \sigma^2) \ldots f(y_n; \mu_n, \sigma^2) = (2\pi)^{-\frac{n}{2}} \sigma^{-n} e^{-\frac{1}{2\sigma^2} \Sigma(y_i - \mu_i)^2}. \tag{A4.6}$$

Now, let's write this equation in terms of τ instead of σ^2. The likelihood can be written:

$$P(\text{data} \mid \mu_1, \mu_n, \tau) = (2\pi)^{-\frac{n}{2}} \tau^{\frac{n}{2}} e^{-\frac{1}{2}\tau \Sigma(y_i - \mu_i)^2}. \tag{A4.7}$$

Next, we know that the mean for each datapoint is given by $b_0 + b_1 x_i$. Thus, the likelihood for the simple linear regression problem is:

$$P(\text{data} \mid b_0, b_1, \tau) = (2\pi)^{-\frac{n}{2}} \tau^{\frac{n}{2}} e^{-\frac{1}{2}\tau \Sigma(y_i - (b_0 + b_1 x_i))^2}. \tag{A4.8}$$

This likelihood applies for all three parameters. As you'll soon see, we will write this as $P(\text{data} \mid b_0)$ when we compute the posterior distribution of b_0 given data, b_1 and τ, $P(\text{data} \mid b_1)$ when we compute the posterior distribution of b_1 given data b_0 and τ, and $P(\text{data} \mid \tau)$ when we compute the posterior distribution of τ given data b_0 and b_1. In other words:

$$P(data \mid b_0) = P(data \mid b_1) = P(data \mid \tau) = (2\pi)^{-\frac{n}{2}} \tau^{\frac{n}{2}} e^{-\frac{1}{2}\tau \Sigma(y_i - (b_0 + b_1 x_i))^2}. \tag{A4.9}$$

The remaining portions of this appendix are dedicated to showing the conjugate solutions for each of the three parameters in turn.

The b_0 parameter

The goal now is to focus only on the unknown parameter b_0. We assume that both b_1 and τ are known. (You may recall that at each MCMC trial, we draw a proposal from the posterior distribution for a parameter of interest – here, b_0, conditional on the values for the remaining parameters in that trial.)

The prior distribution

The prior distribution for the unknown parameter, b_0 provides weights of belief in alternative values of the intercept, b_0. The prior distribution is a normal distribution defined by hyperparameters for mean and precision, μ_0 and τ_0.
As shown in Appendix 3, we may write:

$$P(b_0) = \frac{\sqrt{\tau_0}}{\sqrt{2\pi}} e^{-\frac{\tau_0}{2}(b_0 - \mu_0)^2}. \tag{A4.10}$$

The posterior distribution for b_0 will be another normal distribution, with a mean of

$$\mu_{0[\text{posterior}]} = \frac{\tau_0 \mu_0 + \tau \Sigma (\boldsymbol{y_i} - \boldsymbol{b_1 x_i})}{\tau_0 + n\tau} \tag{A4.11}$$

and precision of:

$$\tau_{0[\text{posterior}]} = \tau_0 + n\tau \tag{A4.12}$$

We will next show how Bayes' Theorem provides the posterior hyperparameters.

Bayes' Theorem

We start with Bayes' Theorem. Here, we write it in the most general form for one parameter called θ to review the general form.

$$P(\theta \,|\, \text{data}) = \frac{P(\text{data} \,|\, \theta) * P(\theta)}{\int_{-\infty}^{\infty} P(\text{data} \,|\, \theta) * P(\theta) d\theta}. \tag{A4.13}$$

The prior distribution is shown in blue, and the likelihood is shown in red. For b_0, Bayes' Theorem looks like this:

$$P(b_0 \,|\, \text{data}) = \frac{P(\text{data} \,|\, b_0) * P(b_0)}{\int_{-\infty}^{\infty} P(\text{data} \,|\, b_0) * P(b_0) db_0}. \tag{A4.14}$$

In the equation, $P(b_0 \,|\, \text{data})$ is referring to a specific value of b_0. In the denominator, we are integrating over all possible values that b_0 can assume ($-\infty$ to $+\infty$). To avoid confusion in the derivation, we let u be a variable that takes on these possible values:

$$P(b_0 \,|\, \text{data}) = \frac{P(\text{data} \,|\, b_0) * P(b_0)}{\int_{-\infty}^{\infty} P(\text{data} \,|\, u) * P(u) du}. \tag{A4.15}$$

Let's focus on the numerator only, and replace each term with their respective functions for this simple linear regression problem:

$$P(b_0) * P(\text{data} \,|\, b_0) = (2\pi)^{-\frac{1}{2}} \tau_0^{\frac{1}{2}} e^{-\frac{\tau_0}{2}(b_0 - \mu_0)^2} * (2\pi)^{-\frac{n}{2}} \tau^{\frac{n}{2}} e^{-\frac{1}{2}\tau \Sigma (y_i - (b_0 + b_1 x_i))^2}. \tag{A4.16}$$

We can simplify (and lose our color-coding) as:

$$P(b_0) * P(\text{data} \,|\, b_0) = (2\pi)^{-\frac{n+1}{2}} \tau_0^{\frac{1}{2}} \tau^{\frac{n}{2}} e^{-\frac{1}{2}\tau_0(b_0 - \mu_0)^2 - \frac{1}{2}\tau \Sigma (y_i - b_0 - b_1 x_i)^2}. \tag{A4.17}$$

Now, we notice that the unknown parameter, b_0, occurs in the exponent. **We will focus our attention there**. First, we will expand the terms in the exponent:

$$P(b_0) * P(\text{data} \mid b_0) = (2\pi)^{-\frac{n+1}{2}} \tau_0^{\frac{1}{2}} \tau^{\frac{n}{2}} e^{-\frac{1}{2}\tau_0[b_0^2 - 2b_0\mu_0 + \mu_0^2] - \frac{1}{2}\tau[nb_0^2 - 2b_0\Sigma(y_i - b_1 x_i) + \Sigma(y_i - b_1 x_i)^2]}. \qquad \text{(A4.18)}$$

Next, we combine terms in the exponent involving b_0^2 and b_0:

$$P(b_0) * P(\text{data} \mid b_0) = (2\pi)^{-\frac{n+1}{2}} \tau_0^{\frac{1}{2}} \tau^{\frac{n}{2}} e^{-\frac{1}{2}[\tau_0 + n\tau]b_0^2 - \frac{1}{2}[-2\tau_0\mu_0 - 2\tau\Sigma(y_i - b_1 x_i)]b_0 - \frac{1}{2}\tau_0\mu_0^2 - \frac{1}{2}\tau\Sigma(y_i - b_1 x_i)^2}. \qquad \text{(A4.19)}$$

Next, we will add and subtract a term to the exponent, which is adding 0 to the exponent but will make our next steps easier. This allows us to complete a square for b_0. The term is:

$$\frac{-\frac{1}{2}[\tau_0\mu_0 + \tau\Sigma(y_i - b_1 x_i)]^2}{\tau_0 + n\tau} \qquad \text{(A4.20)}$$

$P(b_0) * P(\text{data} \mid b_0)$

$$= (2\pi)^{-\frac{n+1}{2}} \tau_0^{\frac{1}{2}} \tau^{\frac{n}{2}} e^{-\frac{1}{2}[\tau_0 + n\tau]b_0^2 + [\tau_0\mu_0 + \tau\Sigma(y_i - b_1 x_i)]b_0 + \frac{-\frac{1}{2}[\tau_0\mu_0 + \tau\Sigma(y_i - b_1 x_i)]^2}{\tau_0 + n\tau} - \frac{-\frac{1}{2}[\tau_0\mu_0 + \tau\Sigma(y_i - b_1 x_i)]^2}{\tau_0 + n\tau} - \frac{1}{2}\tau_0\mu_0^2 - \frac{1}{2}\tau\Sigma(y_i - b_1 x_i)^2} . \qquad \text{(A4.21)}$$

The first three terms in the exponent can be re-expressed (completing the square) as:

$$-\frac{1}{2} * (\tau_0 + n\tau) \left[b_0 - \frac{\left(\tau_0\mu_0 + \tau\Sigma(y_i - b_1 x_i)\right)}{\tau_0 + n\tau} \right]^2 . \qquad \text{(A4.22)}$$

The last three terms in the exponent don't depend on b_0. Let's simplify them as C.

Remember that we have been focusing on the numerator of Bayes' Theorem only. We can carry out these same steps for the denominator but won't show them because they are essentially identical.

Plugging in our equations into Bayes' Theorem, we have:

$$P(b_0 \mid \text{data}) = \frac{(2\pi)^{-\frac{n+1}{2}} \tau_0^{\frac{1}{2}} \tau^{\frac{n}{2}} e^{-\frac{1}{2}(\tau_0 + n\tau)[b_0 - \frac{(\tau_0\mu_0 + \tau\Sigma(y_i - b_1 x_i))}{\tau_0 + n\tau}]^2 + C}}{\int_{-\infty}^{\infty} (2\pi)^{-\frac{n+1}{2}} \tau_0^{\frac{1}{2}} \tau^{\frac{n}{2}} e^{-\frac{1}{2}(\tau_0 + n\tau)[u - \frac{(\tau_0\mu_0 + \tau\Sigma(y_i - b_1 x_i))}{\tau_0 + n\tau}]^2 + C} \, du} . \qquad \text{(A4.23)}$$

Various terms cancel out from the numerator and denominator, which leaves:

$$P(b_0 \mid \text{data}) = \frac{e^{-\frac{1}{2}(\tau_0 + n\tau)[b_0 - \frac{(\tau_0\mu_0 + \tau\Sigma(y_i - b_1 x_i))}{\tau_0 + n\tau}]^2}}{\int_{-\infty}^{\infty} e^{\frac{1}{2}(\tau_0 + n\tau)[u - \frac{(\tau_0\mu_0 + \tau\Sigma(y_i - b_1 x_i))}{\tau_0 + n\tau}]^2} \, du} . \qquad \text{(A4.24)}$$

Next, we multiply both the numerator and denominator by the term:

$$\frac{\sqrt{\tau_0 + n\tau}}{\sqrt{2\pi}} . \qquad \text{(A4.25)}$$

This gives:

$$P(b_0 \mid \text{data}) = \frac{\frac{\sqrt{\tau_0 + n\tau}}{\sqrt{2\pi}} e^{-\frac{1}{2}(\tau_0 + n\tau)[b_0 - \frac{(\tau_0\mu_0 + \tau\Sigma(y_i - b_1 x_i))}{\tau_0 + n\tau}]^2}}{\int_{-\infty}^{\infty} \frac{\sqrt{\tau_0 + n\tau}}{\sqrt{2\pi}} e^{-\frac{1}{2}(\tau_0 + n\tau)[u - \frac{(\tau_0\mu_0 + \tau\Sigma(y_i - b_1 x_i))}{\tau_0 + n\tau}]^2} \, du} . \qquad \text{(A4.26)}$$

The denominator is the integral of a pdf—a normal distribution. Thus, the denominator is equal to 1. This leaves us with:

$$P(b_0 \mid \text{data}) = \frac{\sqrt{\tau_0 + n\tau}}{\sqrt{2\pi}} e^{\frac{1}{2}(\tau_0 \mid n\tau)[b_0 \; \frac{(\tau_0\mu_0 + \tau\Sigma(y_i - b_1 x_i))}{\tau_0 + n\tau}]^2}. \tag{A4.27}$$

Here is the normal pdf for our prior distribution for b_0 again:

$$P(b_0) = \frac{\sqrt{\tau_0}}{\sqrt{2\pi}} e^{-\frac{\tau_0}{2}(b_0 - \mu_0)^2}. \tag{A4.28}$$

Thus, the posterior distribution of b_0 is a normal distribution with mean and precision hyperparameters that are defined by:

$$\mu_{0\text{posterior}} = \frac{\tau_0\mu_0 + \tau\Sigma(y_i - b_1 x_i)}{\tau_0 + n\tau} \tag{A4.29}$$

$$\tau_{0\text{posterior}} = \tau_0 + n\tau. \tag{A4.30}$$

Match these results with Equations A4.11 and A4.12.

The b_1 parameter

We now focus only on the conjugate solutions for the unknown parameter b_1 in a simple linear regression problem. We assume that both b_0 and τ are known. (You may recall that at each MCMC trial, we draw a proposal from the posterior distribution for a parameter of interest—here, b_1, conditional on the values for the remaining parameters in that trial.)

The prior distribution

The prior distribution for the unknown parameter, b_1 provides weights of belief in alternative values of the slope, b_1. The prior distribution is a normal distribution defined by hyperparameters μ_1 and τ_1.

Recall from Equation A4.9 that we may write this normal distribution as:

$$P(b_1) = \frac{\sqrt{\tau_1}}{\sqrt{2\pi}} e^{-\frac{\tau_1}{2}(b_1 - \mu_1)^2}. \tag{A4.31}$$

The posterior distribution for b_1 will be another normal distribution, with a mean hyperparameter equal to

$$\mu_{1[\text{posterior}]} = \frac{\tau_1\mu_1 + \tau\Sigma x_i(y_i - b_0)}{\tau_1 + \tau\Sigma x_i^2} \tag{A4.32}$$

and precision hyperparameter equal to

$$\tau_{1[\text{posterior}]} = \tau_1 + \tau\Sigma x_i^2 \tag{A4.33}$$

We will next show how Bayes' Theorem provides these posterior hyperparameters.

Bayes' Theorem

We start with Bayes' Theorem. Here, we write it in the most general form for one parameter called θ to review the general form:

$$P(\theta \mid \text{data}) = \frac{P(\text{data} \mid \theta) * P(\theta)}{\int P(\text{data} \mid \theta) * P(\theta) d\theta}.$$

(A4.34)

For b_1, Bayes' Theorem looks like this:

$$P(b_1 \mid \text{data}) = \frac{P(\text{data} \mid b_1) * P(b_1)}{\int_{-\infty}^{\infty} P(\text{data} \mid b_1) * P(b_1) db_1}.$$

(A4.35)

In the equation, $P(b_1 \mid \text{data})$ is referring to a specific value of b_1. In the denominator, we are integrating over all possible values that b_1 can assume ($-\infty$ to $+\infty$). To avoid confusion in the derivation, we let u be a variable that takes on these possible values:

$$P(b_1 \mid \text{data}) = \frac{P(\text{data} \mid b_1) * P(b_1)}{\int_{-\infty}^{\infty} P(\text{data} \mid u) * P(u) du}.$$

(A4.36)

The prior distribution is shown in blue, and the likelihood is shown in red.
Let's focus on the numerator only and replace each term with their respective functions for this simple linear regression problem:

$$P(b_1) * P(\text{data} \mid b_1) = P(b_1) = \frac{\sqrt{\tau_1}}{\sqrt{2\pi}} e^{-\frac{\tau_1}{2}(b_1-\mu_1)^2} * (2\pi)^{-\frac{n}{2}} \tau^{\frac{n}{2}} e^{-\frac{1}{2}\tau\Sigma(y_i-(b_0+b_1x_i))^2}$$

(A4.37)

We can simplify (and lose our color-coding) as:

$$P(b_1) * P(\text{data} \mid b_1) = (2\pi)^{-\frac{n+1}{2}} \tau_1^{\frac{1}{2}} \tau^{\frac{n}{2}} e^{-\frac{1}{2}\tau_1(b_1-\mu_1)^2 - \frac{1}{2}\tau\Sigma((y_i-b_0)-b_1x_i)^2}$$

(A4.38)

Now, we notice that the unknown parameter, b_1, occurs in the exponent. **We will focus our attention there**. First, we will expand the terms in the exponent:

$$P(b_1) * P(\text{data} \mid b_1) = (2\pi)^{-\frac{n+1}{2}} \tau_1^{\frac{1}{2}} \tau^{\frac{n}{2}} e^{-\frac{1}{2}\tau_1[b_1^2-2b_1\mu_1+\mu_1^2] - \frac{1}{2}\tau[b_1^2\Sigma x_i^2-2b_1\Sigma x_i(y_i-b_0)+\Sigma(y_i-b_0)^2]}$$

(A4.39)

Next, we combine terms in the exponent involving b_1^2 and b_1:

$$P(b_1) * P(\text{data} \mid b_1) = (2\pi)^{-\frac{n+1}{2}} \tau_0^{\frac{1}{2}} \tau^{\frac{n}{2}} e^{-\frac{1}{2}[\tau_1+\tau\Sigma x_i^2]b_1^2-\frac{1}{2}[-2\tau_1\mu_1-2\tau\Sigma x_i(y_i-b_0)]b_1-\frac{1}{2}\tau_1\mu_1^2-\frac{1}{2}\tau\Sigma(y_i-b_0)^2}$$

(A4.40)

Next, we will add and subtract a term to the exponent, which is adding 0 to the exponent but will make our next steps easier. This allows us to complete a square for b_1. The term is:

$$\frac{-\frac{1}{2}[\tau_1\mu_1 + \tau\Sigma x_i(y_i-b_0)]^2}{\tau_1 + \tau\Sigma x_i^2}$$

(A4.41)

$P(b_1)*P(\text{data}|b_1)$

$$= (2\pi)^{-\frac{n+1}{2}} \tau_1^{\frac{1}{2}} \tau^{\frac{n}{2}} e^{-\frac{1}{2}[\tau_1+\tau\Sigma x_i^2]b_1^2-\frac{1}{2}[-2\tau_1\mu_1-2\tau\Sigma x_i(y_i-b_0)]b_1+\frac{-\frac{1}{2}[\tau_1\mu_1+\tau\Sigma x_i(y_i-b_0)]^2}{\tau_1+\tau\Sigma x_i^2}-\frac{-\frac{1}{2}[\tau_1\mu_1+\tau\Sigma x_i(y_i-b_0)]^2}{\tau_1+\tau\Sigma x_i^2}-\frac{1}{2}\tau_1\mu_1^2-\frac{1}{2}\tau\Sigma(y_i-b_0)^2}$$

(A4.42)

The *first* three terms in the exponent can be re-expressed (completing the square) as:

$$-\frac{1}{2}[\tau_1 + \tau\Sigma x_i^2]\left[b_1 - \frac{\tau_1\mu_1 + \tau\Sigma x_i(y_i-b_0)}{\tau_1 + \tau\Sigma x_i^2}\right]^2$$

(A4.43)

The *last* three terms in the exponent don't depend on b_1. Let's simplify them as C. We are now ready to plug these into Bayes' Theorem. In this case, it is:

$$P(b_1 \mid \text{data}) = \frac{(2\pi)^{-\frac{n+1}{2}} \tau_0^{\frac{1}{2}} \tau^{\frac{n}{2}} e^{-\frac{1}{2}[\tau_1 + \tau\Sigma x_i^2]\left[b_1 - \frac{\tau_1\mu_1 + \tau\Sigma x_i(y_i - b_0)}{\tau_1 + \tau\Sigma x_i^2}\right]^2 + C}}{\displaystyle\int_{-\infty}^{\infty} (2\pi)^{-\frac{n+1}{2}} \tau_0^{\frac{1}{2}} \tau^{\frac{n}{2}} e^{-\frac{1}{2}[\tau_1 + \tau\Sigma x_i^2]\left[u - \frac{\tau_1\mu_1 + \tau\Sigma x_i(y_i - b_0)}{\tau_1 + \tau\Sigma x_i^2}\right]^2 + C} \, du}$$

(A4.44)

Various terms cancel out from the numerator and denominator, which leaves:

$$P(b_1 \mid \text{data}) = \frac{e^{-\frac{1}{2}[\tau_1 + \tau\Sigma x_i^2]\left[b_1 - \frac{\tau_1\mu_1 + \tau\Sigma x_i(y_i - b_0)}{\tau_1 + \tau\Sigma x_i^2}\right]^2}}{\displaystyle\int_{-\infty}^{\infty} e^{-\frac{1}{2}[\tau_1 + \tau\Sigma x_i^2]\left[u - \frac{\tau_1\mu_1 + \tau\Sigma x_i(y_i - b_0)}{\tau_1 + \tau\Sigma x_i^2}\right]^2} \, du}$$

(A4.45)

Next, we multiply both the numerator and denominator by the term:

$$\frac{\sqrt{\tau_1 + \tau\Sigma x_i^2}}{\sqrt{2\pi}}$$

(A4.46)

This gives:

$$P(b_1 \mid \text{data}) = \frac{\frac{\sqrt{\tau_1 + \tau\Sigma x_i^2}}{\sqrt{2\pi}} e^{-\frac{1}{2}[\tau_1 + \tau\Sigma x_i^2]\left[b_1 - \frac{\tau_1\mu_1 + \tau\Sigma x_i(y_i - b_0)}{\tau_1 + \tau\Sigma x_i^2}\right]^2}}{\displaystyle\int_{-\infty}^{\infty} \frac{\sqrt{\tau_1 + \tau\Sigma x_i^2}}{\sqrt{2\pi}} e^{-\frac{1}{2}[\tau_1 + \tau\Sigma x_i^2]\left[u - \frac{\tau_1\mu_1 + \tau\Sigma x_i(y_i - b_0)}{\tau_1 + \tau\Sigma x_i^2}\right]^2} \, du}$$

(A4.47)

The denominator is the integral of a pdf—a normal distribution. Thus, the denominator is equal to 1. Thus, we have:

$$P(b_1 \mid \text{data}) = \frac{\sqrt{\tau_1 + \tau\Sigma x_i^2}}{\sqrt{2\pi}} e^{-\frac{1}{2}[\tau_1 + \tau\Sigma x_i^2]\left[b_1 - \frac{\tau_1\mu_1 + \tau\Sigma x_i(y_i - b_0)}{\tau_1 + \tau\Sigma x_i^2}\right]^2}$$

(A4.48)

Here is the normal pdf again for our prior distribution for the unknown parameter, b_1, defined by hyperparameters τ_1 and μ_1 (see Equation A4.28):

$$P(b_1) = \frac{\sqrt{\tau_1}}{\sqrt{2\pi}} e^{-\frac{\tau_1}{2}(b_1 - \mu_1)^2}.$$

(A4.49)

Now look for the corresponding terms in Equation A4.43. You can see that the posterior distribution of b_1 is a normal distribution with hyperparameters defined by:

$$\mu_{1 posterior} = \frac{\tau_1\mu_1 + \tau\Sigma x_i(y_i - b_0)}{\tau_1 + \tau\Sigma x_i^2}$$

(A4.50)

$$\tau_{1 posterior} = \tau_1 + \tau\Sigma x_i^2.$$

(A4.51)

Match these results with Equations A4.32 and A4.33.

The τ parameter

We now focus only on the conjugate solutions for the unknown parameter τ in a simple linear regression problem. We assume that both b_0 and b_1 are known. (You may recall that at each MCMC trial, we draw a proposal from the posterior distribution for a parameter of interest—here, τ, conditional on the values for the remaining parameters in that trial.)

The prior distribution

The prior distribution for the unknown parameter, τ, provides weights of belief in alternative values of the precision, τ. The prior distribution is a gamma distribution, and you may recall that the gamma pdf can be written:

$$g(x; \alpha, \beta) = \frac{\beta^\alpha x^{\alpha-1} e^{-\beta x}}{\Gamma(\alpha)}, \quad 0 \leq x \leq \infty. \tag{A4.52}$$

Remember that the gamma distribution is controlled by two parameters: α (also called the "shape" parameter) and β (also called the "rate" parameter). Figure A4.2 shows a few alternative gamma distributions.

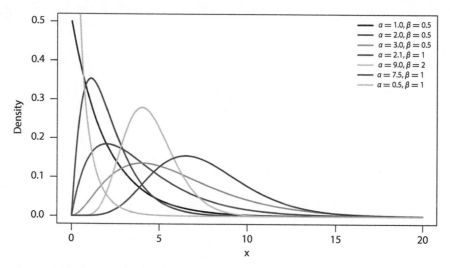

Figure A4.2 Gamma distributions

The prior distribution is a gamma distribution defined by hyperparameters α_0 and β_0:

$$P(\tau) = \frac{\beta_0^{\alpha_0} \tau^{\alpha_0-1} e^{-\beta_0 \tau}}{\Gamma(\alpha_0)} \quad 0 \leq \tau \leq \infty. \tag{A4.53}$$

The posterior distribution for τ will be another gamma distribution, with a hyperparameter equal to

$$\alpha_{[\text{posterior}]} = \alpha_0 + \frac{n}{2} \tag{A4.54}$$

$$\beta_{[\text{posterior}]} = \beta_0 + \frac{\Sigma_{i=1}^n \left(y_i - (b_0 + b_1 x_i) \right)^2}{2}. \tag{A4.55}$$

We will next show how Bayes' Theorem provides these posterior hyperparameters.

Bayes' Theorem

We start with Bayes' Theorem. Here, we write it in the most general form for one parameter called θ to review the general form:

$$P(\theta\,|\,\text{data}) = \frac{P(\text{data}\,|\,\theta) * P(\theta)}{\int P(\text{data}\,|\,\theta) * P(\theta)d\theta}. \tag{A4.56}$$

For τ, Bayes' Theorem looks like this:

$$P(\tau\,|\,\text{data}) = \frac{P(\text{data}\,|\,\tau) * P(\tau)}{\int_0^\infty P(\text{data}\,|\,\tau) * P(\tau)d\tau}. \tag{A4.57}$$

In the equation, $P(\tau\,|\,\text{data})$ is referring to a specific value of τ. In the denominator, we are integrating over all possible values that τ can assume (0 to $+\infty$). To avoid confusion in the derivation, we let u be a variable that takes on these possible values:

$$P(\tau\,|\,\text{data}) = \frac{P(\text{data}\,|\,\tau) * P(\tau)}{\int_0^\infty P(\text{data}\,|\,u) * P(u)du}. \tag{A4.58}$$

The prior distribution is shown in blue, and the likelihood is shown in red. Let's focus on the numerator only and replace each term with their respective functions for this simple linear regression problem:

$$P(\tau) * P(\text{data}\,|\,\tau) = \frac{\beta_0^{\alpha_0}\tau^{\alpha_0-1}e^{-\beta_0\tau}}{\Gamma(\alpha_0)} * (2\pi)^{-\frac{n}{2}}\tau^{\frac{n}{2}}e^{-\frac{1}{2}\tau\Sigma(y_i-(b_0-b_1x_i))^2} \tag{A4.59}$$

For convenience, we'll combine terms and re-express this as:

$$P(\text{data}\,|\,\tau) * P(\tau) = (2\pi)^{-\frac{n}{2}}\tau^{\alpha_0+\frac{n}{2}-1}\beta_0^{\alpha_0}e^{-\beta_0\tau-\frac{\tau}{2}\Sigma(y_i-b_0-b_1x_i)^2}/\Gamma(\alpha_0) \tag{A4.60}$$

We are now ready to plug these into Bayes' Theorem. In this case, it is:

$$P(\tau|\text{data}) = \frac{(2\pi)^{-\frac{n}{2}}\tau^{\alpha_0+\frac{n}{2}-1}\beta_0^{\alpha_0}e^{-\tau[\beta_0+\frac{1}{2}\Sigma(y_i-b_0-b_1x_i)^2]}/\Gamma(\alpha_0)}{\int_0^\infty (2\pi)^{-\frac{n}{2}}u^{\alpha_0+\frac{n}{2}-1}\beta_0^{\alpha_0}e^{-u[\beta_0+\frac{1}{2}\Sigma(y_i-b_0-b_1x_i)^2]}/\Gamma(\alpha_0)du} \tag{A4.61}$$

The terms, $(2\pi)^{-\frac{n}{2}}$ and $\beta_0^{\alpha_0}$ and $\Gamma(\alpha_0)$ cancel from the numerator and denominator. Next, we'll multiply both the numerator and denominator by a term:

$$\frac{[\beta_0 + \frac{1}{2}\Sigma(y_i-b_0-b_1x_i)^2]^{\alpha_0+\frac{n}{2}}}{\Gamma(\alpha_0 + \frac{n}{2})} \tag{A4.62}$$

This gives us:

$$P(\tau\,|\,\text{data}) = \frac{\frac{\tau^{\alpha_0+\frac{n}{2}-1}}{\Gamma(\alpha_0+\frac{n}{2})}\left[\beta_0 + \frac{1}{2}\Sigma(y_i-b_0-b_1x_i)^2\right]^{\alpha_0+\frac{n}{2}}e^{-(\beta_0+\frac{1}{2}\Sigma(y_i-b_0-b_1x_i)^2)\tau}}{\int_0^\infty \frac{u^{\alpha_0+\frac{n}{2}-1}}{\Gamma(\alpha_0+\frac{n}{2})}\left[\beta_0 + \frac{1}{2}\Sigma(y_i-b_0-b_1x_i)^2\right]^{\alpha_0+\frac{n}{2}}e^{-(\beta_0+\frac{1}{2}\Sigma(y_i-b_0-b_1x_i)^2)u}du} \tag{A4.63}$$

The denominator is the integral of a pdf—a gamma distribution. Thus, the denominator is equal to 1.

With some minor rearrangement, we now have:

$$P(\tau \mid \text{data}) = \frac{\left[\beta_0 + \frac{1}{2}\Sigma(y_i - b_0 - b_1 x_i)^2\right]^{\alpha_0 + \frac{n}{2}} \tau^{\alpha_0 + \frac{n}{2} - 1} e^{-\left(\beta_0 + \frac{1}{2}\Sigma(y_i - b_0 - b_1 x_i)^2\right)\tau}}{\Gamma\left(\alpha_0 + \frac{n}{2}\right)} \qquad (A4.64)$$

Here is the gamma pdf again for our prior distribution for the unknown parameter, τ, defined by hyperparameters α_0 and β_0 (see Equation A4.53):

$$P(\tau) = \frac{\beta_0^{\alpha_0} \tau^{\alpha_0 - 1} e^{-\beta_0 \tau}}{\Gamma(\alpha_0)}, \quad 0 \le x \le \infty. \qquad (A4.65)$$

You can see that the posterior distribution of τ is a gamma distribution with hyperparameters defined by:

$$\alpha_{[\text{posterior}]} = \alpha_0 + \frac{n}{2} \qquad (A4.66)$$

$$\beta_{[\text{posterior}]} = \beta_0 + \frac{\sum_{i=1}^n (y_i - b_0 - b_1 x_i)^2}{2}. \qquad (A4.67)$$

Match these results with Equations A4.54 and A4.55.

The Standardization of Regression Data

In Chapter 17 we introduced MCMC as a method for developing a simple linear regression model. In simple linear regression, there is a dependent variable, Y, and an independent or predictor variable, X. It is "simple" because there is just one predictor variable.

Here is the statistical model we wish to build, where y_i is **success** in Survivor (number of days) and x_i is **years** of formal education:

$$y_i = b_0 + b_1 x_i + \epsilon_i \qquad i = 1, \dots, n \tag{A5.1}$$

We presume that each, Y_i follows a normal distribution whose mean, μ_i, is equal to $b_0 + b_1 x_i$ and whose standard deviation is σ. We can write:

$$Y_i \sim \mathrm{N}(b_0 + b_1 x_i, \sigma^2) \tag{A5.2}$$

For a given datapoint x_i, y_i, arises from a normal distribution:

- The mean of this distribution is the parameter, b_0, plus a second parameter, b_1, multiplied by the level of the predictor variable, x_i.
- The standard deviation of this distribution is σ. The precision of this distribution is τ, which is $\frac{1}{\sigma^2}$.

Thus, there are three unknown parameters to estimate in this linear model. In a Bayesian analysis, a prior distribution for each parameter is specified. Then, MCMC approaches with Gibbs Sampling can be used to estimate the posterior distributions of b_0, b_1, and τ. In Chapter 17, we demonstrated how MCMC approaches with Gibbs sampling can be used to estimate the parameters. We mentioned that datasets may be standardized to Z scores prior to an MCMC analysis. The reason for this is that an MCMC analysis often produces a joint posterior distribution in which the parameters are highly correlated. Standardization of data can relieve some of this correlation.

Here, we aim to illustrate how to standardize variables in a dataset, and then how to back-transform them to the original scale for a simple linear regression analysis.

Standardizing datasets

To illustrate the process, we start with our original **grit** dataset (see Table A5.1), focusing only on the columns **Success** and **Years** (as in Chapter 17):

Here, the dependent variable is **success**, and the predictor variable is **years**. The original dataset is highlighted in gray.

Table A5.1

	Success	Z_Success	Years	Z_Years
1	33.48	−0.39	12.00	−0.67
2	42.53	0.38	14.00	−0.22
3	48.53	0.89	18.00	0.67
4	30.21	−0.67	10.00	−1.12
5	38.76	0.06	13.00	−0.45
6	38.59	0.04	22.00	1.57
7	52.93	1.26	17.00	0.45
8	32.65	−0.46	15.00	0.00
9	52.42	1.22	16.00	0.22
10	22.22	−1.35	9.00	−1.34
11	41.40	0.28	19.00	0.89
12	16.28	−1.86	8.00	−1.57
13	40.83	0.23	20.00	1.12
14	24.43	−1.16	11.00	−0.89
15	56.38	1.55	21.00	1.34
Mean	38.11	0	15.00	0
Std	11.76	1	4.47	1

To standardize the dataset column by column, we find the mean and standard deviation for each column. Then, the standardized Z score for success for a given individual, $z_{success_i}$, is computed as:

$$z_{success_i} = \frac{(y_i - \bar{y})}{s_y}. \tag{A5.3}$$

For example, the mean **success**, $\bar{y}$, is 33.11 days, and the standard deviation, s_y, is 11.76. The first record in Table A5.1 is a contestant that lasted 33.48 days in a Survivor contest. The Z score for this contestant is:

$$z_{success_1} = \frac{(y_1 - \bar{y})}{s_y} = \frac{33.48 - 38.11}{11.76} = -0.39. \tag{A5.4}$$

Similarly, the Z score for the first contestant for the years variable can be computed as:

$$z_{years_1} = \frac{(x_1 - \bar{x})}{s_x} = \frac{12 - 15}{4.47} = -0.67. \tag{A5.5}$$

Now that the data are standardized, the MCMC can be run, and the resulting joint posterior distribution of the three parameters will hopefully have less correlation. The "challenge" is that the parameters are now on this new scale, whose meaning is not as apparent. For instance, b_1 no longer is the unit increase that a contestant will last on Survivor with each unit increase in years of formal education.

Back-transforming standardized coefficients to the original scale

The main challenge now is how to convert the b_0 and b_1 estimates to their original scale. For simple linear regression, the solution is straightforward.

- Let's let b_0^* represent the standardized coefficient for b_0. This is equal to 0. Let b_1^* represent the standardized coefficient for b_1.
- Let s_y represent the standard deviation of the y vector. Let's let s_x represent the standard deviation of the x vector.

We begin by writing the equation to predict the standardized response score (success) for contestant i as a function of the standardized coefficients and standardized predictor score (years). Let's let the predicted value of z for individual i be expressed as $\hat{z}_{y_i}$:

$$\hat{z}_{y_i} = \frac{\hat{y}_i - \bar{y}}{s_y} = b_0^* + b_1^* \frac{(x_i - \bar{x})}{s_x}. \tag{A5.6}$$

Next, multiply through by s_y:

$$\hat{y}_i - \bar{y} = 0 + b_1^* \frac{s_y}{s_x}(x_i - \bar{x}). \tag{A5.7}$$

Now, we isolate $\hat{y}_i$ by adding $\bar{y}$ to both sides and then add a bit of color-coding to help us see the pattern:

$$\hat{y}_i = 0 + \bar{y} - b_1^* \frac{s_y}{s_x}\bar{x} + b_1^* \frac{s_y}{s_x} x_i. \tag{A5.8}$$

This gives us the linear form that we seek:

$$\hat{y}_i = b_0 + b_1 x_i. \tag{A5.9}$$

Thus, the back-transformed estimate of b_0 is:

$$b_0 = \bar{y} - b_1^* \frac{s_y}{s_x}\bar{x}. \tag{A5.10}$$

And the back-transformed estimate of b_1 is:

$$b_1 = b_1^* \frac{s_y}{s_x}. \tag{A5.11}$$

These coefficients are now on the original scale. In an MCMC analysis, b_0^* and b_1^* are generated in each trial. Similarly, the back-transformed estimates are computed for each trial.

Bibliography

J. Albert. *Bayesian Computation with R*. Springer New York, 2009. DOI: 10.1007/978-0-387-92298-0. http://doi.org/10.1007%2F978-0-387-92298-0.

M. Allaby. *Oxford Dictionary of Zoology*. Oxford University Press, 2014.

C. Andrieu, N. D. Freitas, A. Doucet, et al. "An introduction to MCMC for machine learning." *Machine Learning* 50.1–2 (2003): 5–43.

T. Bayes and R. Price. "An Essay towards solving a Problem in the Doctrine of Chance. By the late Rev. Mr. Bayes, communicated by Mr. Price, in a letter to John Canton, A. M. F. R. S." *Philosophical Transactions of the Royal Society of London* 53 (1763): 370–418.

J. O. Berger, D. R. Insua, and F. Ruggeri. "Bayesian robustness." In: *Robust Bayesian Analysis*. Ed. by D. R. Insua and F. Ruggeri. Springer New York, 2000, pp. 1–32. DOI: 10.1007/978-1-4612-1306-2_1. http://doi.org/10.1007%2F978-1-4612-1306-2_1.

J. Bernardo and A. Smith. *Bayesian Theory*. John Wiley, 1994.

J. Bernoulli. *Opera Jacobi Bernoullii*. Genevæ, Sumptibus Hæredum Cramer & Fratrum Philibert, 1774.

B. Bolker. *Ecological Models and Data in R*. Princeton University Press, 2008.

K. P. Burnham and D. R. Anderson, ed. *Model Selection and Multimodel Inference: A Practical Information–Theoretic Approach*. Springer New York, 2004. DOI: 10.1007/b97636. http://doi.org/10.1007%2Fb97636.

C. Clapham and J. Nicholson. *The Concise Oxford Dictionary of Mathematics* (fifth edition). Oxford University Press, 2014.

M. Conroy and J. Peterson. *Decision Making in Natural Resource Management: A Structured, Adaptive Approach*. Wiley-Blackwell, 2013. http://www.wiley.com/WileyCDA/WileyTitle/productCd-0470671742.html.

A. L. Duckworth, C. Peterson, M. D. Matthews, et al. "Grit: Perseverance and passion for long-term goals." *Journal of Personality and Social Psychology* 92.6 (2007): 1087–101. DOI: 10.1037/0022-3514.92.6.1087. http://doi.org/10.1037%2F0022-3514.92.6.1087.

M. Elliot, I. Fairweather, W. Olsen, et al. *A Dictionary of Social Research Methods*. Oxford University Press, 2016.

B. Everitt. *The Cambridge Dictionary of Statistics*. Cambridge University Press, 1998.

C. F. Gauss. *Werke*. Cambridge Library Collection: Mathematics. Cambridge University Press, 2011 [1863–1933].

A. E. Gelfand and A. F. M. Smith. "Sampling-based approaches to calculating marginal densities." *Journal of the American Statistical Association* 85.410 (1990): 398–409. DOI: 10.2307/2289776. http://doi.org/10.2307%2F2289776.

A. Gelman, J. B. Carlin, H. S. Stern, and D. B. Rubin. *Bayesian Data Analysis*. Chapman & Hall, 2004.

S. Geman and D. Geman. "Stochastic relaxation, Gibbs distributions, and the Bayesian restoration of images." In: *Readings in Computer Vision*. Ed. by M. A. Fischler and O. Firschein. Elsevier, 1987, pp. 564–84. DOI: 10.1016/b978-0-08-051581-6.50057-x. http://doi.org/10.1016%2Fb978-0-08-051581-6.50057-x.

W. R. Gilks, A. Thomas, and D. J. Spiegelhalter. "A language and program for complex Bayesian modelling." *The Statistician* 43.1 (1994): 169–77. DOI: 10.2307/2348941. http://doi.org/10.2307%2F2348941.

P. Goodwin and G. Wright. *Decision Analysis for Management Judgment* (fourth edition). Wiley and Sons, 2009. http://bcs.wiley.com/he-bcs/Books?action=index&itemId=0470714395&bcsId=5047.

P. Goodwin and G. Wright. *Decision Analysis for Management Judgment* (fifth edition). John Wiley & Sons, 2014.

S. Greenland. "Bayesian perspectives for epidemiological research: I. Foundations and basic methods." *International Journal of Epidemiology* 35.3 (2006): 765–74.

R. Gregory, L. Failing, M. Harstone, et al. *Structured Decision Making: A Practical Guide to Environmental Management Choices*. Wiley and Sons, 2012. DOI: 10.1002/9781444398557. http://onlinelibrary.wiley.com/book/10.1002/9781444398557.

J. S. Hammond, R. L. Keeney, and H. Raiffa. *Smart Choices: A Practical Guide to Making Better Decisions*. Harvard Business Review Press, 1999.

K. Hastings. "Monte Carlo sampling methods using Markov chains and their applications." *Biometrika* 57.1 (1970): 97–109.

N. T. Hobbs and M. B. Hooten. *Bayesian Models: A Statistical Primer for Ecologists*. Princeton University Press, 2015. DOI: 10.1515/9781400866557. http://doi.org/10.1515%2F9781400866557.

M. B. Hooten and N. T. Hobbs. "A guide to Bayesian model selection for ecologists." *Ecological Monographs* 85.1 (2015): 3–28. DOI: 10.1890/14-0661.1. http://doi.org/10.1890%2F14-0661.1.

M. Kéry. *Introduction to WinBUGS for Ecologists*. Elsevier, 2010. DOI: 10.1016/c2009-0-30639-x. http://doi.org/10.1016%2Fc2009-0-30639-x.

J. Kruschke. *Doing Bayesian Data Analysis: A Tutorial with R, JAGS, and Stan*. Elsevier, 2015. DOI: 10.1016/c2012-0-00477-2. http://doi.org/10.1016%2Fc2012-0-00477-2.

D. Lane. *Online Statistics Education: A Multimedia Course of Study*. http://onlinestatbook.com/, 2011.

J. M. Last, ed. *A Dictionary of Public Health* (first edition). Oxford University Press, 2007.

W. Link and R. Barker. *Bayesian Inference*. Elsevier, 2010. DOI: 10.1016/c2009-0-01674-2. http://doi.org/10.1016%2Fc2009-0-01674-2.

J. Mayer, K. Khairy, and J. Howard. "Drawing an elephant with four complex parameters." *American Journal of Physics* 78.6 (2010): 648–9. DOI: 10.1119/1.3254017. http://doi.org/10.1119%2F1.3254017.

R. K. McCann, B. G. Marcot, and R. Ellis. "Bayesian belief networks: Applications in ecology and natural resource management." *Canadian Journal of Forest Research* 36.12 (2006): 3053–62. DOI: 10.1139/x06-238. http://doi.org/10.1139%2Fx06-238.

M. A. McCarthy. *Bayesian Methods for Ecology*. Cambridge University Press, 2007. DOI: 10.1017/cbo9780511802454. http://doi.org/10.1017%2Fcbo9780511802454.

S. B. McGrayne. The Theory That Would Not Die: How Bayes' Rule Cracked the Enigma Code, Hunted Down Russian Submarines, and Emerged Triumphant from Two Centuries of Controversy. Yale University Press, 2011.

N. Metropolis, A. W. Rosenbluth, M. N. Rosenbluth, et al. "Equation of state calculations by fast computing machines." *J. Chem. Phys.* 21.6 (1953): 1089–92. DOI: 10.1063/1.1699114. http://dx.doi.org/10.1063/1.1699114.

F. Mosteller and D. L. Wallace. *Inference and Disputed Authorship: The Federalist*. Series in Behavioral Science: Quantitative Methods. Addison-Wesley, 1964.

J. Pearl. *Probabilistic Reasoning in Intelligent Systems*. Elsevier, 1988. DOI: 10.1016/c2009-0-27609-4. http://doi.org/10.1016%2Fc2009-0-27609-4.

M. Plummer. "JAGS: A program for analysis of Bayesian graphical models using Gibbs sampling." In: *Proceedings of the 3rd International Workshop on Distributed Statistical Computing (DSC 2003)*. Ed. by K. Hornik, F. Leisch, and A. Zeileis. Technische Universität Wien, 2003, pp. 564–84.

J. W. Pratt, H. Raiffa, and R. Schlaifer. *Introduction to Statistical Decision Theory*. Massachusetts Institute of Technology, 1995.

H. Raiffa and R. Schlaifer. *Applied Statistical Decision Theory*. Division of Research, Graduate School of Business Administration, Harvard University, 1961.

C. R. Rao. *Statistics and Truth: Putting Chance to Work*. World Scientific, 1997.

J. A. Royle and M. Kery. *Applied Hierarchical Modeling in Ecology*. Elsevier, 2016. DOI: 10.1016/c2013-0-19160-x. http://doi.org/10.1016%2Fc2013-0-19160-x.

R. Schlaifer. "Greek theories of slavery from Homer to Aristotle." *Harvard Studies in Classical Philology* 47 (1936): 165–204.

A. F. M. Smith and A. E. Gelfand. "Bayesian statistics without tears: A sampling–resampling perspective." *The American Statistician* 46.2 (1992): 84–8. DOI: 10.2307/2684170. http://doi.org/10.2307%2F2684170.

D. J. Spiegelhalter, N. G. Best, B. P. Carlin, et al. "Bayesian measures of model complexity and fit." *Journal of the Royal Statistical Society: Series B (Statistical Methodology)* 64.4 (2002): 583–639. DOI: 10.1111/1467-9868.00353. http://doi.org/10.1111%2F1467-9868.00353.

A. Stevenson, ed. *Oxford Dictionary of English*. Oxford University Press, 2010. DOI: 10.1093/acref/9780199571123.001.0001. http://doi.org/10.1093%2Facref%2F9780199571123.001.0001.

J. V. Stone. *Bayes' Rule: a Tutorial Introduction to Bayesian Analysis*. Sebtel Press, 2014.

G. Upton and I. Cook. *A Dictionary of Statistics* (third edition). Oxford University Press, 2014.

G. Upton and I. Cook. *Oxford Dictionary Plus Science and Technology*. Oxford University Press, 2016. DOI: 10.1093/acref/9780191826726.001.0001.

J. Venn. "On the diagrammatic and mechanical representation of propositions and reasonings." *The London, Edinburgh, and Dublin Philosophical Magazine and Journal of Science* 10.59 (1880): 1–18.

J. Wei. "Least squares fitting of an elephant." *Chemtech* 5.2 (1975): 128–9.

P. J. Williams and M. B. Hooten. "Combining statistical inference and decisions in ecology." *Ecological Applications* 26.6 (2016): 1930–42. DOI: 10.1890/15-1593.1. http://doi.org/10.1890%2F15-1593.1.

P. G. Wodehouse. *My Man Jeeves*. George Newnes, 1919.

M. Zhu and A. Y. Lu. "The counter-intuitive non-informative prior for the Bernoulli family." *Journal of Statistics Education* 12.2 (2004): 1–10.

Hyperlinks Accessed August 2017

Preface

- Rational Bayes: http://yudkowsky.net/rational/bayes
- Oxford Dictionary of Statistics: http://www.oxfordreference.com/view/10.1093/acref/9780199679188.001.0001/acref-9780199679188
- Wolfram Mathematics: http://www.wolfram.com/
- Online Statistics Education: An Interactive Multimedia Course for Study: http://onlinestatbook.com/2/index.html
- Wikipedia: http://www.wikipedia.org/
- Encyclopedia Britannica: http://www.britannica.com/

Chapter 1: Introduction to Probability

- Gerolamo Cardano: http://www.britannica.com/biography/Girolamo-Cardano
- Pierre de Fermat: http://www.britannica.com/biography/Pierre-de-Fermat
- Blaise Pascal: http://www.britannica.com/biography/Blaise-Pascal
- Definition of a set: http://www.britannica.com/topic/set-mathematics-and-logic
- Definition of sample space: http://mathworld.wolfram.com/SampleSpace.html
- Definition of Law of Large Numbers: http://mathworld.wolfram.com/LawofLargeNumbers.html
- Probability distribution: http://www.oxfordreference.com/search?source=%2F10.1093%2Facref%2F9780199679188.001.0001%2Facref-9780199679188&q=probability+distribution
- Event: http://www.oxfordreference.com/view/10.1093/acref/9780199679188.001.0001/acref-9780199679188-e-1433#

Chapter 2: Joint, Marginal, and Conditional Probability

- Definition of set: http://mathworld.wolfram.com/Set.html
- Definition of mutually exclusive: http://mathworld.wolfram.com/MutuallyExclusiveEvents.html
- Venn diagram: http://mathworld.wolfram.com/VennDiagram.html
- John Venn: http://www.britannica.com/biography/John-Venn
- MacTutor History of Mathematics Archive: http://www-history.mcs.st-andrews.ac.uk/Biographies/Venn.html
- Morton's toe: http://en.wikipedia.org/wiki/Morton%27s_toe
- Definition of mutually exclusive: http://mathworld.wolfram.com/MutuallyExclusiveEvents.html
- Definition of joint probability: http://www.oxfordreference.com/view/10.1093/acref/9780199679188.001.0001/acref-9780199679188-e-847?rskey=BIh8lb&result=1

- Definition of conditional probability: http://mathworld.wolfram.com/Conditional Probability.html
- Kalid Azad: http://www.betterexplained.com

Chapter 3: Bayes' Theorem

- Bayes' Theorem: http://mathworld.wolfram.com/BayesTheorem.html
- Definition of conditional probability: http://mathworld.wolfram.com/Conditional Probability.html
- Breast cancer example: http://yudkowsky.net/rational/bayes

Chapter 4: Bayesian Inference

- Definition of science: http://spaceplace.nasa.gov/science/en/
- Definition of science: http://en.wikipedia.org/wiki/Science
- Scientific method: http://www.britannica.com/science/scientific-method
- Definition of hypothesis: http://www.britannica.com/topic/scientific-hypothesis
- Definition of scientific theory: http://www.dictionary.com/browse/scientific-theory
- Definition of deductive reasoning: http://www.oxfordreference.com/view/10.1093/oi/authority.20110803095706311
- Deductive versus inductive inference quote from livescience: http://www.livescience.com/21569-deduction-vs-induction.html
- Link to the word verify: http://en.wikipedia.org/wiki/Verification_and_validation)
- Link to the word falsify: http://www.britannica.com/topic/criterion-of-falsifiability
- Theory That Would Not Die lecture: http://www.youtube.com/watch?v=8oD6eBkjF9o
- Pierre-Simon Laplace: http://en.wikipedia.org/wiki/Pierre-Simon_Laplace
- Definition of Bayesian inference: http://www.oxfordreference.com/view/10.1093/acref/9780199679188.001.0001/acref-9780199679188-e-135?rskey=CHC1xV&result=1
- Definition of Bayesian inference: http://en.wikipedia.org/wiki/Bayesian_inference
- Definition of posterior probability: http://en.wikipedia.org/wiki/Posterior_probability
- Definition of infer: http://www.merriam-webster.com/dictionary/infer

Chapter 5: The Author Problem

- Frederick Mosteller: http://ww2.amstat.org/about/statisticiansinhistory/index.cfm?fuseaction=biosinfo&BioID=10
- David Wallace: http://www.stat.uchicago.edu/faculty/emeriti/wallace/index.shtml
- Federalist Papers: http://www.britannica.com/topic/Federalist-papers
- Alexander Hamilton: http://www.britannica.com/biography/Alexander-Hamilton-United-States-statesman
- James Madison: http://www.britannica.com/biography/James-Madison
- Federalist Paper 54: http://www.congress.gov/resources/display/content/The+Federalist+Papers#TheFederalistPapers-54
- Additional information on Federalist Paper 54: http://en.wikipedia.org/wiki/Federalist_No._54
- Definition of likelihood: http://www.dictionary.com/browse/likelihood?s=t
- Definition of likelihood: http://mathworld.wolfram.com/Likelihood.html

Chapter 6: The Birthday Problem

- Absent Treatment: http://americanliterature.com/author/p-g-wodehouse/short-story/absent-treatment
- Yudkowski priors: http://yudkowsky.net/rational/bayes
- Definition of non-informative and informative prior: http://support.sas.com/documentation/cdl/en/statug/63033/HTML/default/viewer.htm#statug_introbayes_sect004.htm
- Definition of non-informative and informative prior: http://en.wikipedia.org/wiki/Prior_probability
- Link to the word robust: http://www.dictionary.com/browse/robust
- Zhu and Lu link: http://ww2.amstat.org/publications/jse/v12n2/zhu.pdf
- Link to robust Bayesian analysis: http://en.wikipedia.org/wiki/Robust_Bayesian_analysis

Chapter 7: The Portrait Problem

- IMS Bulletin: http://bulletin.imstat.org/
- Challenge: http://www.york.ac.uk/depts/maths/histstat/bayespic.htm
- History of Life Insurance: http://www.amazon.com/History-life-insurance-formative-years/dp/B00085BVQY
- Definition of a non-conformist: http://www.britannica.com/topic/Nonconformists
- Description of England's Act of Uniformity: http://www.parliament.uk/about/living-heritage/transformingsociety/private-lives/religion/collections/common-prayer/act-of-uniformity-1662/
- Dr. Bellhouse article: http://www2.isye.gatech.edu/~brani/isyebayes/bank/bayesbiog.pdf
- Answer to why men wear wigs: http://boston1775.blogspot.com
- Link to Anne Clark's page: http://www.uvm.edu/~religion/?Page=clark.php

Chapter 8: Probability Mass Functions

- Definition of variable: http://www.oxfordreference.com/view/10.1093/acref/9780199679188.001.0001/acref-9780199679188-e-1703?rskey=cgYSYj&result=4
- Definition of variable: http://www.oxfordreference.com/view/10.1093/acref/9780199679188.001.0001/acref-9780199679188-e-1703?rskey=cgYSYj&result=4
- Definition of a variable: http://www.britannica.com/topic/variable-mathematics-and-logic
- Definition of random variable: http://www.oxfordreference.com/view/10.1093/acref/9780199679188.001.0001/acref-9780199679188-e-1351?rskey=LxzeOh&result=1
- Definition of random variable: http://en.wikipedia.org/wiki/Random_variable
- Definition of probability theory: http://www.britannica.com/topic/probability-theory
- Definition of probability distribution: http://www.oxfordreference.com/view/10.1093/acref/9780199679188.001.0001/acref-9780199679188-e-1295?rskey=D26yAE&result=1
- Definition of probability mass function:http://www.britannica.com/science/statistics/Random-variables-and-probability-distributions#ref367430
- Definition of probability mass function: http://en.wikipedia.org/wiki/Probability_mass_function

- Link to Jakob Bernoulli: http://www.britannica.com/biography/Jakob-Bernoulli
- Definition of parameter: http://www.oxfordreference.com/view/10.1093/acref/ 9780199679188.001.0001/acref-9780199679188-e-1206?rskey=f6caGK&result=2
- Definition of parameter: http://en.wikipedia.org/wiki/Parameter
- Link to negative binomial distribution: http://mathworld.wolfram.com/Negative BinomialDistribution.html
- Link to Bernoulli distribution: http://mathworld.wolfram.com/Bernoulli Distribution.html
- link to Poisson distribution: http://mathworld.wolfram.com/PoissonDistribution. html
- Link to discrete uniform distribution: http://mathworld.wolfram.com/Discrete UniformDistribution.html
- Link to geometric distribution: http://mathworld.wolfram.com/Geometric Distribution.html
- Link to hypergeometric distribution: http://mathworld.wolfram.com/Hypergeo metricDistribution.html
- Definition of Bernoulli distribution: http://www.oxfordreference.com/view/10. 1093/acref/9780199541454.001.0001/acref-9780199541454-e-149
- Definition of Bernoulli distribution: http://en.wikipedia.org/wiki/Bernoulli_ distribution
- Definition of likelihood: http://www.dictionary.com/browse/likelihood?s=t
- Link to John Kruschke's blog: http://doingbayesiandataanalysis.blogspot.com.au/ 2012/05/graphical-model-diagrams-in-doing.html

Chapter 9: Probability Density Functions

- Definition of pmf: http://www.britannica.com/science/statistics/Random-variables-and-probability-distributions#ref367430
- Definition of random variable: http://www.oxfordreference.com/view/10.1093/ acref/9780199679188.001.0001/acref-9780199679188-e-1351?rskey=5IndNY&result=2
- Bacteria growth example: http://en.wikipedia.org/wiki/Probability_density_function
- Uniform or rectangular distribution: http://mathworld.wolfram.com/Uniform Distribution.html
- Online Statistics Education ebook: http://onlinestatbook.com/2/normal_distri bution/intro.html
- Carl Friedrich Gauss: http://www.britannica.com/biography/Carl-Friedrich-Gauss
- Integral symbol: http://en.wikipedia.org/wiki/Integral_symbol
- Definition of integral: http://mathworld.wolfram.com/Integral.html
- Gaussian integral: http://mathworld.wolfram.com/GaussianIntegral.html
- Riemann Sum: http://mathworld.wolfram.com/RiemannSum.html
- Khan Academy: http://www.khanacademy.org/math/
- Better Explained: http://betterexplained.com/calculus/
- How to Enjoy Calculus: http://www.youtube.com/watch?v=GSV-AuGOjsg
- Normal distribution: http://mathworld.wolfram.com/NormalDistribution.html
- Log normal distribution: http://mathworld.wolfram.com/LogNormalDistribution. html
- Beta distribution: http://mathworld.wolfram.com/BetaDistribution.html
- Gamma distribution: http://mathworld.wolfram.com/GammaDistribution.html
- Exponential distribution: http://mathworld.wolfram.com/ExponentialDistribution .html

- Weibull distribution: http://mathworld.wolfram.com/WeibullDistribution.html
- Cauchy distribution: http://mathworld.wolfram.com/CauchyDistribution.html
- Definition of likelihood function: http://www.oxfordreference.com/view/10.1093/acref/9780199679591.001.0001/acref-9780199679591-e-1671?rskey=uGoV2J&result=1
- Definition of closed-form expression: http://en.wikipedia.org/wiki/Closed-form_expression
- Definition of conjugate prior: http://www.oxfordreference.com/view/10.1093/acref/9780199679188.001.0001/acref-9780199679188-e-135?rskey=HNKSuc&result=5
- Dirichlet distribution: http://en.wikipedia.org/wiki/Dirichlet_distribution

Chapter 10: The White House Problem: The Beta-Binomial Conjugate

- Link to bet: http://voices.washingtonpost.com/dcsportsbog/2009/07/could_shaq_get_into_the_white.html?wprss=dcsportsbog
- Binomial pmf link: http://mathworld.wolfram.com/BinomialDistribution.html
- Bernoulli distribution link: http://mathworld.wolfram.com/BernoulliDistribution.html
- Beta distribution description: http://www.oxfordreference.com/view/10.1093/acref/9780199679188.001.0001/acref-9780199679188-e-158?rskey=DESoj3&result=1
- Beta distribution description: http://en.wikipedia.org/wiki/Beta_distribution
- Wolfram calculator link: http://www.wolframalpha.com/
- Zhu and Lu link: http://ww2.amstat.org/publications/jse/v12n2/zhu.pdf
- Link to Howard Raiffa: http://en.wikipedia.org/wiki/Howard_Raiffa
- Link to Robert Schlaifer: http://en.wikipedia.org/wiki/Robert_Schlaifer
- Bayesian conjugate table: http://en.wikipedia.org/wiki/Conjugate_prior
- Conjugate prior definition: http://www.oxfordreference.com/view/10.1093/acref/9780199679188.001.0001/acref-9780199679188-e-135?rskey=HNKSuc&result=5
- Beta distribution link: http://mathworld.wolfram.com/BetaDistribution.html
- Credible interval definition: http://www.oxfordreference.com/view/10.1093/acref/9780191816826.001.0001/acref-9780191816826-e-0087?rskey=cmncUS&result=1
- Credible intervals: http://en.wikipedia.org/wiki/Credible_interval
- Result of bet: http://voices.washingtonpost.com/dcsportsbog/2009/07/shaq_denied_entrance_by_the_wh.html

Chapter 11: The Shark Attack Problem: The Gamma-Poisson Conjugate

- Link to shark attack post: http://www.sciencedaily.com/releases/2001/08/010823084028.htm
- Description of Poisson distribution: http://www.oxfordreference.com/view/10.1093/acref/9780199684274.001.0001/acref-9780199684274-e-6897?rskey=NlK26q&result=5
- Simeon Poisson link: http://www.britannica.com/biography/Simeon-Denis-Poisson
- Euler's number link: http://www.oxfordreference.com/view/10.1093/acref/9780199679591.001.0001/acref-9780199679591-e-1026?rskey=hJEPVP&result=1
- Gamma distribution link: http://mathworld.wolfram.com/GammaDistribution.html
- Gamma distribution link: http://www.oxfordreference.com/view/10.1093/acref/9780199679188.001.0001/acref-9780199679188-e-649?rskey=lJwBFp&result=3

- Gamma distribution link: http://mathworld.wolfram.com/GammaDistribution.html
- Gamma function: http://mathworld.wolfram.com/GammaFunction.html
- Gamma function online calculator: http://www.wolframalpha.com/input/?i=gamma+function
- Definition of conjugate prior: http://www.oxfordreference.com/view/10.1093/acref/9780199679188.001.0001/acref-9780199679188-e-135?rskey=HNKSuc&result=5
- Wikipedia table of conjugate priors: http://en.wikipedia.org/wiki/Conjugate_prior
- Log-normal distribution link: http://mathworld.wolfram.com/LogNormalDistribution.html

Chapter 12: The Maple Syrup Problem: The Normal-Normal Conjugate

- Link to syrup movie: http://entertainment.time.com/2013/09/26/jason-segel-will-star-in-canadian-syrup-heist-film/
- Maple syrup link: http://www.britannica.com/topic/maple-syrup
- Article about cartel: http://business.time.com/2012/12/24/why-does-canada-have-a-maple-syrup-cartel/
- Link to Vermont maple: http://vermontmaple.org/
- Wikipedia conjugate prior table: http://en.wikipedia.org/wiki/Conjugate_prior
- Link to inverse gamma: http://reference.wolfram.com/language/ref/InverseGammaDistribution.html
- Link to inverse chi square: http://reference.wolfram.com/language/ref/InverseChiSquareDistribution.html
- Link to gamma distribution: http://mathworld.wolfram.com/GammaDistribution.html

Chapter 13: The Shark Attack Problem Revisited: MCMC with the Metropolis Algorithm

- Link to Poisson distribution: http://mathworld.wolfram.com/PoissonDistribution.html
- Link to gamma distribution: http://mathworld.wolfram.com/GammaDistribution.html
- Link to Metropolis algorithm: http://en.wikipedia.org/wiki/Equation_of_State_Calculations_by_Fast_Computing_Machines
- Definition of algorithm: http://www.britannica.com/topic/algorithm
- Link to Nicholas Metropolis: http://en.wikipedia.org/wiki/Nicholas_Metropolis
- Link to Andrieu paper: http://www.cs.princeton.edu/courses/archive/spr06/cos598C/papers/AndrieuFreitasDoucetJordan2003.pdf
- Link to Monte Carlo casino: http://www.britannica.com/place/Monte-Carlo-resort-Monaco
- Link to Monte Carlo method: http://www.britannica.com/science/Monte-Carlo-method
- Link to Markov Chain: http://www.oxfordreference.com/view/10.1093/acref/9780191793158.001.0001/acref-9780191793158-e-6731?rskey=Ggi5ev&result=3
- Link to Andrey Markov: http://www.britannica.com/biography/Andrey-Andreyevich-Markov

- Link to Wolfram MCMC demo: http://demonstrations.wolfram.com/ MarkovChainMonteCarloSimulationUsingTheMetropolisAlgorithm/
- Link to calculator: http://www.wolframalpha.com/
- Link to rules of logs: http://www.britannica.com/topic/logarithm

Chapter 14: MCMC Diagnostic Approaches

- SAS Institute link: http://support.sas.com/documentation/cdl/en/statug/63962/ HTML/default/viewer.htm#statug_mcmc_sect024.htm
- Definition of heuristic: http://www.dictionary.com/browse/heuristic
- How your computer stores numbers: http://www.britannica.com/topic/numeral #toc233819
- Converting numbers to binary: http://www.wolframalpha.com/examples/Number Bases.html
- Link to numerical overflow: http://en.wikipedia.org/wiki/Integer_overflow

Chapter 15: The White House Problem Revisited: MCMC with the Metropolis-Hastings Algorithm

- Link to Metropolis algorithm: http://en.wikipedia.org/wiki/Equation_of_State_ Calculations_by_Fast_Computing_Machines
- Link to Keith Hastings: http://www.probability.ca/hastings/
- Normal pdf: http://mathworld.wolfram.com/NormalDistribution.html
- Link to beta pdf: http://mathworld.wolfram.com/BetaDistribution.html
- Link to Metropolis–Hastings algorithm: http://support.sas.com/documentation/ cdl/en/statug/63033/HTML/default/viewer.htm#statug_introbayes_sect007.htm
- Independence Metropolis–Hastings Sampler: http://support.sas.com/documen tation/cdl/en/statug/63033/HTML/default/viewer.htm#statug_introbayes_sect007. htm
- Link to Monte Carlo resort: http://www.britannica.com/place/Monte-Carlo-resort- Monaco
- Definition of Monte Carlo methods: http://www.britannica.com/science/Monte- Carlo-method
- Link to Andrey Markov: http://www.britannica.com/biography/Andrey- Andreyevich-Markov
- Link to Markov Chain: http://www.oxfordreference.com/view/10.1093/acref/ 9780191793158.001.0001/acref-9780191793158-e-6731?rskey=Ggi5ev&result=3
- Link to beta distribution (to show table): http://en.wikipedia.org/wiki/Beta_ distribution
- Link to Hastings entry: http://www.oxfordreference.com/view/10.1093/acref/ 9780199679188.001.0001/acref-9780199679188-e-1977?rskey=IyPQN9&result=1

Chapter 16: The Maple Syrup Problem Revisited: MCMC with Gibbs Sampling

- SAS link to Gibbs sampler: http://support.sas.com/documentation/cdl/cn/statug/ 63033/HTML/default/viewer.htm#statug_introbayes_sect007.htm
- Definition of algorithm: http://www.merriam-webster.com/dictionary/algorithm

- Wikipedia figure of joint and marginal distributions: http://en.wikipedia.org/wiki/Marginal_distribution
- Link to gamma distribution: http://en.wikipedia.org/wiki/Gamma_distribution
- Link to Josiah Gibbs: http://www.britannica.com/biography/J-Willard-Gibbs
- Link to Stuart Gemen: http://en.wikipedia.org/wiki/Stuart_Geman
- Link to Donald Gemen: http://en.wikipedia.org/wiki/Donald_Geman
- Link to Theory That Would Not Die: http://www.mcgrayne.com/the_theory_that_would_not_die__how_bayes__rule_cracked_the_enigma_code__hunted_d_107493.htm
- Link to Geman and Geman paper: http://www.ncbi.nlm.nih.gov/pubmed/22499653
- Link to BUGS: http://www.mrc-bsu.cam.ac.uk/software/bugs/
- Link to JAGS: http://mcmc-jags.sourceforge.net/

Chapter 17: The Survivor Problem

- Definition of function: http://www.oxfordreference.com/view/10.1093/acref/9780191826726.001.0001/acref-9780191826726-e-468?rskey=kC5yBF&result=11
- Definition of linear function: http://www.oxfordreference.com/view/10.1093/acref/9780199679591.001.0001/acref-9780199679591-e-1692?rskey=N38sCm&result=1
- Definition of variable: http://www.britannica.com/topic/variable-mathematics-and-logic
- Definition of error term: http://www.britannica.com/science/statistics/Experimental-design#ref367485
- Additional link to statistical model: http://en.wikipedia.org/wiki/Statistical_model
- Definition of science: http://spaceplace.nasa.gov/science/en/
- Definition of science: http://en.wikipedia.org/wiki/Science
- Definition of hypothesis: http://www.britannica.com/topic/scientific-hypothesis
- Definition of scientific theory: http://www.dictionary.com/browse/scientific-theory
- Definition of deductive reasoning: http://www.oxfordreference.com/view/10.1093/oi/authority.20110803095706311
- Wassertheil-Smoller quotes: http://www.livescience.com/21569-deduction-vs-induction.html
- Link to verification and validation: http://en.wikipedia.org/wiki/Verification_and_validation
- Definition of falsification: http://www.britannica.com/topic/criterion-of-falsifiability
- Definition of statistical inference: http://www.oxfordreference.com/view/10.1093/acref/9780199679188.001.0001/acref-9780199679188-e-1557?rskey=oryjck&result=1
- Angela Duckworth: http://angeladuckworth.com/
- Grit scale: http://angeladuckworth.com/grit-scale/
- Grit paper: http://www.dropbox.com/s/0y545gn2withb5e/DuckworthPetersonMatthewsKelly_2007_PerseveranceandPassion.pdf?dl=0
- Description of Survivor show: http://en.wikipedia.org/wiki/Survivor_(franchise)
- Definition of statistical model: http://en.wikipedia.org/wiki/Statistical_model
- Link to binomial distribution: http://reference.wolfram.com/language/ref/BinomialDistribution.html
- Link to normal distribution: http://reference.wolfram.com/language/ref/NormalDistribution.html
- Link to gamma distribution: http://reference.wolfram.com/language/ref/GammaDistribution.html

- Wikipedia table of conjugate shortcuts: http://en.wikipedia.org/wiki/Conjugate_prior
- Alan Gelfand homepage: http://www2.stat.duke.edu/~alan/
- Adrian Smith bio: http://en.wikipedia.org/wiki/Adrian_Smith_(statistician)
- Link to Bayesian Statistics without Tears: http://www.jstor.org/stable/2684170?seq=1#page_scan_tab_contents
- Link to proper gamma distribution: http://en.wikipedia.org/wiki/Gamma_distribution
- Link to posterior predictive distribution: http://www.oxfordreference.com/view/10.1093/acref/9780191816826.001.0001/acref-9780191816826-e-0031?rskey=wjRLiA&result=1
- Link to posterior predictive distribution: http://en.wikipedia.org/wiki/Posterior_predictive_distribution
- Link to goodness of fit definition: http://www.oxfordreference.com/view/10.1093/acref/9780199679188.001.0001/acref-9780199679188-e-687?rskey=B6T6sC&result=6

Chapter 18: The Survivor Problem Continued: Introduction to Bayesian Model Selection

- Link to Angela Duckworth: http://angeladuckworth.com/
- Link to SSE: http://www.oxfordreference.com/view/10.1093/oi/authority.20111013151335747?rskey=W9nlte&result=9
- Laws of logarithms: http://www.britannica.com/topic/logarithm
- Occam's razor: http://www.britannica.com/topic/Occams-razor
- John von Neumann: http://www.britannica.com/biography/John-von-Neumann
- Fitting an elephant: http://demonstrations.wolfram.com/FittingAnElephant/
- Link to DIC: http://support.sas.com/documentation/cdl/en/statug/63347/HTML/default/viewer.htm#statug_introbayes_sect009.htm#statug.introbayes.bayesdic
- Link to WinBUGS: http://www.mrc-bsu.cam.ac.uk/software/bugs/the-bugs-project-winbugs/

Chapter 19: The Lorax Problem: Introduction to Bayesian Networks

- Summary of The Lorax: http://en.wikipedia.org/wiki/The_Lorax
- Definition of sustainable forest management: http://en.wikipedia.org/wiki/Sustainable_forest_management
- Link to sustainability: http://www.britannica.com/topic/sustainability
- Example of Bayesian network: http://en.wikipedia.org/wiki/Bayesian_network
- Netica: http://www.norsys.com/tutorials/netica/nt_toc_A.htm
- Influence diagram: http://www.lumina.com/technology/influence-diagrams/
- Definition of sample space and event: http://www.oxfordreference.com/view/10.1093/acref/9780199679188.001.0001/acref-9780199679188-e-1433#
- DAG: http://www.oxfordreference.com/view/10.1093/acref/9780191816826.001.0001/acref-9780191816826-e-0168?rskey=gb73bl&result=2
- Dictionary definition of marginal: http://www.dictionary.com/browse/marginal
- Wikipedia definition of marginal probability: http://en.wikipedia.org/wiki/Marginal_distribution
- Bruce Marcot: http://www.fs.fed.us/research/people/profile.php?alias=bmarcot
- Chain rule in probability: http://en.wikipedia.org/wiki/Chain_rule_(probability)
- Machine learning: http://www.britannica.com/technology/machine-learning

- Judea Pearl: http://www.britannica.com/biography/Judea-Pearl
- Definition of utility: http://www.oxfordreference.com/view/10.1093/acref/978019 9679591.001.0001/acref-9780199679591-e-2939?rskey=bRAwhh&result=9

Chapter 20: The Once-ler Problem: Introduction to Decision Trees

- Link to summary of The Lorax: http://en.wikipedia.org/wiki/The_Lorax
- Definition of decision tree: http://www.oxfordreference.com/view/10.1093/acref/ 9780199541454.001.0001/acref-9780199541454-e-436
- Yogi Berra: http://www.britannica.com/biography/Yogi-Berra
- Expected value: http://www.oxfordreference.com/view/10.1093/acref/978019954 1454.001.0001/acref-9780199541454-e-567
- Utility: http://www.oxfordreference.com/view/10.1093/acref/9780199679591.001. 0001/acref-9780199679591-e-2939?rskey=bRAwhh&result=9
- Elicitation: http://www.dictionary.com/browse/elicitation?s=t
- Binomial probability mass function: http://reference.wolfram.com/language/ref/ BinomialDistribution.html
- Expected value of sample information: http://en.wikipedia.org/wiki/Expected_ value_of_sample_information
- Yogi quotes: http://ftw.usatoday.com/2015/09/the-50-greatest-yogi-berra-quotes

Name Index

Subject Index